QUANTUM MECHANICS DEMYSTIFIED

Demystified Series

Advanced Statistics Demystified
Algebra Demystified
Anatomy Demystified
asp.net Demystified
Astronomy Demystified
Biology Demystified
Business Calculus Demystified
Business Statistics Demystified
C++ Demystified
Calculus Demystified
Chemistry Demystified
College Algebra Demystified
Data Structures Demystified
Databases Demystified
Differential Equations Demystified
Digital Electronics Demystified
Earth Science Demystified
Electricity Demystified
Electronics Demystified
Environmental Science Demystified
Everyday Math Demystified
Genetics Demystified
Geometry Demystified
Home Networking Demystified
Investing Demystified
Java Demystified
JavaScript Demystified
Linear Algebra Demystified
Macroeconomics Demystified
Math Proofs Demystified
Math Word Problems Demystified
Medical Terminology Demystified
Meteorology Demystified
Microbiology Demystified
OOP Demystified
Options Demystified
Organic Chemistry Demystified
Personal Computing Demystified
Pharmacology Demystified
Physics Demystified
Physiology Demystified
Pre-Algebra Demystified
Precalculus Demystified
Probability Demystified
Project Management Demystified
Quality Management Demystified
Quantum Mechanics Demystified
Relativity Demystified
Robotics Demystified
Six Sigma Demystified
sql Demystified
Statistics Demystified
Trigonometry Demystified
uml Demystified
Visual Basic 2005 Demystified
Visual C# 2005 Demystified
xml Demystified

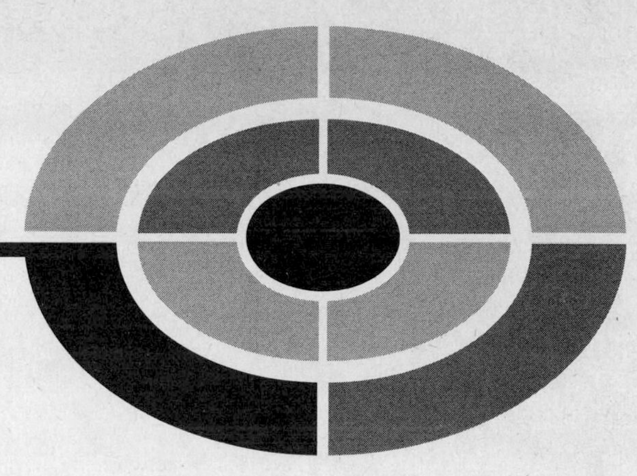

QUANTUM MECHANICS DEMYSTIFIED

DAVID McMAHON

McGRAW-HILL
New York Chicago San Francisco Lisbon London
Madrid Mexico City Milan New Delhi San Juan
Seoul Singapore Sydney Toronto

The McGraw-Hill Companies

Library of Congress Cataloging-in-Publication Data

McMahon, David (David M.)
Quantum mechanics demystified / David McMahon.
p. cm.
Includes bibliographical references and index.
ISBN 0-07-145546-9 (acid-free paper)
1. Quantum theory. I. Title.

QC174.12.M379 2005
530.12—dc22 2005050547

5 6 7 8 9 0 DOC/DOC 0 1 0 9

ISBN 0-07-145546-9

The sponsoring editor for this book was Judy Bass and the production supervisor was Pamela A. Pelton. It was set in Times Roman by Fine Composition. The art director for the cover was Margaret Webster-Shapiro; the cover designer was Handel Low.

Printed and bound by RR Donnelley.

CONTENTS

PREFACE

Quantum mechanics, which by its very nature is highly mathematical (and therefore extremely abstract), is one of the most difficult areas of physics to master. In these pages we hope to help pierce the veil of obscurity by demonstrating, with explicit examples, how to *do* quantum mechanics. This book is divided into three main parts.

After a brief historical review, we cover the basics of quantum theory from the perspective of wave mechanics. This includes a discussion of the wavefunction, the probability interpretation, operators, and the Schrödinger equation. We then consider simple one-dimensional scattering and bound state problems.

In the second part of the book we cover the mathematical foundations needed to do quantum mechanics from a more modern perspective. We review the necessary elements of matrix mechanics and linear algebra, such as finding eigenvalues and eigenvectors, computing the trace of a matrix, and finding out if a matrix is Hermitian or unitary. We then cover Dirac notation and Hilbert spaces. The postulates of quantum mechanics are then formalized and illustrated with examples. In the chapters that cover these topics, we attempt to "demystify" quantum mechanics by providing a large number of solved examples.

The final part of the book provides an illustration of the mathematical foundations of quantum theory with three important cases that are typically taught in a first semester course: angular momentum and spin, the harmonic oscillator, and an introduction to the physics of the hydrogen atom. Other topics covered at some level with examples include the density operator, the Bloch vector, and two-state systems.

Unfortunately, due to the large amount of space that explicitly solved examples from quantum mechanics require, it is not possible to include everything about the theory in a volume of this size. As a result we hope to prepare a second volume

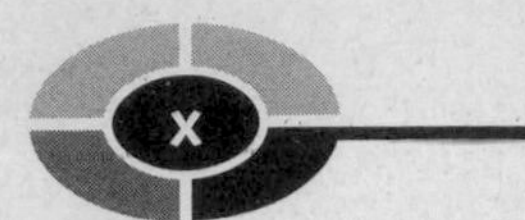

to cover advanced topics from non-relativistic quantum theory such as scattering, identical particles, addition of angular momentum, higher Z atoms, and the WKB approximation.

There is no getting around the mathematical background necessary to learn quantum mechanics. The reader should know calculus, how to solve ordinary and partial differential equations, and have some exposure to matrices/linear algebra and at least a basic working knowledge of complex numbers and vectors. Some knowledge of basic probability is also helpful. While this mathematical background is extensive, it is our hope that the book will help "demystify" quantum theory for those who are interested in self-study or for those from different backgrounds such as chemistry, computer science, or engineering, who would like to learn something about quantum mechanics.

ACKNOWLEDGMENTS

Thanks to Daniel M. Topa of Wavefront Sciences in Albuquerque, New Mexico, Sonja Daffer of Imperial College, London, and Bryan Eastin of the University of New Mexico, for review of the manuscript.

Historical Review

In this chapter we very briefly sketch out four of the main ideas that led to the development of quantum theory. These are Planck's solution to the blackbody radiation problem, Einstein's explanation of the photoelectric effect, the Bohr model of the atom, and the de Broglie wavelength of material particles.

Blackbody Radiation and Planck's Formula

A *blackbody* is an object that is a perfect absorber of radiation. In the ideal case, it absorbs all of the light that falls on it, no light is reflected by it, and no light passes through it. While such an object doesn't reflect any light, if we heat up a blackbody, it can *radiate* light. The study of this radiated light generated a bit of controversy in the late 19th century. Specifically, there was a problem explaining the *spectrum* of the thermal radiation emitted from a blackbody.

Simply put, a spectrum is a plot, at fixed temperature, of the amount of light emitted at each wavelength (or if we choose at each frequency). A plot of the amount of light (specifically, the energy density) emitted versus wavelength looks something like the curve in Fig. 1-1.

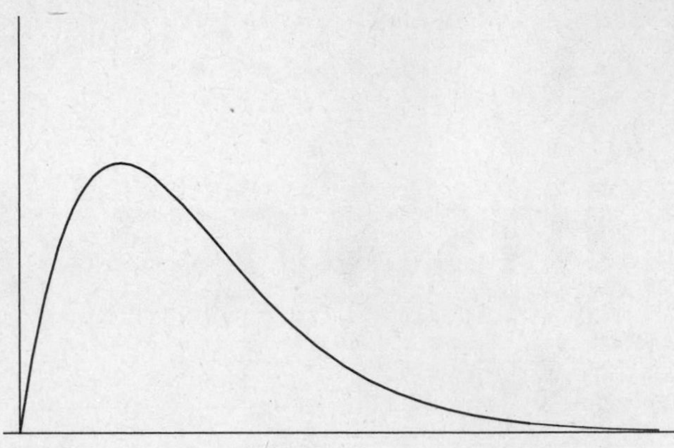

Fig. 1-1

As the temperature is increased, more light is emitted at higher frequencies. This means that the peak in this plot would shift more to the right. Classical theory was not able to explain the high frequency behavior of blackbody emission. Spectra like the one shown here were found experimentally.

An attempt to explain these results using classical theory was codified in the *Rayleigh-Jeans formula,* which is an expression that attempts to give us the energy density $u(\nu, T)$ of radiation in the cavity, where ν is frequency and T is the temperature. Qualitatively, it is formed as a product of two quantities:

$$u = \left(\begin{array}{c} \text{number degrees of} \\ \text{freedom for frequency } \nu \end{array} \right) \times \left(\begin{array}{c} \text{average energy per} \\ \text{degree of freedom} \end{array} \right)$$

Using classical physics, the average energy per degree of freedom can be calculated in the following way. Let's call the energy E, Boltzmann's constant k, and the temperature T. The average energy $\overline{E}$ is given by:

$$\overline{E} = \frac{\displaystyle\int_0^\infty E e^{-E/kT} dE}{\displaystyle\int_0^\infty e^{-E/kT} dE}$$

Both of these integrals are easy to do. The integral in the denominator can be done immediately by using the substitution $y = -E/(kT)$:

$$\int_0^\infty e^{-E/kT}\,dE = kT\int_{-\infty}^0 e^y dy = kTe^y\Big|_{-\infty}^0 = kT$$

In the numerator, we use integration by parts. The integration by parts formula is:

$$\int u\,dv = uv - \int v\,du$$

We let $u = E$, then $du = dE$. Using the previous result, $dv = e^{-E/kT}$ and so $v = -kTe^{-E/kT}$. We then have:

$$\int_0^\infty Ee^{-E/kT}\,dE = kTe^{-E/kT}\Big|_0^\infty + kT\int_0^\infty e^{-E/kT}\,dE = kTe^{-E/kT}\,E\Big|_0^\infty + (kT)^2$$

Now:

$$\lim_{E\to\infty} e^{-E/kT} = 0$$

And so the evaluation at the upper limit of $kTe^{-E/kT}E$ vanishes. Also, as $E \to 0$, this term clearly vanishes and so:

$$\int_0^\infty Ee^{-E/kT}\,dE = (kT)^2$$

And so we find that:

$$\overline{E} = \frac{(kT)^2}{kT} = kT$$

The other term in the Rayleigh-Jeans formula is the number of degrees of freedom per frequency. Using classical theory, the number of degrees of freedom was found to be:

$$\frac{8\pi\nu^2}{c^3}$$

All together the Rayleigh-Jeans formula tells us that the energy density is:

$$u(\nu, T) = \frac{8\pi\nu^2}{c^3}kT$$

You can see from this formula that as ν gets large, its going to blow sky-high. Worse—if you integrate over all frequencies to get the total energy per unit volume, you will get infinity. The formula only works at low frequencies. Obviously this is not what is observed experimentally, and the prediction that the energy density at high frequencies would go to infinity became known as the "ultraviolet catastrophe" (since ultraviolet is light of high frequency).

Planck fixed the problem by examining the calculation of $\overline{E}$, a calculation that gave us the simple result of kT and seems so reasonable if you've studied thermodynamics. Consider the implicit assumption that is expressed by the way the formula is calculated. The formula is computed using integration, which means that it has been assumed that energy exchange is *continuous*. What if instead, only certain fixed values of energy exchange were allowed?

PLANCK'S RADICAL ASSUMPTION

A practical blackbody is made of a metallic cavity with a small hole through which radiation can escape. Planck made the assumption that an exchange of energy between the electrons in the wall of the cavity and electromagnetic radiation can only occur in discrete amounts. This assumption has an immediate mathematical consequence. The first consequence of this assumption is that the integrals above turn into discrete sums. So when we calculate the average energy per degree of freedom, we must change all integrals to sums:

$$\int \to \sum$$

The second important piece of data that Planck told us, was that energy comes in little bundles, that we will call the basic "quantum of energy." According to Planck, the basic quantum of energy ε is given by:

$$\varepsilon = h\nu$$

where ν is the frequency of the radiation. Furthermore, energy can only come in amounts that are integer multiples of the basic quantum:

$$E = n\varepsilon = nh\nu, \quad n = 0, 1, 2, \ldots$$

The constant $h = 6.62 \times 10^{-34}$ (Joules-seconds) is called *Planck's constant*. It is frequently convenient to use the symbol $\hbar = h/2\pi$.

Incorporating this assumption with the change from integrals to discrete sums, we now have:

$$\overline{E} = \frac{\sum_{n=0}^{\infty} n\varepsilon e^{-n\varepsilon/kT}}{\sum_{n=0}^{\infty} e^{-n\varepsilon/kT}}$$

To evaluate this formula, we recall that a geometric series sums to:

$$\sum_{n=0}^{\infty} ar^n = \frac{a}{1-r}$$

where $|r| < 1$. Returning to the formula for average energy, let's look at the denominator. We set $a = 1$ and let $r = e^{-n\varepsilon/kT}$. Clearly r is always less than one, and so:

$$\sum_{n=0}^{\infty} e^{-n\varepsilon/kT} = \frac{1}{1-e^{-\varepsilon/kT}}$$

In the exercises, you will show that the other term we have can be written as:

$$\sum_{n=0}^{\infty} n\varepsilon e^{-n\varepsilon/kT} = \frac{\varepsilon e^{-\varepsilon/kT}}{(1-e^{-\varepsilon/kT})^2}$$

These results allow us to rewrite the average energy in the following way:

$$\overline{E} = \frac{\sum_{n=0}^{\infty} n\varepsilon e^{-n\varepsilon/kT}}{\sum_{n=0}^{\infty} e^{-n\varepsilon/kT}} = \frac{\frac{\varepsilon e^{-\varepsilon/kT}}{(1-e^{-\varepsilon/kT})^2}}{\frac{1}{1-e^{-\varepsilon/kT}}} = \frac{\varepsilon e^{-\varepsilon/kT}}{(1-e^{-\varepsilon/kT})^2}(1-e^{-\varepsilon/kT})$$

$$= \frac{\varepsilon e^{-\varepsilon/kT}}{1-e^{-\varepsilon/kT}}$$

We can put this in a more familiar form by letting $\varepsilon = h\nu$ and doing some algebraic manipulation:

$$\overline{E} = \frac{h\nu e^{-h\nu/kT}}{1-e^{-h\nu/kT}} = \frac{h\nu}{e^{h\nu/kT}(1-e^{-h\nu/kT})} = \frac{h\nu}{e^{h\nu/kT}-1}$$

To get the complete Planck formula for blackbody radiation, we just substitute this term for kT in the Rayleigh-Jeans law. The exponential in the denominator decays much faster than ν^2. The net result is that the average energy term cuts off any energy density at high frequencies. The complete Planck formula for the energy density of blackbody radiation is:

$$u(\nu, T) = \frac{8\pi\nu^2}{c^3}\frac{h\nu}{e^{h\nu/kT}-1}$$

The Photoelectric Effect

In 1905, Einstein made the radical proposal that light consisted of particles called *photons*. Each photon carries energy:

$$E = h\nu$$

and linear momentum:

$$p = \frac{h}{\lambda}$$

where ν and λ are the frequency and wavelength of the lightwave. Using the relation $c = \nu\lambda$ where c is the speed of light in vacuum, we can rewrite the momentum of a photon as:

$$p = \frac{h}{\lambda} = \frac{h}{c/\nu} = \frac{h}{c(h/E)} = \frac{E}{c}$$

Einstein made this proposal to account for several unexplained features associated with the *photoelectric effect*. This is a process that involves the emission of electrons from a metal when light strikes the surface. The maximum energy of the emitted electrons is found to be:

$$qV_o = E_{\max}$$

where q is the charge of the electron and V_o is the *stopping potential*. Experiment shows that:

1. When light strikes a metal surface, a current flows instantaneously, even for very weak light.
2. At a fixed frequency, the strength of the current is directly proportional to the intensity of the light.
3. The stopping potential V_o, and therefore the maximum energy of the emitted electrons, depends only on the frequency of the light and the type of metal used.
4. Each metal has a characteristic threshold frequency ν_o such that:

$$qV_o = h(\nu - \nu_o)$$

5. The constant h is found to be the same for all metals, and not surprisingly turns out to be the same constant used by Planck in his blackbody derivation.

Each of these experimental ideas can be explained by accepting that light is made up of particles. For example, consider observation 2, which is easy to explain in the photon picture. If the intensity of the light beam is increased, then the number of photons is increased in turn and there are more photons striking the metal surface.

Specifically, suppose we double the intensity of the light. Twice as many photons strike the metal surface and knock out twice as many electrons—making a current that is twice as strong. In the wave picture, however, you would expect that increasing the intensity would increase the energy of the electrons, and not their number. Classical wave theory disagrees with observation.

The ideas of Planck and Einstein can be summarized by the Planck-Einstein relations.

DEFINITION: The Planck-Einstein Relations

The Planck-Einstein relations connect the particle-like properties of energy and momentum to wavelike properties of frequency and wave vector **k**. Recalling that frequency $\nu = \omega/2\pi$

$$E = h\nu = \hbar\omega$$

$$p = \hbar k$$

The Bohr Theory of the Atom

Light again took center stage in 1913 when Bohr worked out the basic structure of the hydrogen atom. He did this by considering the light that atoms emit.

The light emitted by isolated atoms takes the form of a discrete series of lines called spectral lines. It is found that these lines occur at specific frequencies for type of atom. So a sodium atom has a different line spectrum than a hydrogen atom, and a helium atom has yet another spectrum. Think of a spectrum as the fingerprint of each element. It is also found that atoms absorb light at specific, well-defined frequencies as well.

This tells us that like Planck's blackbody oscillators, atoms can exchange energy only in fixed discrete amounts. Neils Bohr noticed this and proposed two radical ideas about the behavior of electrons in atoms.

Bohr Makes Two Key Assumptions About the Atom

1. An electron can only orbit about the nucleus in such a way that the orbit is defined by the relationship:

$$mvr = n\hbar \quad n = 1, 2, \ldots$$

where v is the velocity of the electron, r is the radius of the orbit, and m is the mass of the electron. The presence of n in the formula restricts the angular momentum of the electron to integer multiples of $\hbar$, where the angular momentum is given by:

$$L = n\hbar$$

2. Electrons only radiate during transitions between states. A transition from energy state E_i to energy state E_f is accompanied by the emission of a photon of energy:

$$h\nu = E_i - E_f$$

The Coulomb force between the positively charged nucleus and the negatively charged electron is what keeps the electrons in orbit. Setting this equal to the centrifugal force:

$$\frac{e^2}{r} = \frac{mv^2}{r}$$

Results in the following expressions for the velocity of the electron and the radius of the orbit. We label each quantity with subscript n to conform with assumption (a) above:

$$v_n = \frac{e^2}{n\hbar} \qquad \text{(velocity of electron in orbit } n\text{)}$$

$$r_n = \frac{n^2\hbar^2}{me^2} \qquad \text{(radius of orbit } n\text{)}$$

EXAMPLE 1.1

Derive the energy of an electron in the hydrogen atom using Bohr's formulas.

SOLUTION

We start by recalling that the

$$\text{total energy} = \text{kinetic energy} + \text{potential energy} = T + V$$

For an electron moving in the Coloumb potential of a proton, the potential is just

$$V = -\frac{e^2}{r}$$

Using the formula for the radius of orbit n this becomes:

$$V_n = -\frac{e^2}{r_n} = -\frac{e^2}{n^2\hbar^2}me^2 = -\frac{me^4}{n^2\hbar^2}$$

For the kinetic energy, we obtain:

$$T = \frac{1}{2}mv_n^2 = \frac{1}{2}m\left(\frac{e^2}{n\hbar}\right)^2 = \frac{me^4}{2n^2\hbar^2}$$

The total energy of an electron in orbit n is therefore:

$$E_n = T_n + V_n = \frac{me^4}{2n^2\hbar^2} - \frac{me^4}{n^2\hbar^2} = -\frac{me^4}{2n^2\hbar^2} = -\frac{2\pi^2 me^4}{2n^2h^2}$$

EXAMPLE 1.2 e.g.

Derive a relation that predicts the frequencies of the line spectra of hydrogen.

SOLUTION 

Bohr proposed that the frequency of a photon emitted by an electron in the hydrogen atom was related to transitions of energy states as:

$$h\nu = E_i - E_f$$

The energy of state n is:

$$E_n = -\frac{2\pi^2 me^4}{n^2h^2}$$

Therefore:

$$E_i - E_f = -\frac{2\pi^2 me^4}{n_i^2h^2} + \frac{2\pi^2 me^4}{n_f^2h^2} = \frac{2\pi^2 me^4}{h^2}\left(\frac{1}{n_f^2} - \frac{1}{n_i^2}\right)$$

Putting this together with Bohr's proposal we find the frequency is:

$$\nu = \frac{E_i - E_f}{h} = \frac{2\pi^2 me^4}{h^3}\left(\frac{1}{n_f^2} - \frac{1}{n_i^2}\right)$$

This formula can be used to predict the line spectra of hydrogen.

de Broglie's Hypothesis

In 1923 Louis de Broglie proposed that the Planck-Einstein relations should be extended to material particles. A particle with energy E is associated with a wave of frequency $\omega = E/\hbar$. In addition, momentum is related to the wave vector via $\mathbf{p} = \hbar\mathbf{k}$. Applying these simple relations to material particles like electrons, de Broglie proposed that a material particle moving with momentum p has a wavelength:

$$\lambda = \frac{h}{p}$$

If a particle of mass m is moving with a nonrelativistic energy E, we can write:

$$\lambda = \frac{h}{\sqrt{2mE}}$$

EXAMPLE 1.3

A thermal neutron has a speed v that corresponds to room temperature $T = 300$ K. What is the wavelength of a thermal neutron?

SOLUTION

At temperature T average energy is:

$$\overline{E} = \frac{3}{2}kT$$

where k is Boltzmann's constant. By equating the kinetic energy to this quantity with $T = 300$ K, we can find the momentum of the neutron:

$$\frac{p^2}{2m_n} = \frac{3}{2}kT$$

Using de Broglie's relation we obtain the wavelength of the thermal neutron:

$$\lambda = \frac{h}{p} = \frac{h}{\sqrt{3m_n kT}} = \frac{6.63 \times 10^{-34}}{\sqrt{3(1.67 \times 10^{-27})(1.38 \times 10^{-23})(300)}} = 1.4 \text{ Å}$$

Quiz

1. Making the following definition:

$$f(\varepsilon) = \sum_{n=0}^{\infty} e^{-n\varepsilon/kT}$$

Write the following series in terms of $f'(\varepsilon)$:

$$g(\varepsilon) = \sum_{n=1}^{\infty} n\varepsilon e^{-n\varepsilon/kT}$$

Then use the geometric series result to show that g can be written in the form:

$$\varepsilon + 2\varepsilon^2 + 3\varepsilon^3 + \cdots = \frac{\varepsilon}{(1-\varepsilon)^2}$$

2. The lowest energy of an electron in the hydrogen atom occurs for $n = 1$ and is called the ground state. Show that the ground state energy is -13.6 eV.

3. Using the formula for quantized orbits, show that the ground state radius is 0.529×10^{-8} cm. This is known as the *Bohr radius*.

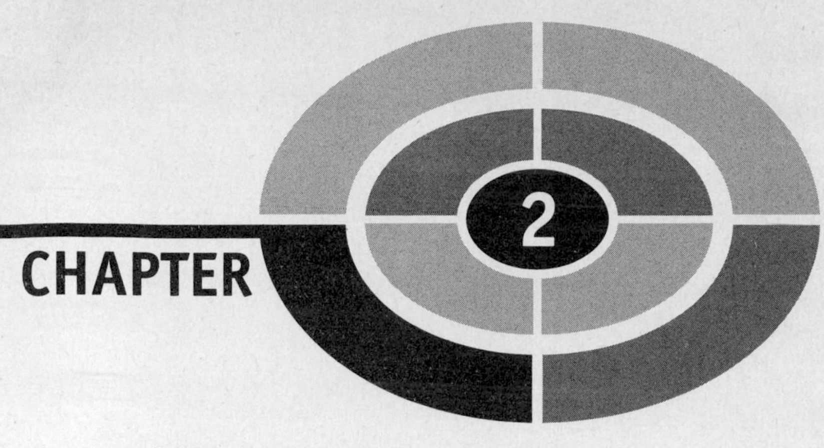

Basic Developments

In quantum mechanics, information about the state of a particle is described by a *wavefunction*. This function is usually denoted by $\psi(x, t)$. The equation that describes its time evolution is called the *Schrödinger equation*.

The Schrödinger Equation

The behavior of a particle of mass m subject to a potential $V(x, t)$ is described by the following partial differential equation:

$$i\hbar \frac{\partial \psi(x, t)}{\partial t} = -\frac{\hbar^2}{2m} \frac{\partial^2 \psi(x, t)}{\partial x^2} + V(x, t)\psi(x, t)$$

where $\psi(x, t)$ is called the *wavefunction*. The wavefunction contains information about where the particle is located, its square being a probability density. A wavefunction must be "well behaved," in other words it should be defined and continuous everywhere. In addition it must be square-integrable, meaning:

$$\int_{-\infty}^{\infty} |\psi(x, t)|^2 \, dx < \infty$$

Summary: Properties of a Valid Wavefunction

A wavefunction ψ must:

- Be single valued
- Be continuous
- Be differentiable
- Be square integrable

EXAMPLE 2.1

Let two functions ψ and Φ be defined for $0 \leq x < \infty$. Explain why $\psi(x) = x$ cannot be a wavefunction but $\Phi(x) = e^{-x^2}$ could be a valid wavefunction.

SOLUTION

Both functions are continuous and defined on the interval of interest. They are both single valued and differentiable. However, consider the integral of x:

$$\int_{0}^{\infty} |\psi(x)|^2 \, dx = \int_{0}^{\infty} x^2 dx = \left. \frac{x^3}{3} \right|_0^{\infty} = \infty$$

Given that, $\psi(x) = x$ is not square integrable over this range it cannot be a valid wavefunction. On the other hand:

$$\int_{0}^{\infty} |\Phi(x)|^2 \, dx = \int_{0}^{\infty} e^{-2x^2} dx = \sqrt{\frac{\pi}{8}}$$

Therefore, $\Phi(x) = e^{-x^2}$ is square integrable—so we can say it's well behaved over the interval of interest. This makes it a valid candidate wavefunction. A plot Figure 2-1 shows that this function is smooth and decays rather quickly:

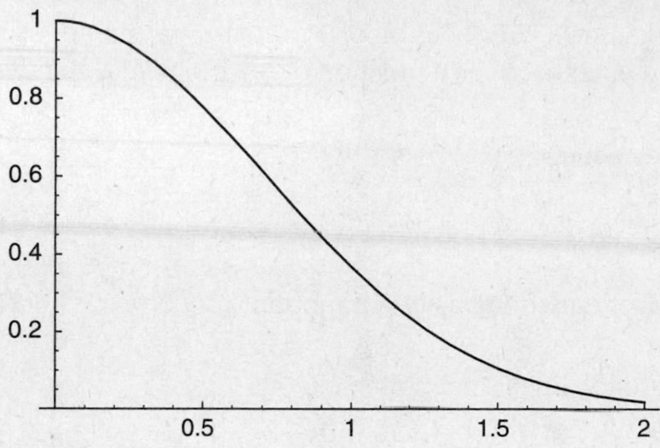

Fig. 2-1

Definition: The Probability Interpretation of the Wavefunction

At time t, the probability of finding the particle within the interval x and $x + dx$ is given by the *square* of the wavefunction. Calling this probability $dP(x, t)$, we write:

$$dP(x, t) = |\psi(x, t)|^2 \, dx$$

The square is given by $|\psi(x, t)|^2$ as opposed to $\psi(x, t)^2$ because in general, the wavefunction can be *complex*. In these dimensions, the Schrödinger equation is readily generalized to:

$$i\hbar \frac{\partial \psi(r, t)}{\partial t} = -\frac{\hbar^2}{2m} \nabla^2 \psi(r, t) + V(r, t)\psi(r, t)$$

where ∇^2 is the Laplacian operator which can be written as

$$\frac{\partial^2}{\partial x^2} + \frac{\partial^2}{\partial y^2} + \frac{\partial^2}{\partial z^2}$$

in Cartesian coordinates. The probability of finding the particle then becomes the probability of locating it within a volume $d^3r = dxdydz$:

$$dP(r,t) = |\psi(r,t)|^2 \, d^3r$$

For the time being we will focus on one-dimensional situations. In most cases of interest, the potential V is a function of position only, and so we can write:

$$i\hbar \frac{\partial \psi(x,t)}{\partial t} = -\frac{\hbar^2}{2m} \frac{\partial^2 \psi(x,t)}{\partial x^2} + V(x)\psi(x,t)$$

The Schrödinger equation has two important properties. These are:

1. The equation is linear and homogeneous
2. The equation is first order with respect to time—meaning that the state of a system at some initial time t_o determines its behavior for all future times.

An important consequence of the first property is that the *superposition* principle holds. This means that if $\psi_1(x,t), \psi_2(x,t), \ldots, \psi_n(x,t)$ are solutions of the Schrödinger equation, then the linear combination of these functions:

$$\psi = C_1\psi_1(x,t) + C_2\psi_2(x,t) + \ldots + C_n\psi_n(x,t) = \sum_{i=1}^{n} C_i\psi_i(x,t)$$

is also a solution. We will see that this superposition property is of critical importance in quantum theory.

Solutions to the Schrödinger equation when a particle is subject to a time-independent potential $V(x)$ can be found using separation of variables. This is done by writing the wavefunction in the form:

$$\psi(x,t) = \Phi(x) f(t)$$

Since the Schrödinger equation is first order in time derivative, this leads to a simple solution to the time dependent part of the wavefunction.

$$f(t) = e^{-iEt/\hbar}$$

where E is the energy.

EXAMPLE 2.2

Consider a particle trapped in a well with potential given by:

$$V(x) = \begin{cases} 0 & 0 \le x \le a \\ \infty & \text{otherwise} \end{cases}$$

Show that $\psi(x,t) = A\sin(kx)\exp(iEt/\hbar)$ solves the Schrödinger equation provided that

$$E = \frac{\hbar^2 k^2}{2m}$$

SOLUTION

The potential is infinite at $x = 0$ and a, therefore the particle can never be found outside of this range. So we only need to consider the Schrödinger equation inside the well, where $V = 0$. With this condition the Schrödinger equation takes the form:

$$i\hbar\frac{\partial\psi(x,t)}{\partial t} = -\frac{\hbar^2}{2m}\frac{\partial^2\psi(x,t)}{\partial x^2}$$

Setting $\Psi(x,t) = A\sin(kx)\exp(iEt/\hbar)$, we consider the left side of the Schrödinger equation first:

$$\begin{aligned} i\hbar\frac{\partial\psi(x,t)}{\partial t} &= i\hbar\frac{\partial}{\partial t}(A\sin(kx)\exp(-iEt/\hbar)) \\ &= i\hbar(-iE/\hbar)A\sin(kx)\exp(-iEt/\hbar) \\ &= E(A\sin(kx)\exp(-iEt/\hbar)) = E\psi \end{aligned}$$

Now consider the derivative with respect to x:

$$\frac{\partial}{\partial x}\psi = \frac{\partial}{\partial x}[A\sin(kx)\exp(-iEt/\hbar)] = kA\cos(kx)\exp(-iEt/\hbar)$$

$$\begin{aligned} \rightarrow -\frac{\hbar^2}{2m}\frac{\hbar^2\psi(x,t)}{\partial x^2} &= -\frac{\hbar^2}{2m}\frac{\partial}{\partial x}[kA\cos(kx)\exp(-iEt/\hbar)] \\ &= -\frac{\hbar^2}{2m}[-k^2A\sin(kx)\exp(-iEt/\hbar)] = \frac{\hbar^2}{2m}k^2\psi \end{aligned}$$

Using

$$i\hbar\frac{\partial\psi(x,t)}{\partial t} = -\frac{\hbar^2}{2m}\frac{\partial^2\psi(x,t)}{\partial x^2}$$

we equate both terms, finding that:

$$E\psi = \frac{\hbar^2}{2m}k^2\psi$$

And so we conclude that the Schrödinger equation is satisfied if

$$E = \frac{\hbar^2 k^2}{2m}$$

Solving the Schrödinger Equation

We have seen that when the potential is time-independent the solution to the Schrödinger equation is given by:

$$\psi(x, t) = \Phi(x) \exp(-iEt/\hbar)$$

The spatial part of the wavefunction, $\Phi(x)$, satisfies the *time-independent Schrödinger equation.*

Definition: The Time-Independent Schrödinger Equation

Let $\Psi(x, t) = \Phi(x) \exp(-iEt/\hbar)$ be a solution to the Schrödinger equation with time-independent potential $V = V(x)$. The spatial part of the wavefunction $\Phi(x)$ satisfies:

$$-\frac{\hbar^2}{2m}\frac{\partial^2 \Phi(x)}{\partial x^2} + V(x)\Phi(x) = E\Phi(x)$$

where E is the energy of the particle. This equation is known as the time-independent Schrödinger equation.

Solutions that can be written as $\Psi(x, t) = \Phi(x) \exp(-iE\hbar t)$ are called *stationary*.

Definition: Stationary State

A solution $\Psi(x, t) = \Phi(x) \exp(-iEt/\hbar)$ to the Schrödinger equation is called *stationary* because the probability density does not depend on time:

$$\begin{aligned}
|\psi(x, t)|^2 &= \psi^*(x, t)\psi(x, t) \\
&= (\Phi(x) \exp(-iEt/\hbar))^* \, \Phi(x) \exp(-iEt/\hbar) \\
&= \Phi^*(x) \exp(iEt/\hbar)\Phi(x) \exp(-iEt/\hbar) \\
&= \Phi^*(x)\Phi(x)
\end{aligned}$$

We now consider an example where we are given the wavefunction. If we know the form of the wavefunction, it is easy to find the potential so that the Schrödinger equation is satisfied.

EXAMPLE 2.3 e.g.

Suppose $\Psi(x,t) = A(x - x^3)e^{-iEt/\hbar}$. Find $V(x)$ such that the Schrödinger equation is satisfied.

SOLUTION

The wavefunction is written as a product:

$$\Psi(x,t) = \Phi(x)\exp(-iEt/\hbar)$$

Therefore it is not necessary to work with the full Schrödinger equation. Recalling the time-independent Schrödinger equation:

$$-\frac{\hbar^2}{2m}\frac{\partial^2\Phi(x)}{\partial x^2} + V(x)\Phi(x) = E\Phi(x)$$

We set $\Phi(x) = A(x - x^3)$ and solve to find V. The right-hand side is simply:

$$E\Phi(x) = EA(x - x^3)$$

To find the form of the left-side of the equation, we begin by computing the first derivative:

$$\frac{\partial\Phi(x)}{\partial x} = \frac{\partial}{\partial x}\left[A(x - x^3)\right] = A(1 - 3x^2)$$

For the second derivative, we obtain:

$$\frac{\partial^2\Phi(x)}{\partial x^2} = \frac{\partial}{\partial x}\left(A(1 - 3x^2)\right) = -6Ax, \Rightarrow -\frac{\hbar^2}{2m}\frac{\partial^2\Phi(x)}{\partial x^2} = \frac{\hbar^2}{2m}6Ax$$

Putting this in the left-side of the time-independent Schrödinger equation and equating this to $EA(x - x^3)$ gives:

$$\frac{\hbar^2}{2m}6Ax + V(x)A(x - x^3) = EA(x - x^3)$$

Now subtract $(\hbar^2/2m)6Ax$ from both sides:

$$V(x)A(x - x^3) = EA(x - x^3) - \frac{\hbar^2}{2m}6Ax$$

Dividing both sides by $A(x - x^3)$ gives us the potential:

$$V(x) = E - \frac{\hbar^2}{2m}\frac{6x}{(x - x^3)}$$

Most of the time, we are given a specific potential and asked to find the form of the wavefunction. In many cases this involves solving a *boundary value problem*. A wavefunction must be continuous and defined everywhere, so the wavefunction must match up at boundaries. The process of applying boundary conditions to find a solution to a differential equation is no doubt familiar. We show how this is done in quantum mechanics with a problem we introduced earlier, the infinite square-well.

e.g. **EXAMPLE 2.4**

A particle of mass m is trapped in a one dimensional box with a potential described by:

$$V(x) = \begin{cases} 0 & 0 \le x \le a \\ \infty & \text{otherwise} \end{cases}$$

Solve the Schrödinger equation for this potential.

✔ **SOLUTION**

Inside the box, the potential V is zero and so the Schrödinger equation takes the form:

$$i\hbar \frac{\partial \psi(x,t)}{\partial t} = -\frac{\hbar^2}{2m}\frac{\partial^2 \psi(x,t)}{\partial x^2}$$

We already showed that the solution to this problem is separable. Therefore, we take:

$$\psi(x,t) = \psi(x)e^{-iEt/\hbar}$$

The time-independent equation with zero potential is:

$$E\psi(x) = -\frac{\hbar^2}{2m}\frac{d^2\psi}{dx^2}$$

Dividing through by $-\hbar^2/2m$ and moving all terms to one side, we have:

$$\frac{d^2\psi}{dx^2} + \frac{2mE}{\hbar^2}\psi = 0$$

If we set $k^2 = 2mE/\hbar^2$, we obtain:

$$\frac{d^2\psi}{dx^2} + k^2\psi = 0$$

The solution of this equation is:

$$\psi(x) = A\sin(kx) + B\cos(kx)$$

To determine the constants A and B, we apply the boundary conditions. Since the potential is infinite at $x = 0$ and $x = a$, the wavefunction is zero outside of the well. A wavefunction must be defined *and* continuous everywhere. This tells us that any solution we find inside the well must match up to what is outside at the boundaries.

At $x = 0$, the solution becomes:

$$\Psi(0) = A\sin(0) + B\cos(0) = B$$

The requirement, that $\Psi(0) = 0$ tells us that:

$$B = 0, \Rightarrow \Psi(x) = A\sin(kx)$$

The wavefunction must also vanish at the other end of the well. This means we must also enforce the condition $\Psi(a) = 0$:

$$\Psi(a) = A\sin(ka) = 0$$

This can only be true if $A = 0$ or if $\sin(ka) = 0$. The first possibility is not of any interest; a wavefunction that is zero everywhere is the same as saying there is no particle present. So we pursue the latter possibility. Now $\sin(ka) = 0$ if:

$$ka = n\pi$$

where $n = 1, 2, \ldots$. So we set $k = (n\pi)/a$ and the wavefunction can be written as:

$$\psi(x) = A\sin\left(\frac{n\pi}{a}x\right)$$

The constant A is determined by *normalizing* the wavefunction, a technique we will discuss in the next section. For now we will just carry it along.

Earlier, we defined the constant k in terms of the particle's energy E. Let's use this definition to write down the form of the energy in terms of $(n\pi)/a$:

$$k^2 = \frac{2mE}{\hbar^2} \Rightarrow E = \frac{k^2\hbar^2}{2m} = \frac{n^2\pi^2\hbar^2}{2ma^2}$$

If $n = 0$, the wavefunction would vanish indicating there is no particle in the well. Therefore the lowest possible energy a particle can have is found by setting $n = 1$:

$$E_1 = \frac{\pi^2\hbar^2}{2ma^2}$$

The lowest energy is called the *ground state* energy.

Definition: Ground State

The ground state is the state with the lowest energy a system can assume.

For the infinite square well, all other energies are given in terms of integer multiples of the ground state energy. Here are sample plots (Figures 2-2, 2-3, 2-4) of the first three wavefunctions with $a = 3$:

$$\psi_1(x) = A \sin\left(\frac{\pi}{a}x\right), \text{ with energy } E_1 = \frac{\pi^2\hbar^2}{2ma^2}:$$

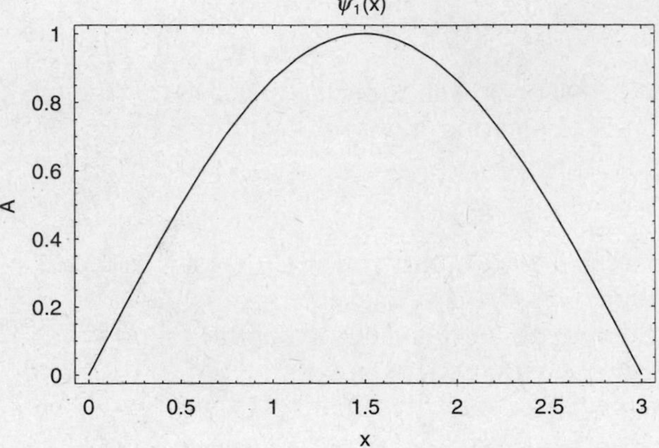

Fig. 2-2

$$\psi_2(x) = A \sin\left(\frac{2\pi}{a}x\right), \text{ with energy } E_2 = \frac{4\pi^2\hbar^2}{2ma_2} = 4E_1$$

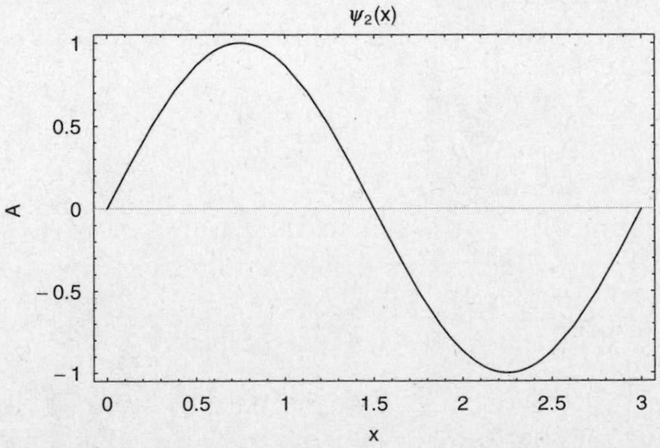

Fig. 2-3

$$\psi_3(x) = A \sin\left(\frac{3x}{a}x\right), \text{ with energy } E_3 = \frac{9\pi^2\hbar^2}{2ma^2} = 9E_1$$

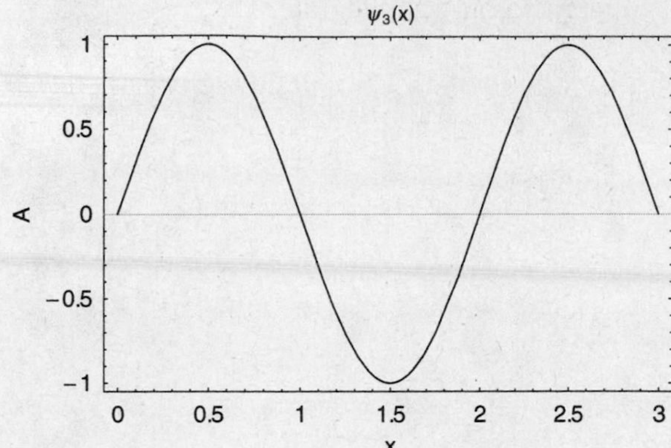

Fig. 2-4

As you might have guessed, in general $E_n = n^2 E_1$.

The square of the wavefunction gives the probability density. We can say that the intensity of the wave at a given point is proportional to the likelihood of finding the particle there. Here is a plot in Figure 2-5 of $|\Psi_3(x)|^2$:

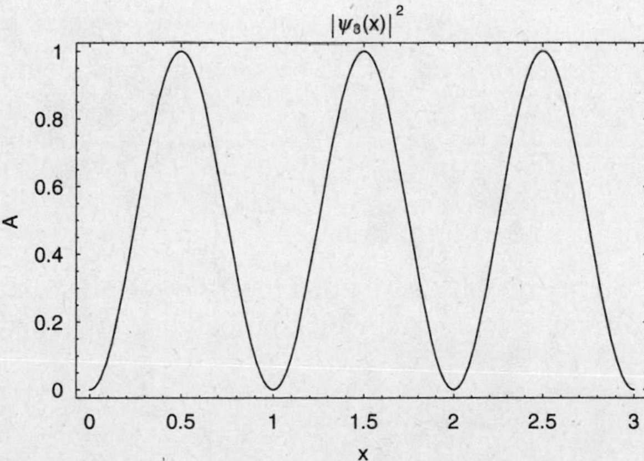

Fig. 2-5

From the plot we can surmise that the particle is most likely to be found at $x = 0.5$, 1.5, and 2.5, but is not ever going to be found at $x = 1$, $x = 2$, or at the boundaries.

The Probability Interpretation and Normalization

The probability interpretation tells us that $|\Psi(x,t)|^2 dx$ gives the probability for finding the particle between x and $x + dx$. To find the probability that a particle is located within a given region, we integrate.

Definition: Finding the Probability a Particle is Located in the Region

$$a \leq x \leq b$$

The probability p that a particle is located within $a \leq x \leq b$ is:

$$p = \int_a^b |\psi(x,t)|^2 \, dx$$

It is common to denote a probability density $|\Psi(x,t)|^2$ as $\rho(x,t)$.

e.g. EXAMPLE 2.5

Suppose that a certain probability distribution is given by $\rho(x) = \frac{9}{4}\frac{1}{x^3}$ for $1 \leq x \leq 3$. Find the probability that $\frac{5}{2} \leq x \leq 3$.

✔ SOLUTION

$$p = \frac{9}{4}\int_{\frac{5}{2}}^{3} \frac{1}{x^3} dx = \frac{9}{4}\frac{x^{-2}}{(-2)}\bigg|_{\frac{5}{2}}^{3} = -\frac{9}{8}\left(\frac{1}{9} - \frac{4}{25}\right) = \frac{11}{200} = 0.055$$

If this were a wavefunction, there would be about a 5.5 percent chance of finding the particle between $5/2 \leq x \leq 3$.

Note: Probabilities Must Sum to 1

The total probability for any distribution must sum to unity. If the probability distribution is discrete with n individual probabilities p_i, this means that:

$$\sum_i p_i = 1$$

For a continuous probability distribution $\rho(x)$, the fact that probabilities must sum to unity means that:

$$\int_{-\infty}^{\infty} \rho(x) dx = 1$$

In quantum mechanics, this condition means that the particle is located *somewhere* in space with certainty

$$\int_{-\infty}^{\infty} |\psi(x,t)|^2\, dx = 1$$

As we saw in the solution to the square well, a wavefunction that solves the Schrödinger equation is determined up to an unknown constant. In that case, the constant is called the *normalization constant* and we find the value of that constant by *normalizing* the wavefunction.

Definition: Normalizing the Wavefunction

When a wavefunction that solves the Schrödinger equation is multiplied by an undetermined constant A, we normalize the wavefunction by solving:

$$\frac{1}{A^2} = \int_{-\infty}^{\infty} |\psi(x,t)|^2\, dx$$

The normalized wavefunction is then $A\Psi(x,t)$.

EXAMPLE 2.6

The wave function for a particle confined to $0 \le x \le a$ in the ground state was found to be:

$$\Psi(x) = A \sin(\pi x/a)$$

where A is the normalization constant. Find A and determine the probability that the particle is found in the interval $\frac{a}{2} \le x \le \frac{3a}{4}$.

SOLUTION

Normalization means that:

$$\int_{-\infty}^{\infty} |\psi|^2\, dx = 1$$

The wavefunction is zero outside of the interval, $0 \le x \le a$, therefore we only need to consider

$$\int_0^a |\psi|^2\, dx = \int_0^a A^2 \sin^2\left(\frac{\pi x}{a}\right) dx = A^2 \int_0^a \sin^2\left(\frac{\pi x}{a}\right) dx$$

We use the trigometric identity $\sin^2 u = (1 - \cos 2u)/2$ to rewrite the integrand:

$$A^2 \int_0^a \sin^2\left(\frac{\pi x}{a}\right) dx = A^2 \int_0^a \frac{1 - \cos\left(\frac{2\pi x}{a}\right)}{2} dx$$

$$= \frac{A^2}{2} \int_0^a dx - \frac{A^2}{2} \int_0^a \cos\left(\frac{2\pi x}{a}\right) dx$$

The first term can be integrated immediately:

$$\frac{A^2}{2}\int_0^a dx = \frac{A^2}{2}x\Big|_0^a = A^2\frac{a}{2}$$

For the second term, let $u = (2\pi x)/a, \Rightarrow du = (2\pi)/a\ dx$ and:

$$\int_0^a \cos\left(\frac{2\pi x}{a}\right) dx = \frac{a}{2\pi}\int_0^{2\pi} \cos(u)du = \frac{a}{2\pi}\sin(u)\Big|_0^{2\pi}$$

$$\frac{a}{2\pi}\left[\sin(2\pi) - \sin(0)\right] = 0$$

And so, only the first term contributes and we have:

$$\int_0^a |\psi|^2\, dx = \frac{A^2}{2}a = 1$$

Solving for the normalization constant A, we find:

$$A = \sqrt{\frac{2}{a}}$$

and the normalized wavefunction is:

$$\psi(x) = \sqrt{\frac{2}{a}}\sin(\pi x/a)$$

The probability that the particle is found in the interval $a/2 \le x \le 3a/4$ is given by:

$$P\left(\frac{a}{2} \le x \le \frac{3a}{4}\right) = \int_{\frac{a}{2}}^{\frac{3a}{4}} |\psi(x)|^2\, dx = \int_{\frac{a}{2}}^{\frac{3a}{4}} \left(\frac{2}{a}\right)\sin^2\left(\frac{\pi x}{a}\right) dx$$

$$= \left(\frac{2}{a}\right)\int_{\frac{a}{2}}^{\frac{3a}{4}} \frac{1-\cos(2\pi x)}{2} dx$$

$$= \frac{1}{a}\int_{\frac{a}{2}}^{\frac{3a}{4}} dx - \frac{1}{a}\int_{\frac{a}{2}}^{\frac{3a}{4}} \cos\left(\frac{2\pi x}{a}\right) dx$$

$$= \frac{1}{a}x\Big|_{\frac{a}{2}}^{\frac{3a}{4}} - \frac{1}{2\pi}\sin\left(\frac{2\pi x}{a}\right)\Big|_{\frac{a}{2}}^{\frac{3a}{4}}$$

$$= \frac{1}{a}\left[\frac{3a}{4} - \frac{a}{2}\right] - \frac{1}{2\pi}\left[\sin\left(\frac{6\pi}{4}\right) - \sin(\pi)\right]$$

$$= \frac{1}{a}\left[\frac{a}{4}\right] - \frac{1}{2\pi}\sin\left(\frac{3\pi}{2}\right)$$

$$= \frac{1}{4} + \frac{1}{2\pi} = \frac{\pi + 2}{4\pi} = 0.41$$

EXAMPLE 2.7

Find an A and B so that:

$$\Phi(x) = \begin{cases} A & \text{for } 0 \le x \le a \\ Bx & \text{for } a \le x \le b. \end{cases}$$

is normalized.

SOLUTION

$$\int_{-\infty}^{\infty} |\Phi(x)|^2\, dx = \int_0^a A^2 dx + \int_a^b B^2 x^2 dx$$

$$= A^2 x\big|_0^a + \frac{B^2(x^3)}{3}\big|_a^b = A^2 a + \frac{B^2(b^3 - a^3)}{3}$$

Using $\int_{-\infty}^{\infty} |\Phi(x)|^2\, dx = 1$, we obtain:

$$A^2 a + \frac{B^2(b^3 - a^3)}{3} = 1, \Rightarrow A^2 = \left(\frac{1}{a}\right)\left(1 - \frac{B^2(b^3 - a^3)}{3}\right)$$

As long as $\int_{-\infty}^{\infty} |\Phi(x)|^2\, dx = 1$ is satisfied, we are free to arbitrarily choose one of the constants as long as it's not zero. So we set $B = 1$:

$$A^2 = \left(\frac{1}{a}\right)\left(1 - \frac{(b^3 - a^3)}{3}\right), \Rightarrow A = \sqrt{\frac{1}{a}\left(1 - \frac{(b^3 - a^3)}{3}\right)}$$

EXAMPLE 2.8

Normalize the wavefunction

$$\psi(x) = \frac{C}{x^2 + a^2}$$

SOLUTION

We start by finding the square of the wavefunction:

$$|\psi(x)|^2 = \frac{C^2}{(x^2 + a^2)^2}$$

To compute $\int |\psi(x)|^2 \, dx$, we will need two integrals which can be found in integration tables. These are:

$$\int \frac{du}{u^2 + a^2} = \frac{1}{a} \tan^{-1} \frac{u}{a} \text{ and } \int \frac{du}{(u^2 + a^2)^2} = \frac{1}{2a^2} \left(\frac{u}{u^2 + a^2} \right) + \int \frac{du}{u^2 + a^2}$$

We begin by using the second of these:

$$\Rightarrow \int_{-\infty}^{\infty} |\psi(x)|^2 \, dx = C^2 \int_{-\infty}^{\infty} \frac{dx}{(x^2 + a^2)^2}$$

$$= \frac{C^2}{2a^2} \left(\left. \frac{x}{x^2 + a^2} \right|_{-\infty}^{\infty} + \int_{-\infty}^{\infty} \frac{dx}{x^2 + a^2} \right)$$

Where the first term is to be evaluated at $\pm\infty$. Consider the limit as $x \to \infty$:

$$\lim_{x \to \infty} \frac{x}{x^2 + a^2} = \lim_{x \to \infty} \frac{1}{2x} = 0$$

where L'Hopitals rule was used. Similarly, for the lower limit we find:

$$\lim_{x \to -\infty} \frac{x}{x^2 + a^2} = \lim_{x \to -\infty} \frac{1}{2x} = 0$$

So we can discard the first term altogether. Now we can use $\int (du)/(u^2 + a^2) = (1/a) \tan^{-1}(u/a)$ for the remaining piece.

$$\int_{-\infty}^{\infty} |\psi(x)|^2 \, dx = \frac{C^2}{2a^2} \int_{-\infty}^{\infty} \frac{dx}{x^2 + a^2} = \frac{C^2}{2a^2} \left(\frac{1}{a} \right) \tan^{-1} \left. \frac{x}{a} \right|_{-\infty}^{\infty}$$

$$= \lim_{u \to \infty} \frac{C^2}{2a^3} \left[\tan^{-1}(u) - \tan^{-1}(-u) \right]$$

$$= \frac{C^2}{2a^3} \left[\frac{\pi}{2} - \left(-\frac{\pi}{2} \right) \right] = \frac{C^2}{2a^3} \pi$$

Again we recall that the normalization condition is:

$$\int_{-\infty}^{\infty} |\psi(x)|^2 \, dx = 1$$

therefore, setting our result equal to one we find that:

$$\frac{C^2}{2a^3}\pi = 1 \Rightarrow C = \sqrt{\frac{2a^3}{\pi}}$$

And so, the normalized wavefunction is:

$$\psi(x) = \sqrt{\frac{2a^3}{\pi}} \frac{1}{x^2 + a^2}$$

In the next examples, we consider the normalization of some Gaussian functions. Three frequently seen integrals we will use are:

$$\int_{-\infty}^{\infty} e^{-z^2} dz = \sqrt{\pi}$$

$$\int_{-\infty}^{\infty} z^{2n} e^{-z^2} dz = \sqrt{\pi} \frac{1.3.5 \ldots (2n-1)}{2^n}, \qquad n = 1, 2 \ldots$$

$$\int_{-\infty}^{\infty} z e^{-z^2} dz = 0$$

We also introduce the *error function*:

$$\operatorname{erf}(z) = \frac{2}{\sqrt{\pi}} \int_0^z e^{-u^2} du$$

EXAMPLE 2.9

e.g.

$\psi(x) = Ae^{-\lambda(x-x_0)^2}$. Find A such that $\Psi(x)$ is normalized. The constants λ and x_0 are real.

SOLUTION

$$|\psi(x)|^2 = A^2 e^{-2\lambda(x-x_0)^2} \Rightarrow \int_{-\infty}^{\infty} |\psi(x)|^2\,dx$$
$$= \int_{-\infty}^{\infty} A^2 e^{-2\lambda(x-x_0)^2} dx$$
$$= A^2 \int_{-\infty}^{\infty} e^{-2\lambda(x-x_0)^2} dx$$

We can perform this integral by using the substitution $z^2 = 2\lambda(x - x_0)^2$. Then:

$$z = \sqrt{2\lambda}(x - x_0),\, dz = \sqrt{2\lambda} dx \Rightarrow A^2\psi(x) = Ae^{-\lambda(x-x_0)^2}$$

Using $\int_{-\infty}^{\infty} |\psi(x)|^2\,dx = 1$, we find that $A = (2\lambda/\pi)^{\frac{1}{4}}$.

EXAMPLE 2.10

$$\psi(x,t) = (A_1 e^{-x^2/a} + A_2 x e^{-x^2/b})e^{-ict}$$

for $-\infty < x < \infty$. (a) Write the normalization condition for A_1 and A_2. (b) Normalize $\Psi(x)$ for $A_1 = A_2$ and $a = 32$, $b = 8$. (c) Find the probability that the particle is found in the region $0 < x < 32$.

SOLUTION

(a)

$$\psi^*(x,t) = (A_1 e^{-x^2/a} + A_2 x e^{-x^2/b})e^{ict}, \Rightarrow$$
$$|\psi(x,t)|^2 = \psi^*(x,t)\psi(x,t)$$
$$= [(A_1 e^{-x^2/a} + A_2 x e^{-x^2/b})\, e^{ict}]\left[\left(A_1 e^{-x^2/a} + A_2 x e^{-x^2/b}\right) e^{-ict}\right]$$
$$= A_1^2 e^{-2x^2/a} + 2A_1 A_2 x e^{-x^2\left(\frac{1}{a}+\frac{1}{b}\right)} + A_2^2 x^2 e^{-2x^2/b}$$

The normalization condition is $\int_{-\infty}^{\infty} |\psi(x,t)|^2\,dx = 1$. Now since $\int_{-\infty}^{\infty} ze^{-z^2} dz = 0$, the middle term vanishes, and we have:

$$\int_{-\infty}^{\infty} |\psi(x,t)|^2\,dx = \int_{-\infty}^{\infty} A_1^2\, e^{-2x^2/a}\,dx + \int_{-\infty}^{\infty} A_2^2 x^2 e^{-2x^2/b}\,dx$$

Looking at the first term, we use the substitution technique from calculus to get it in the form

$$\int_{-\infty}^{\infty} e^{-z^2} dz.$$

Let $z^2 = (2x^2)/a$. Then $z = \sqrt{2/a}x$, $dz = \sqrt{2/a}dx$, or $dx = \sqrt{a/2}dz$ Therefore we can write the integral as:

$$\int_{-\infty}^{\infty} A_1^2 e^{-2x^2/a} dx = A_1^2 \int_{-\infty}^{\infty} e^{-2x^2/a} dx$$

$$= A_1^2 \int_{-\infty}^{\infty} e^{-z^2} \sqrt{\frac{a}{2}} dz$$

$$= A_1^2 \sqrt{\frac{a}{2}} \int_{-\infty}^{\infty} e^{-z^2} dz = A_1^2 \sqrt{\frac{\pi a}{2}}$$

Turning to the second term, we use again use a substitution. This time we want the form:

$$\int_{-\infty}^{\infty} z^{2n} e^{-z^2} dz$$

As before let $z^2 = (2x^2)/b$, then:

$$z = \sqrt{\frac{2}{b}}x, \ dz = \sqrt{\frac{2}{b}}dx, \text{ or } dx = \sqrt{\frac{b}{2}}dz$$

Also

$$x^2 = \frac{b}{2} z^2$$

So we can rewrite the second integral in the following way:

$$\int_{-\infty}^{\infty} A_2^2 x^2 e^{-2x^2/b} dx = A_2^2 \int_{-\infty}^{\infty} \frac{b}{2} z^2 e^{-z^2} \sqrt{\frac{b}{2}}\, dz$$

$$= A_2^2 \left(\frac{b}{2}\right)^{3/2} \int_{-\infty}^{\infty} z^2 e^{-z^2}\, dz$$

Using $\int_{-\infty}^{\infty} z^{2n} e^{-z^2} dz = \sqrt{\pi}(1 \cdot 3 \cdot 5 \ldots (2n-1))/(2^n)$ for $n = 1$, we obtain:

$$\int_{-\infty}^{\infty} A_2^2 x^2 e^{-2x^2/b} dx = A_2^2 \left(\frac{b}{2}\right)^{3/2} \frac{\sqrt{\pi}}{2}$$

Putting these results together we obtain the normalization condition:

$$1 = \int_{-\infty}^{\infty} |\psi(x,t)|^2\,dx = \int_{-\infty}^{\infty} A_1^2 e^{-2x^2/a} dx + \int_{-\infty}^{\infty} A_2^2 x^2 e^{-2x^2/b} dx$$

$$= A_1^2 \sqrt{\frac{\pi a}{2}} + A_2^2 \left(\frac{b}{2}\right)^{3/2} \frac{\sqrt{\pi}}{2}$$

(b) Now we let $A_1 = A_2$ and $a = 32$, $b = 8$. The normalization condition becomes:

$$1 = A_1^2 \left(\sqrt{\frac{\pi a}{2}} + \left(\frac{b}{2}\right)^{3/2} \frac{\sqrt{\pi}}{2} \right) \Rightarrow A_1 = A_2 = \frac{1}{\sqrt{8\sqrt{\pi}}}$$

$$= A_1^2 \left(\sqrt{\frac{\pi 32}{2}} + \left(\frac{8}{2}\right)^{3/2} \frac{\sqrt{\pi}}{2} \right)$$

$$= \sqrt{\pi} A_1^2 \left(\sqrt{16} + (4)^{3/2} \frac{1}{2} \right)$$

$$= \sqrt{\pi} A_1^2 \left(4 + \frac{8}{2} \right) = 8\sqrt{\pi} A_1^2$$

And so, the normalized wavefunction in this case is:

$$\psi(x,t) = \frac{1}{\sqrt{8\sqrt{\pi}}} \left(e^{\frac{-x^2}{32}} + x e^{\frac{-x^2}{8}} \right) e^{-ict}$$

(c) The probability that the particle is found between $0 < x < 32$ is:

$$P(0 < x < 32) = \int_0^{32} |\psi(x,t)|^2\,dx = \frac{1}{8\sqrt{\pi}} \int_0^{32} \left(e^{-x^2/32} + x e^{-x^2/8} \right)^2 dx$$

$$= \frac{1}{8\sqrt{\pi}} \int_0^3 2e^{-x^2/16} dx + \frac{2}{8\sqrt{\pi}} \int_0^3 2xeA_1 e^{-3x^2/16} dx$$

$$+ \frac{1}{8\sqrt{\pi}} \int_0^3 2x^2 e^{-x^2/4} dx$$

To perform these integrals, we consider the error function. This is given by:

$$\text{erf}(z) = \frac{2}{\sqrt{\pi}} \int_0^z e^{-u^2} du$$

We use substitution techniques to put the integrals into this form. It turns out that erf(32) $\approx$ 1.

$$\int_0^{32} e^{-x^2/16} dx = 4\int_0^{32} e^{-u^2} du = 4\frac{\sqrt{\pi}}{2} erf(32) = 2\sqrt{\pi}$$

For the second integral, we use integration by parts. Let $u = x$, then $du = dx$. If $dV = e^{-3x^2/16}dx$, then

$$V = \int e^{-3x^2/16} dx$$

Now let a dummy variable $s = (\sqrt{3}/4)x$. This allows us to put this in the form of erf(32):

$$V = \int e^{-3x^2/16} dx = 2\sqrt{\frac{\pi}{3}}\text{erf}\left(\frac{\sqrt{3}x}{4}\right)$$

And so, with $u = x$, using integration by parts, $\int udV = uV - \int Vdu$ we obtain:

$$\int_0^{32} xe^{-3x^2/16} dx = (x)2\sqrt{\frac{\pi}{3}} erf\left(\frac{\sqrt{3}x}{4}\right) - 2\sqrt{\frac{\pi}{3}}\int_0^{32} \text{erf}\left(\frac{\sqrt{3}x}{4}\right) dx$$

These terms can be evaluated numerically. The integration is between the limits $x = 0$ and $x = 32$. Now $erf(0) = 0$, and so the first term is:

$$(32)2\sqrt{\frac{\pi}{3}}\text{erf}\left(\frac{\sqrt{3}32}{4}\right) = 64\sqrt{\frac{\pi}{3}}\text{erf}\left(8\sqrt{3}\right) \approx 65.5$$

The second term is:

$$2\sqrt{\frac{\pi}{3}}\int_0^{32} \text{erf}\left(\frac{\sqrt{3}x}{4}\right) dx$$

$$= 2\sqrt{\frac{\pi}{3}}\left(\frac{4}{\sqrt{3\pi}} + \frac{4}{192\sqrt{3\pi}} + 32\text{erf}\left[8\sqrt{3}\right]\right) \approx 62.8$$

$$\Rightarrow \int_0^{32} xe^{-3x^2/16} dx = 65.5 - 62.8 = 2.7$$

Now the final term, which we evaluate numerically, is:

$$\int_0^{32} x^2 e^{-x^2/4} dx = -\frac{64}{256} + 2\sqrt{\pi}\,\text{erf}\,[16] \approx 3.54$$

Pulling all of these results together, the probability that the particle is found between $0 < x < 32$ is:

$$\begin{aligned} P\,(0 < x < 32) &= \frac{1}{8\sqrt{\pi}} \int_0^{32} e^{-x^2/16} dx + \frac{2}{8\sqrt{\pi}} \int_0^{32} x e^{-3x^2/16} dx \\ &\quad + \frac{1}{8\sqrt{\pi}} \int_0^{32} x^2 e^{-x^2/4} dx \\ &= \frac{1}{8\sqrt{\pi}} 2\sqrt{\pi} + \frac{2}{8\sqrt{\pi}}(2.7) + \frac{1}{8\sqrt{\pi}}(3.54) = 0.88 \end{aligned}$$

Conclusion: for $\psi(x, t) = \left(1/\left(\sqrt{8\sqrt{\pi}}\right)\right)\left(e^{-x^2/32} + x e^{-x^2/8}\right) e^{-ict}$ defined for $-\infty < x < \infty$, there is an 88% chance that we will find the particle between $0 < x < 32$.

e.g. **EXAMPLE 2.11**

Let $\psi(x) = Ae^{-|x|/2a} e^{i(x-x_0)}$. Find the constant A by normalizing the wavefunction.

✔ **SOLUTION**

First we compute:

$$\begin{aligned} \psi^*(x) &= A^* e^{-\frac{|x|}{2a}} e^{-i(x-x_0)}, \Rightarrow \\ \psi^*(x)\psi(x) &= |A|^2\, e^{-\frac{|x|}{2a}} e^{-i(x-x_0)} e^{-\frac{|x|}{2a}} e^{i(x-x_0)} \\ &= |A|^2\, e^{-\frac{|x|}{a}} \left[e^{-i(x-x_0)} e^{i(x-x_0)}\right] \\ &= |A|^2\, e^{-\frac{|x|}{a}} \end{aligned}$$

To integrate $e^{-x/a}$ think about the fact that it's defined in terms of the absolute value function. For $|x| < 0$, this term is $e^{x/a}$, while for $x > 0$ it's $e^{-x/a}$.

So we split the integral into two parts:

$$\int_{-\infty}^{\infty} |\psi(x, t)|^2\, dx = A^2 \int_{-\infty}^{0} e^{\frac{x}{a}} dx + A^2 \int_0^{\infty} e^{\frac{-x}{a}} dx$$

Let's look at the first term (the second can be calculated in the same way modulo a minus sign). Let $u = x/a$, then $du = dx/a$. And so:

$$\int_{-\infty}^{0} e^{\frac{x}{a}} dx = a \int e^{u} du = a e^{\frac{x}{a}} \Big|_{-\infty}^{0} = a\left[e^{0} - e^{-\infty}\right] = a\,[1 - 0] = a$$

Application of the same technique to the second term also gives a, and so:

$$\int_{-\infty}^{\infty} |\psi(x,t)|^2\, dx = A^2 \int_{-\infty}^{0} e^{\frac{x}{a}} dx + A^2 \int_{0}^{\infty} e^{\frac{-x}{a}} dx = A^2 a + A^2 a = 2A^2 a$$

Using the normalization condition:

$$\int_{-\infty}^{\infty} |\psi(x,t)|^2\, dx = 1$$

We obtain:

$$A = \frac{1}{\sqrt{2a}}$$

The normalized wavefunction is then:

$$\psi(x) = \frac{1}{\sqrt{2a}} e^{-\frac{|x|}{2a}} e^{i(x - x_0)}$$

Expansion of the Wavefunction and Finding Coefficients

Earlier we noted that the *superposition* principle holds. Consider the superposition of stationary states $\psi_1(x,t), \psi_2(x,t), \ldots, \psi_n(x,t)$ which are solutions of the Schrödinger equation for a given potential V. Since these are stationary states, they can be written as:

$$\psi_n(x,t) = \Phi_n(x) \exp\left(-i E_n t/\hbar\right)$$

At time $t = 0$, any wavefunction Ψ can be written as a linear combination of these states:

$$\psi(x,0) = \sum C_n \Phi(x)$$

If we set $E = \hbar\omega$, the time evolution of this state is then:

$$\psi(x,t) = \sum C_n \Phi_n(x) e^{-i\omega_n t}$$

Since any function Ψ can be expanded in terms of the Φ_n, we say that the Φ_n are a set of *basis functions*. This is analogous to the way a vector can be expanded in terms of the basis vectors $\{i, j, k\}$. Sometimes we say that the set Φ_n is *complete*—this is another way of saying that any wavefunction can be expressed as a linear combination of them.

Definition: State Collapse

Suppose that a state Ψ is given by:

$$\psi(x,t) = \sum_{n=1}^{\infty} C_n \Phi_n(x) e^{-i\omega_n t}$$

A measurement of the energy is made and found to be $E_i = \hbar\omega_i$. The state of the system immediately after measurement is $\Phi_i(x)$, that is:

$$\Psi(x,t) \xrightarrow{\text{measurement finds } E_i} \Phi_i(x)$$

A second measurement of the energy conducted *immediately* after the first will find $E = \hbar\omega_i$ with certainty. However, if the system is left alone, the wavefunction will "spread out" as dictated by the time evolution of the Schrödinger equation and will again become a superposition of states.

To find the constants C_n in the expansion of Ψ, we use the *inner product*.

Definition: Inner Product

The *inner product* of two wave functions $\Psi(x)$ and $\Phi(x)$ is defined by:

$$(\Phi, \Psi) = \int \Phi^*(x)\Psi(x)dx$$

The square of (Φ, Ψ) tells us is the probability that a measurement will find the system in state $\Phi(x)$, *given* that it is originally in the state $\Psi(x)$.

Basis states are orthogonal. That is:

$$\int \Phi_m^*(x)\Phi_n(x)dx = 0 \quad \text{if } m \neq n$$

If we have normalized the basis states Φ_n, then they are *orthonormal*, which means that:

$$\int \Phi_m^*(x)\Phi_n(x)dx = \begin{cases} 0 & \text{if } m \neq n \\ 1 & \text{if } m = n \end{cases}$$

A shorthand for this type of relationship is provided by the Kronecker delta function:

$$\delta_{mn} = \begin{cases} 0 & \text{if } m \neq n \\ 1 & \text{if } m = n \end{cases}$$

This allows us to write:

$$\int \Phi_m^*(x)\Phi_n(x)dx = \delta_{mn}$$

The fact that basis states are orthogonal allows us to calculate the expansion coefficients using inner products.

Definition: Calculating a Coefficient of Expansion

If a state $\psi(x, 0)$ is written as a summation of basis functions $\Phi_n(x)$, we find the nth coefficient of the expansion c_n by computing the inner product of $\Phi_n(x)$ with $\psi(x, 0)$. That is:

$$C_n = (\Phi_n(x), \psi(x, 0)) = \int \Phi_n^*(x)\psi(x, 0)dx$$

Notice that:

$$\int \Phi_n^*(x)\psi(x, 0)dx = \int \Phi_n^*(x) \sum c_m \Phi_m(x)dx$$

$$= \sum c_m \int \Phi_n^*(x)\Phi_m(x)dx = \sum c_m \delta_{mn} = c_n$$

Definition: The Meaning of the Expansion Coefficient

If a state is written as $\psi(x, t) = \sum C_n \Phi_n(x)e^{-i\omega_n t}$, the modulus squared of the expansion coefficient C_n is the probability of finding the system in state $\Phi_n(x)$, i.e.:

$$\text{Probability system is in state } \Phi_n(x) = |C_n|^2$$

Another way to put this is, if a state $\psi(x, t) = \sum C_n \Phi_n(x)e^{-i\omega_n t}$ and we measure the energy, what is the probability we find $E_n = \hbar\omega_n$? The answer is $|C_n|^2$. Since the coefficients C_n define probabilities, it must be true that:

$$\sum |C_n|^2 = 1$$

EXAMPLE 2.12

A particle of mass m is trapped in a one-dimensional box of width a. The wavefunction is known to be:

$$\psi(x) = \frac{i}{2}\sqrt{\frac{2}{a}}\sin\left(\frac{\pi x}{a}\right) + \sqrt{\frac{1}{a}}\sin\left(\frac{3\pi x}{a}\right) - \frac{1}{2}\sqrt{\frac{2}{a}}\sin\left(\frac{4\pi x}{a}\right)$$

If the energy is measured, what are the possible results and what is the probability of obtaining each result? What is the most probable energy for this state?

SOLUTION

We begin by recalling that the nth excited state of a particle in a one-dimensional box is described by the wavefunction:

$$\Phi_n(x) = \sqrt{\frac{2}{a}}\sin\left(\frac{n\pi x}{a}\right), \text{ with energy } E_n = \frac{n^2\hbar^2\pi^2}{2ma^2}$$

Table 2-1 gives the first few wavefunctions and their associated energies.

Table 2-1

n	$\Phi_n(x)$	E_n
1	$\sqrt{\frac{2}{a}}\sin\left(\frac{\pi x}{a}\right)$	$\frac{\hbar^2\pi^2}{2ma^2}$
2	$\sqrt{\frac{2}{a}}\sin\left(\frac{2\pi x}{a}\right)$	$\frac{4\hbar^2\pi^2}{2ma^2}$
3	$\sqrt{\frac{2}{a}}\sin\left(\frac{3\pi x}{a}\right)$	$\frac{9\hbar^2\pi^2}{2ma^2}$
4	$\sqrt{\frac{2}{a}}\sin\left(\frac{4\pi x}{a}\right)$	$\frac{16\hbar^2\pi^2}{2ma^2}$

Noting that all of the Φ_n multiplied by the constant $\sqrt{2/a}$, we rewrite the wavefunction as given so that all three terms look this way:

$$\psi(x) = \frac{i}{2}\sqrt{\frac{2}{a}}\sin\left(\frac{nx}{a}\right) + \sqrt{\frac{1}{a}}\sin\left(\frac{3\pi x}{a}\right) - \frac{1}{2}\sqrt{\frac{2}{a}}\sin\left(\frac{4\pi x}{a}\right)$$

$$= \frac{i}{2}\sqrt{\frac{2}{a}}\sin\left(\frac{\pi x}{a}\right) + \sqrt{\frac{2}{2}}\sqrt{\frac{1}{a}}\sin\left(\frac{3\pi x}{a}\right) - \frac{1}{2}\sqrt{\frac{2}{a}}\sin\left(\frac{4\pi x}{a}\right)$$

$$= \frac{i}{2}\sqrt{\frac{2}{a}}\sin\left(\frac{\pi x}{a}\right) + \frac{1}{\sqrt{2}}\sqrt{\frac{2}{a}}\sin\left(\frac{3\pi x}{a}\right) - \frac{1}{2}\sqrt{\frac{2}{a}}\sin\left(\frac{4\pi x}{a}\right)$$

Now we compare each term to the table, allowing us to write this as:

$$\psi(x) = \frac{i}{2}\Phi_1(x) + \frac{1}{\sqrt{2}}\Phi_3(x) - \frac{1}{2}\Phi_4(x)$$

Since the wavefunction is written in the form $\psi(x) = \sum c_n \Phi_n(x)$, we see that the coefficients of the expansion areshown in Table 2-1:

Table 2-2

n	C_n	Associated Basis Function	Associated Energy
1	$\frac{i}{2}$	$\Phi_1(x) = \sqrt{\frac{2}{a}}\sin(\pi x/a)$	$\frac{\hbar^2\pi^2}{2ma^2}$
2	0	$\Phi_2(x) = \sqrt{\frac{2}{a}}\sin(2\pi x/a)$	$\frac{2\hbar^2\pi^2}{ma^2}$
3	$\frac{1}{\sqrt{2}}$	$\Phi_3(x) = \sqrt{\frac{2}{a}}\sin(3\pi x/a)$	$\frac{9\hbar^2\pi^2}{2ma^2}$
4	$-\frac{1}{2}$	$\Phi_4(x) = \sqrt{\frac{2}{a}}\sin(4\pi x/a)$	$\frac{8\hbar^2\pi^2}{ma^2}$

Since c_2 is not present in the expansion, there is no chance of seeing the energy $E = (2\hbar^2\pi^2)/(2ma^2)$. The square of the other coefficient terms gives the probability of measuring each energy. So the possible energies that can be measured with their respective probabilities are:

$$E_1 = \frac{\hbar^2\pi^2}{2ma^2},\ P(E_1) = |c_1|^2 = c_1^* c_1 = \left(\frac{-i}{2}\right)\frac{i}{2} = \frac{1}{4} = 0.25$$

$$E_3 = \frac{3\hbar^2\pi^2}{2ma^2},\ P(E_3) = |c_3|^2 = \left(\frac{1}{\sqrt{2}}\right)^2 = \frac{1}{2} = 0.50$$

$$E_4 = \frac{4\hbar^2\pi^2}{2ma^2},\ P(E_4) = |c_4|^2 = \left(\frac{1}{2}\right)^2 = \frac{1}{4} = 0.25$$

$\Rightarrow$the most probable energy is $E_3 = (9\hbar^2\pi^2)/(2ma^2)$

EXAMPLE 2.13

A particle in a one-dimensional box $0 \le x \le a$ is in the state:

$$\psi(x) = \frac{1}{\sqrt{10a}} \sin\left(\frac{\pi x}{a}\right) + A\sqrt{\frac{2}{a}} \sin\left(\frac{2\pi x}{a}\right) + \frac{3}{\sqrt{5a}} \sin\left(\frac{3\pi x}{a}\right)$$

(a) Find A so that $\psi(x)$ is normalized.

(b) What are the possible results of measurements of the energy, and what are the respective probabilities of obtaining each result?

(c) The energy is measured and found to be $(2\pi^2\hbar^2)/(ma^2)$. What is the state of the system immediately after measurement?

SOLUTION

(a) The basis functions for a one-dimensional box are:

$$\Phi_n(x) = \sqrt{\frac{2}{a}} \sin(n\pi x/a), \text{ with inner product } (\Phi_m(x), \Phi_n(x)) = \delta_{mn}$$

First we manipulate $\Psi(x)$ so we can write each term with the coefficient $\sqrt{2/a}$:

$$\begin{aligned}
\psi(x) &= \frac{1}{\sqrt{10a}} \sin\left(\frac{\pi x}{a}\right) + A\sqrt{\frac{2}{a}} \sin\left(\frac{2\pi x}{a}\right) + \frac{3}{\sqrt{5a}} \sin\left(\frac{3\pi x}{a}\right) \\
&= \frac{1}{\sqrt{10a}}\sqrt{\frac{2}{2}} \sin\left(\frac{\pi x}{a}\right) + A\sqrt{\frac{2}{a}} \sin\left(\frac{2\pi x}{a}\right) + \frac{3}{\sqrt{5a}}\sqrt{\frac{2}{2}} \sin\left(\frac{3\pi x}{a}\right) \\
&= \frac{1}{\sqrt{20}}\sqrt{\frac{2}{a}} \sin\left(\frac{\pi x}{a}\right) + A\sqrt{\frac{2}{a}} \sin\left(\frac{2\pi x}{a}\right) + \frac{3}{\sqrt{10}}\sqrt{\frac{2}{a}} \sin\left(\frac{3\pi x}{a}\right) \\
&= \frac{1}{\sqrt{20}}\Phi_1(x) + A\Phi_2(x) + \frac{3}{\sqrt{10}}\Phi_3(x)
\end{aligned}$$

Now we compute the inner product $(\psi(x), \psi(x))$. We must have $(\psi(x), \psi(x)) = 1$ for the state to be normalized. Since $(\Phi_m(x), \Phi_n(x)) = \delta_{mn}$, we can drop all terms where $m \neq n$:

$$(\psi(x), \psi(x)) = \left(\frac{1}{\sqrt{20}}\Phi_1(x) + A\Phi_2(x) + \frac{3}{\sqrt{10}}\Phi_3(x), \right.$$
$$\left. \frac{1}{\sqrt{20}}\Phi_1(x) + A\Phi_2(x) + \frac{3}{\sqrt{10}}\Phi_3(x)\right)$$

$$= \frac{1}{20}(\Phi_1(x), \Phi_1(x)) + A^2(\Phi_2(x), \Phi_2(x)) + \frac{9}{10}(\Phi_3(x), \Phi_3(x))$$

$$= \frac{1}{20} + A^2 + \frac{9}{10} = \frac{19}{20} + A^2$$

Setting this equal to 1 and subtracting 19/20 from both sides:

$$(\psi(x), \psi(x)) = 1 = \frac{19}{20} + A^2 \Rightarrow$$

$$A^2 = 1 - \frac{19}{20} = \frac{1}{20}, \Rightarrow A = \frac{1}{\sqrt{20}}$$

So the normalized wavefunction is:

$$\psi(x) = \frac{1}{\sqrt{20}}\Phi_1(x) + \frac{1}{\sqrt{20}}\Phi_2(x) + \frac{3}{\sqrt{10}}\Phi_3(x)$$

(b) In the one-dimensional box or infinite square well, the possible result of a measurement of energy for $\Phi_n(x)$ is:

$$E_n = \frac{n^2\pi^2\hbar^2}{2ma^2}$$

To determine the possible results of a measurement of the energy for a wavefunction expanded in basis states of the Hamiltonian, we look at each state in the expansion. If the wavefunction is normalized the squared modulus of the coefficient multiplying each state gives the probability of obtaining the given measurement. Looking at $\Psi(x)$, we see that the possible results of a measurement of the energy and their probabilities are:

Table 2-3

State	Energy measurement	Probability
$\Phi_1(x)$	$\frac{\pi^2\hbar^2}{2ma^2}$	$\left(\frac{1}{\sqrt{20}}\right)^2 = \frac{1}{20} = 0.05$
$\Phi_2(x)$	$\frac{4\pi^2\hbar^2}{2ma^2}$	$\left(\frac{1}{\sqrt{20}}\right)^2 = \frac{1}{20} = 0.05$
$\Phi_3(x)$	$\frac{9\pi^2\hbar^2}{2ma^2}$	$\left(\frac{3}{\sqrt{10}}\right)^2 = \frac{9}{10} = 0.90$

(c) If the energy is measured and found to be $(2\pi^2\hbar^2)/(ma^2) = (4\pi^2\hbar^2)/(2ma^2)$, the state of the system immediately after measurement is:

$$\psi(x) = \Phi_2(x) = \sqrt{\frac{2}{a}}\sin\left(\frac{2\pi x}{a}\right)$$

EXAMPLE 2.14

A particle of mass m is confined in a one-dimensional box with dimensions $0 \le x \le a$ and is known to be in the second excited state $(n = 3)$. Suddenly the width of the box is doubled $(a \to 2a)$, without disturbing the state of the particle. If a measurement of the energy is measured, find the probabilities that we find the system in the ground state and the first excited state.

SOLUTION

The stationary states for a particle in a box of length a is:

$$\psi_n(x) = \sqrt{\frac{2}{a}}\sin\frac{n\pi x}{a}$$

The second excited state wavefunction for a particle in a one-dimensional box is $n = 3$. So the wavefunction is:

$$\psi(x) = \sqrt{\frac{2}{a}}\sin\frac{3\pi x}{a}$$

Now let's consider a box of length $2a$. Labeling these states by Φ_n, we just let $(a \to 2a)$:

$$\Phi_n(x) = \sqrt{\frac{1}{a}}\sin\frac{n\pi x}{2a}$$

The energy of each state is also found by letting $a \to 2a$:

$$E_n = \frac{n^2\pi^2\hbar^2}{2ma^2} \quad (\text{ for length } a)$$

$$\Rightarrow E_n = \frac{n^2\pi^2\hbar^2}{2m(2a)^2} = \frac{n^2\pi^2\hbar^2}{8ma^2} \quad (\text{ for length } 2a)$$

Given that the system is in the state $\Psi(x) = \sqrt{2/a}\,\sin((3\pi x)/a)$, we use the inner product (Φ_n, Ψ) to find the probability that the particle is found in state $\Phi_n(x)$

when a measurement is made.

$$(\Phi_n, \psi) = \int \Phi_n^*(x)\psi(x)dx = \int_0^{2a} \left(\sqrt{\frac{1}{a}} \sin \frac{n\pi x}{2a}\right)\left(\sqrt{\frac{2}{a}} \sin \frac{3\pi x}{a}\right) dx$$

$$= \frac{\sqrt{2}}{a} \int_0^a \sin \frac{n\pi x}{2a} \sin \frac{3\pi x}{a} dx$$

To perform this integral, we use the trig identity:

$$\sin A \sin B = \frac{1}{2}\left[\cos(A - B) - \cos(A + B)\right]$$

$$\Rightarrow \sin \frac{n\pi x}{2a} \sin \frac{3\pi x}{a} = \frac{1}{2}\left[\cos\left(\frac{n\pi x}{2a} - \frac{3\pi x}{a}\right) - \cos\left(\frac{n\pi x}{2a} + \frac{3\pi x}{a}\right)\right]$$

$$= \frac{1}{2}\cos\left[\frac{\pi x}{a}\left(\frac{n}{2} - 3\right)\right] - \frac{1}{2}\cos\left[\frac{\pi x}{a}\left(\frac{n}{2} + 3\right)\right]$$

Putting this result into the integral gives:

$$(\Phi_n, \psi) = \frac{\sqrt{2}}{a}\left(\frac{1}{2}\right)\int_0^a \cos\left[\frac{\pi x}{a}\left(\frac{n}{2} - 3\right)\right] - \cos\left[\frac{\pi x}{a}\left(\frac{n}{2} + 3\right)\right] dx$$

Each term is easily evaluated with a substitution. We set $u = ((\pi x)/a)((n/2) - 3)$ in the first integral and $u = ((\pi x)/a)((n/2) + 3)$ in the second. Therefore:

$$\int_0^a \cos\left[\frac{\pi x}{a}\left(\frac{n}{2} - 3\right)\right] dx = \frac{a}{\pi\left(\frac{n}{2} - 3\right)} \sin\left[\pi\left(\frac{n}{2} - 3\right)\right]$$

$$\int_0^a \cos\left[\frac{\pi x}{a}\left(\frac{n}{2} + 3\right)\right] dx = \frac{a}{\pi\left(\frac{n}{2} + 3\right)} \sin\left[\pi\left(\frac{n}{2} + 3\right)\right]$$

If n is even, then we get sin of an integral multiple of π and both terms are zero. Putting all of these results together, we have:

$$(\Phi_n, \psi) = \frac{1}{a\sqrt{2}}\left\{\frac{a}{\pi\left(\frac{n}{2} - 3\right)} \sin\left[\pi\left(\frac{n}{2} - 3\right)\right] - \frac{a}{\pi\left(\frac{n}{2} + 3\right)} \sin\left[\pi\left(\frac{n}{2} + 3\right)\right]\right\}$$

The first excited state has $n = 2$. Since:

$$\sin\left[\pi\left(\frac{2}{2} - 3\right)\right] = \sin\left[\pi(1 - 3)\right] = \sin\left[-2\pi\right] = 0$$

$$\sin\left[\pi\left(\frac{2}{2} + 3\right)\right] = \sin\left[\pi(1 + 3)\right] = \sin\left[4\pi\right] = 0$$

$(\Phi_2, \Psi) = 0$, and so the probability of finding the system in the first excited state vanishes. For the ground state $n = 1$ and we have:

$$\begin{aligned}(\Phi_1, \psi) &= \frac{1}{\pi\sqrt{2}}\left\{\frac{1}{\left(\frac{1}{2}-3\right)}\sin\left[\pi\left(\frac{1}{2}-3\right)\right] - \frac{1}{\left(\frac{1}{2}+3\right)}\sin\left[\pi\left(\frac{1}{2}+3\right)\right]\right\} \\ &= \frac{1}{\pi\sqrt{2}}\left\{-\frac{2}{5}\sin\left[\left(-\frac{5\pi}{2}\right)\right] - \frac{2}{7}\sin\left[\left(\frac{7\pi}{x}\right)\right]\right\} \\ &= \frac{1}{\pi\sqrt{2}}\left\{\frac{2}{5}\sin\left[\left(\frac{5\pi}{2}\right)\right] - \frac{2}{7}\sin\left[\left(\frac{7\pi}{2}\right)\right]\right\} \\ &= \left(\frac{12\sqrt{2}}{35\pi}\right) \\ &= 0.15\end{aligned}$$

The probability is found by *squaring* this number. So the probability of finding the system in the ground state of the newly widened box is:

$$P(n = 1) = (0.15)^2 = 0.024$$

The Phase of a Wavefunction

A phase factor is some function of the form $e^{i\theta}$ that multiplies a wavefunction. If the phase is an overall multiplicative factor it is called a *global phase*. Suppose we have two wavefunctions:

$$\psi(x, t)$$

And:

$$\Phi(x, t) = e^{i\theta}\psi(x, t)$$

Notice that:

$$\begin{aligned}|\Phi(x, t)|^2 &= \Phi^*(x, t)\Phi(x, t) = e^{-i\theta}\psi^*(x, t)e^{i\theta}\psi(x, t) \\ &= e^0\psi^*(x, t)\psi(x, t) = \psi^*(x, t)\psi(x, t) = |\psi(x, t)|^2\end{aligned}$$

Since both states give the *same* physical predictions (i.e., they predict the same probabilities of measurement in terms of the modulus square of the wavefunction) we say that they are physically equivalent. We can go so far as to say that they are the same state.

Now we consider a type of phase called *relative* phase. Suppose that we expand a wavefunction in terms of normalized basis states. For simplicity, we consider:

$$\psi(x,t) = u_1(x,t) + u_2(x,t)$$

Then:

$$\begin{aligned} |\psi(x,t)|^2 &= \psi^*(x,t)\psi(x,t) = \left(u_1^*(x,t) + u_2^*(x,t)\right)\left(u_1(x,t) + u_2(x,t)\right) \\ &= |u_1(x,t)|^2 + |u_2(x,t)|^2 + 2Re\left[u_1(x,t)u_2^*(x,t)\right] \end{aligned}$$

Now suppose that:

$$\Phi(x,t) = u_1(x,t) + e^{i\theta}u_2(x,t)$$

Then:

$$= |u_1(x,t)|^2 + |u_2(x,t)|^2 + e^{-i\theta}u_1(x,t)u_2^*(x,t) + e^{i\theta}u_1^*(x,t)u_2(x,t)$$

in this case, $|\Phi(x,t)|^2 \neq |\psi(x,t)|^2$, and therefore these wavefunctions *do not* represent the same physical state. Each wavefunction will give rise to different probabilities for the results of physical measurements. We summarize these results.

Definition: Global Phase

A *global phase* factor in the form of some function $e^{i\theta}$ that is an overall multiplicative constant for a wavefunction, has *no* physical significance—$\Psi(x)$ and $e^{i\theta}\Psi(x)$ represent the same physical state.

Definition: Local Phase

A *relative phase* factor, in the form of a term $e^{i\theta}$ that multiplies an expansion coefficient in $\Psi(x) = \sum c_n\Phi_n(x)$ *does* change physical predictions.

$$\psi(x) = c_1\Phi_1 + c_2\Phi_2 + \cdots. + c_n\Phi_n + \cdots.$$

and

$$\psi(x) = c_1\Phi_1 + c_2\Phi_2 + \cdots. + e^{i\theta}c_n\Phi_n + \cdots.$$

do not represent the same state.

Operators in Quantum Mechanics

An *operator* is a mathematical instruction to do something to the function that follows. The result is a new function. A simple operator could be to just take the derivative of a function. Suppose we call this operator D. Then:

$$Df(x) = \frac{df}{dx} = g(x)$$

Or in shorthand, we just write;

$$Df(x) = g(x)$$

Consider another operator, which we call X, then multiply a function by x:

$$Xf(x) = xf(x)$$

Operators are often represented by capital letters, sometimes using a caret (i.e. $\hat{X}$).

If given scalars α, β, and functions $\psi(x)$ and (x) an operator A satisfies:

$$A\left[a\varphi(x) + \beta\varphi(x)\right] = aA\varphi(x) + \beta a\varphi(x)$$

we say that the operator A is linear. If the action of an operator on a function $\Phi(x)$ is to multiply that function by some constant:

$$A\varphi(x) = \varphi(x)$$

we say that the constant A is an *eigenvalue* of the operator A, and we call $\Phi(x)$ an eigenfunction of A.

Two operators are of fundamental importance in quantum mechanics. We have already seen one of them. The position operator acts on a wavefunction by multiplying it by the coordinate:

$$X\varphi(x, t) = x\varphi(x, t)$$

In three dimensions, we also have operators Y and Z:

$$Y\psi(x, y, z, t) = y\psi(x, t)$$

$$Z\psi(x, y, z, t) = z\psi(x, t)$$

In quantum mechanics, we also have a momentum operator. The momentum operator is given in terms of differentiation with respect to the position coordinate:

$$P_x\psi = -i\hbar\frac{\partial\psi}{\partial x}$$

Therefore the momentum operator is $P_x = -i\hbar(\partial/\partial x)$. If we are working in three dimensions, we also have momentum operators for the other coordinates:

$$P_y = -i\hbar\frac{\partial}{\partial y}$$

$$P_z = -i\hbar\frac{\partial}{\partial z}$$

The expectation value, or mean of an operator A with respect to the wavefunction $\psi(x)$ is:

$$\langle A\rangle = \int_{-\infty}^{\infty} \psi^*(x)A\psi(x)dx$$

The mean value of an operator is not necessarily a value that can actually be measured. So it may not turn out to be equal to one of the eigenvalues of A.

EXAMPLE 2.15

e.g.

A particle m a one-dimensional box $0 \leq x \leq a$ is in the ground state. Find $\langle x\rangle$ and $\langle p\rangle$.

SOLUTION

✓

The wavefunction is:

$$\psi(x) = \sqrt{\frac{2}{a}}\sin\left(\frac{\pi x}{a}\right)$$

$$\langle x\rangle = \int_{-\infty}^{\infty} \psi^*(x)x\psi(x)dx = \int_0^a \sqrt{\frac{2}{a}}\sin\left(\frac{\pi x}{a}\right)(x)\sqrt{\frac{2}{a}}\sin\left(\frac{\pi x}{a}\right)dx$$

$$= \frac{2}{a}\int_0^a x\left(\sin\left(\frac{\pi x}{a}\right)\right)^2 dx$$

$$= \frac{2}{a}\int_0^a x\left(\frac{1-\cos\left(\frac{2\pi x}{a}\right)}{2}\right)dx$$

$$= \frac{2}{a}\int_0^a xdx - \frac{1}{a}\int_0^a x\cos\left(\frac{2\pi x}{a}\right)dx$$

integrating the first term yields:

$$\int_0^a xdx = x^2/2\big|_0^a = a^2/2$$

The second term can be integrated by parts, giving

$$\int_0^a x\cos\left(\frac{2\pi x}{a}\right)dx = a^2/4\pi^2\cos(2\pi x/a) + a/2\pi x\sin\left(\frac{2\pi x}{a}\right)\Big|_0^a$$
$$= a^2/4\pi\,[\cos 2\pi - \cos 0] = 0$$

And so, we find that:

$$\langle x\rangle = \frac{1}{a}\int_0^a x\,dx = \left(\frac{1}{a}\right)x^2/2\Big|_0^a = a/2$$

To calculate $\langle p\rangle$, we write p as a derivative operator:

$$\langle P\rangle = \int_{-\infty}^{\infty}\psi^*(x)p\psi(x)dx$$
$$= \int_0^a \sqrt{\frac{2}{a}}\sin\left(\frac{\pi x}{a}\right)\left(-ih\frac{d}{dx}\sqrt{\frac{2}{a}}\sin\left(\frac{\pi x}{a}\right)\right)dx$$
$$= -ih\frac{2}{a}\left(\frac{\pi}{a}\right)\int_0^a \sin\frac{\pi x}{a}\cos\frac{\pi x}{a}dx$$

You may already know this integral is zero, but we can calculate it easily so we proceed.

Let $u = \sin\left(\frac{\pi x}{a}\right)$, then $du = \left(\frac{\pi}{a}\right)\cos\left(\frac{\pi x}{a}\right)dx$. This gives:

$$\int u\,du = u^2/2, \Rightarrow$$
$$\int_{-\infty}^{\infty}\sin\left(\frac{\pi x}{a}\right)\cos\left(\frac{\pi x}{a}\right)dx = (1/2)\left(\sin\left(\frac{\pi x}{a}\right)\right)^2\Big|_0^a$$
$$= (1/2)\left[(\sin(\pi))^2 - (\sin(0))^2\right] = 0$$

And so, for the ground state of the particle in a box, we have found:

$$\langle p\rangle = 0$$

EXAMPLE 2.16

e.g.

Let

$$\psi(x) = \left(\frac{2a}{\pi}\right)^{1/4} \exp(-ax^2)$$

Assuming that u is real, find $\langle x^n \rangle$ for arbitrary integers $n > 0$.

SOLUTION

✔

ψ is real so:

$$\psi^*(x) = \psi(x) = \left(\frac{2a}{\pi}\right)^{1/4} \exp(-ax^2)$$

So the expectation value of x^n is given by:

$$\begin{aligned}
\langle x^n \rangle &= \int_{-\infty}^{\infty} \psi^*(x) x^n \psi(x) dx \\
&= \int_{-\infty}^{\infty} \left(\frac{2a}{\pi}\right)^{1/4} \exp(-ax^2) x^n \left(\frac{2a}{\pi}\right)^{1/4} \exp(-ax^2) dx \\
&= \sqrt{\frac{2a}{\pi}} \int_{-\infty}^{\infty} x^n \exp(-2ax^2) dx
\end{aligned}$$

We can do this integral recalling that:

$$\int_{-\infty}^{\infty} z^{2m} e^{-z^2} dz = \sqrt{\pi}\frac{1.3.5\cdots(2m-1)}{2^m}, \quad m = 1, 2\ldots$$

Looking at the integral for $\langle x^n \rangle$, let $z = \sqrt{2a}\, x$, then $dz = \sqrt{2a}\, dx$. For even $n = 2, 4, 6, 8\ldots.$ we obtain:

$$\begin{aligned}
\langle x^n \rangle &= \sqrt{\frac{2a}{\pi}} \int_{-\infty}^{\infty} x^n \exp(-2ax^2) dx \\
&= \frac{1}{\sqrt{\pi}} \int_{-\infty}^{\infty} \left(\frac{z}{\sqrt{2a}}\right)^n \exp(-z^2) dz \\
&= \frac{1}{\sqrt{\pi}} \left(\frac{1}{\sqrt{2a}}\right)^n \int_{-\infty}^{\infty} z^n \exp(-z^2) dz \\
&= \frac{1}{\sqrt{\pi}} \left(\frac{1}{\sqrt{2a}}\right)^n \sqrt{\pi}\frac{1.3.5\cdots(2n-1)}{2^n} \\
&= \left(\sqrt{\frac{2a}{\pi}}\right)^n \frac{1.3.5\cdots(2n-1)}{2^n}
\end{aligned}$$

Earlier we noted that:

$$\int_{-\infty}^{\infty} z e^{-z^2} dz = 0$$

In fact,

$$\int_{-\infty}^{\infty} z^m e^{-z^2} dz = 0$$

for any odd m, therefore $\langle x^n \rangle = 0$ if n is odd.

If a normalized wavefunction is expanded in a basis with known expansion coefficients, we can use this expansion to calculate mean values. Specifically, if a wavefunction has been expanded as:

$$\psi(x, 0) = \sum c_n \phi_n(x),$$

And the basis functions $\phi_n(x)$ are eigenfunctions of operator A with eigenvalues a_n, then the mean of operator A can be found from:

$$\langle A \rangle = \sum a_n |c_n|^2$$

If the following relationship holds:

$$\int \phi^*(x) \left[A\psi(x) \right] dx = \int (\psi(x)A)^* \psi(x) dx$$

we say that the operator A is *Hermitian*. Hermitian operators, which have real eigenvalues, are fundamentally important in quantum mechanics. Quantities that can be measured experimentally, like energy or momentum, are represented by Hermitian operators.

We can also calculate the mean of the square of an operator:

$$\langle A^2 \rangle = \int_{-\infty}^{\infty} \psi^*(x) A^2 \psi(x) dx$$

where A^2 is the operator $(A)(A)$. This leads us to a quantity known as the standard deviation or uncertainty in A:

$$\Delta A = \sqrt{\langle A^2 \rangle - \langle A \rangle^2}$$

This quantity measures the spread of values about the mean for A.

EXAMPLE 2.17

e.g.

$$\psi(x) = A\left(ax - x^2\right) \text{ for } 0 \le x \le a$$

(a) Normalize the wavefunction

(b) Find $\langle x \rangle$, $\langle x^2 \rangle$ and Δx

SOLUTION

(a) The wavefunction is real. So $\psi^*\psi = \psi^2$ and we have:

$$\begin{aligned}
\int \psi^2 dx &= \int_0^a A^2(ax - x^2)^2 dx = A^2 \int_0^a a^2x^2 - 2ax^3 + x^4 dx \\
&= A^2\left[a^2(x^3/3) - ax^4/2 + x^5/5\right]\Big|_0^a \\
&= A^2\left[a^5/3 - a^5/2 + a^5/5\right] \\
&= A^2\left[10a^5/30 - 15a^5/30 + 6a^5/30\right] \\
&= A^2a^5/30, \Rightarrow A = \sqrt{30/a^5}
\end{aligned}$$

(b)

$$\begin{aligned}
\langle x \rangle &= \int x^2\psi^2 dx = (30/a^5)\int_0^a x(ax - x^2)^2 dx \\
&= (30/a^5)\int_0^a a^2x^3 - 2ax^4 + x^5 dx \\
&= (30/a^5)\left[a^2\left(x^4/4\right) - 2a\left(x^5/5\right) + x^6/6\right]\Big|_0^a \\
&= (30/a^5)\left[a^6/4 - 2a^6/5 + a^6/6\right] \\
&= (30/a^5)(a^2/60) = \frac{a}{2}
\end{aligned}$$

$$\begin{aligned}
\langle x^2 \rangle &= \int x^2\psi^2 dx = (30/a^5)\int_0^a x^2\left(ax - x^2\right)^2 dx \\
&= (30/a^5)\int_0^a a^2x^4 - 2ax^5 + x^6 dx \\
&= (30/a^5)\left[a^2\left(x^5/5\right) - 2a\left(x^6 + x^7/7\right)\right]\Big|_0^a \\
&= (30/a^5)\left[a^7/5 - 2a^7/6 + a^7/7\right] \\
&= \frac{2a^2}{7}
\end{aligned}$$

$$\Delta x = \sqrt{\langle x^2 \rangle - \langle x \rangle^2} = \sqrt{\frac{2a^2}{7} - \left(\frac{a}{2}\right)^2} = \sqrt{\frac{2a^2}{7} - \frac{a^2}{4}} = \frac{a}{\sqrt{28}}$$

One very important operator that we will come across frequently is the Hamiltonian operator:

$$H = -\frac{\hbar^2}{2m}\frac{d^2}{dx^2} + V(x,t) = \frac{P^2}{2m} + V(x,t)$$

As you can see this is just one side of the Schrödinger equation. Therefore acting H on a wavefunction gives:

$$H\psi(x,t) = i\hbar\frac{d\psi(x,t)}{dt}$$

When the potential is time independent, and we have the time-independent Schrödinger equation, we arrive at an eigenvalue equation for H:

$$H\psi(x) = E\psi(x)$$

The eigenvalues E of the Hamiltonian are the energies of the system. Or you can say the allowed energies of the system are the eigenvalues of the Hamiltonian operator H. The average or mean energy of a system expanded in the basis states ϕ_n is found from:

$$\langle H \rangle = \sum E_n |c_n|^2 = \sum E_n P_n$$

where p_n, is the probability that energy is measured.

---

e.g. **EXAMPLE 2.18**

Earlier we considered the following state for a particle trapped in a one-dimensional box:

$$\psi(x) = \frac{i}{2}\sqrt{\frac{2}{a}}\sin\left(\frac{\pi x}{a}\right) + \sqrt{\frac{1}{a}}\sin\left(\frac{3\pi x}{a}\right) - \frac{1}{2}\sqrt{\frac{2}{a}}\sin\left(\frac{4\pi x}{a}\right)$$

What is the mean energy for this system?

✔ **SOLUTION**

For this state, we found that the possible energies that could be measured and their respective probabilities were:

$$E_1 = \frac{\hbar^2\pi^2}{2ma^2}, \; P(E_1) = |c_1^* c_1| = \left(\frac{-i}{2}\right)\frac{i}{2} = \frac{1}{4} = 0.25$$

$$E_3 = \frac{9\hbar^2\pi^2}{2ma^2},\ P(E_3) = |c_3|^2 = \left(\frac{1}{\sqrt{2}}\right)^2 = \frac{1}{2} = 0.25$$

$$E_4 = \frac{8\hbar^2\pi^2}{ma^2},\ P(E_4) = |c_4|^2 = \left(\frac{1}{2}\right)^2 = \frac{1}{4} = 0.25$$

So the mean energy is:

$$\langle H\rangle = \sum E_n P_n = \frac{\hbar^2\pi^2}{2ma^2}\left(\frac{1}{4}\right) + \frac{3\hbar^2\pi^2}{2ma^2}\left(\frac{1}{2}\right) + \frac{4\hbar^2\pi^2}{2ma^2}\left(\frac{1}{4}\right)$$

$$= \frac{\hbar^2\pi^2}{2ma^2}\left[\frac{1}{4} + \frac{3}{2} + \frac{4}{4}\right] + \frac{\hbar^2\pi^2}{2ma^2}\left(\frac{11}{4}\right) = \frac{11\hbar^2\pi^2}{8ma^2}$$

Note that the mean energy would never actually be measured for this system.

EXAMPLE 2.19

e.g.

Is the momentum operator p Hermitian?

SOLUTION

✓

We need to show that:

$$\int \phi^*(x)[p\psi(x)]dx = \int (\phi(x)p)^*\psi(x)]dx$$

Using $p\psi = -i\hbar(d\psi/dx)$, we have:

$$\int \phi^*(x)[p\psi^*(x)]dx = \int \phi(x)\left(-i\hbar\frac{d\phi}{dx}\right)dx = i\hbar\int \phi^*(x)\frac{d\phi}{dx}dx$$

Recalling the formula for integration by parts:

$$\int udv = uv - \int vdu$$

We let

$$u = \phi^*(x), \Rightarrow du = \frac{d\phi^*(x)}{dx}dx$$

$$dv = \frac{d\phi}{dx}, \Rightarrow v = \psi(x)$$

The wavefunction must vanish as $x \to \pm\infty$, and so the boundary term $(uv = \phi^*(x)\psi(x))$ vanishes. We are then left with:

$$i\hbar \int \phi^*(x)\frac{d\phi}{dx}dx = i\hbar\left[\phi^*(x)\psi(x)\Big|_{-\infty}^{\infty} - \int \frac{d\phi^*(x)}{dx}\psi(x)dx\right]$$
$$= -i\hbar \int \frac{d\phi^*(x)}{dx}\psi(x)dx$$
$$= \int \left(i\hbar\frac{d\phi}{dx}\right)^* \psi(x)dx$$

We have shown that: $\int \phi^*(x)p\psi(x)dx = \int (\phi(x)p)^*\psi(x)dx$, therefore p is Hermitian.

Momentum and the Uncertainty Principle

The fact that momentum can be expressed as $p = \hbar k$ allows us to define a "momentum space" wavefunction that is related to the position space wavefunction via the Fourier transform. A function $f(x)$ and its Fourier transform. $F(k)$ are related via the relations:

$$f(x) = \frac{1}{\sqrt{2\pi}}\int_{-\infty}^{\infty} F(k)e^{ikx}dk$$
$$F(k) = \frac{1}{\sqrt{2\pi}}\int_{-\infty}^{\infty} f(x)e^{-ikx}dx$$

These relations can be expressed in terms of p with a position space wavefunction $\psi(x)$ and momentum space wavefunction $\Phi(p)$ as:

$$\psi(x) = \frac{1}{\sqrt{2\pi\hbar}}\int_{-\infty}^{\infty} \phi(P)e^{ipx/\hbar}dp$$
$$\phi(p) = \frac{1}{\sqrt{2\pi\hbar}}\int_{-\infty}^{\infty} \psi(x)e^{-ipx/\hbar}dx$$

Parseval's theorem tells us that:

$$\int_{-\infty}^{\infty} f(x)g^*(x)dx = \int_{-\infty}^{\infty} F(k)G^*(k)dk$$
$$\int_{-\infty}^{\infty} |f(x)|^2\, dx = \int_{-\infty}^{\infty} |F(k)|^2\, dk$$

These relations tell us that $\Phi(p)$, like $\psi(x)$, represents a probability density. The function $\Phi(p)$ gives us information about the probability of finding momentum between $a \leq p \leq b$:

$$P(a \leq p \leq b) = \int_a^b |\phi(p)|^2 \, dp$$

Parseval's theorem tells us that if the wavefunction $\psi(x)$ is normalized, then the momentum space wavefunction $\Phi(p)$ is also normalized

$$\int_{-\infty}^{\infty} |\psi(x)|^2 \, dx = 1 \Rightarrow \int_{-\infty}^{\infty} |\phi(p)|^2 \, dp = 1$$

EXAMPLE 2.20

e.g.

Suppose that

$$\psi(x) = \frac{1}{\sqrt{a}} \text{ for } -a \leq x \leq a$$

Find the momentum space wavefunction $\Phi(p)$.

SOLUTION

✔

$$\begin{aligned}
\phi(p) &= \frac{1}{\sqrt{2\pi\hbar}} \int_{-\infty}^{\infty} \psi(x) e^{-ipx/\hbar} dx \\
&= \frac{1}{\sqrt{2\pi\hbar}} \int_{-a}^{a} \frac{1}{\sqrt{a}} e^{-ipx/\hbar} dx \\
&= \frac{1}{\sqrt{2\pi\hbar a}} \int_{-a}^{a} e^{-ipx/\hbar} dx \\
&= \frac{1}{\sqrt{2\pi\hbar}} \left(\frac{\hbar}{-ip} \right) e^{-ipx/\hbar} \Big|_{-a}^{a} \\
&= \frac{1}{\sqrt{2\pi\hbar a}} \frac{2\hbar}{p} \left(\frac{e^{ipa/\hbar} - e^{-ipa/\hbar}}{2i} \right) \\
&= \sqrt{\frac{2a}{\pi\hbar}} \frac{\sin(pa/\hbar)}{(pa/\hbar)} = \frac{2a}{\sqrt{\pi\hbar}} = \text{sinc}(pa/\hbar)
\end{aligned}$$

A plot Figure 2-6 of the so-called sinc function shows that the momentum-space wavefunction, like the position space wavefunction in this case, is also localized:

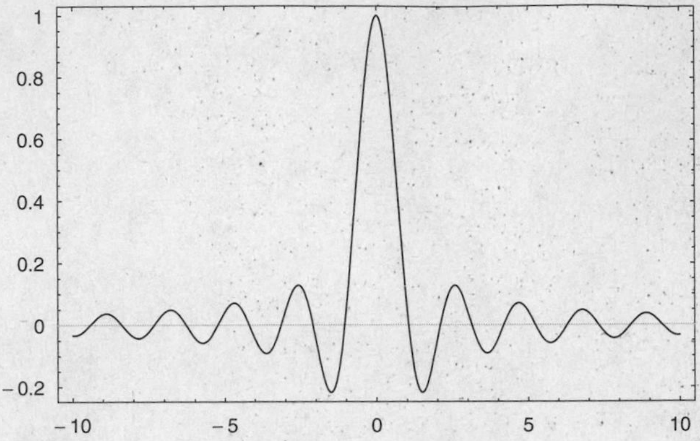

Fig. 2-6

It is a fact of Fourier theory and wave mechanics that the spatial extension of the wave described by $\psi(x)$ and the extension of wavelength described by the Fourier transform $\Phi(p)$ cannot be made arbitrarily small. This observation is described mathematically by the Heisenberg uncertainty principle:

$$\Delta x \Delta p \geq \hbar$$

Or, using $p = \hbar k$, Δx, $\Delta k \geq 1$.

EXAMPLE 2.21

Let

$$\phi(k) = e^{-\frac{a}{b}(k-k_o)^2}$$

Use the Fourier transform to find ψ(x)

SOLUTION

We can find the position space wavefunction from the relation:

$$\psi(x) = \frac{1}{\sqrt{2\pi}} \int_{-\infty}^{\infty} \phi(k) e^{ikx} dk$$

$$= \frac{1}{\sqrt{2\pi}} \int_{-\infty}^{\infty} \exp^{-\frac{a}{b}(k-k_o)^2} e^{ix} dk$$

To perform this daunting integral, we resort to an integral table, where we find that:

$$\int_{-\infty}^{\infty} e^{-\alpha u^2} e^{\beta u} du = \sqrt{\frac{\pi}{\alpha}} e^{\frac{\beta^2}{4\alpha}}$$

Examining the Fourier transform, we let:

$$u = (k - k_o)\sqrt{\frac{a}{b}}$$

Then we have:

$$du = \sqrt{\frac{a}{b}} dk$$

This substitution gives us:

$$\psi(x) = \frac{1}{\sqrt{2\pi}} \int_{-\infty}^{\infty} e^{-\frac{a}{b}(k-k_a)^2} e^{ikx} dk = \frac{1}{\sqrt{2\pi}} \sqrt{\frac{b}{a}} \int_{-\infty}^{\infty} e^{-u^2} e^{i\left(\sqrt{\frac{b}{a}}u+k_a\right)x} du$$

$$= \frac{1}{\sqrt{2\pi}} \sqrt{\frac{b}{a}} e^{ik_a x} \int_{-\infty}^{\infty} e^{-u^2} e^{i\sqrt{\frac{b}{a}}ux} du$$

Now we can make the following identifications with the result from the integral table:

$$\alpha = 1, \beta = i\sqrt{\frac{b}{a}}x, \Rightarrow \int_{-\infty}^{\infty} e^{-u^2} \exp^{\left[i\left(\sqrt{\frac{b}{a}}u\right)^x\right]} du = \sqrt{\pi} \exp^{-\left(\frac{b}{a}\frac{x^2}{4}\right)}$$

And so the position space wavefunction is:

$$\psi(x) = \sqrt{\frac{b}{2a}} e^{ik_o x} e^{-\frac{bx^2}{4a}}$$

EXAMPLE 2.22 *e.g.*

A particle of mass m in a one-dimensional box is found to be in the ground state:

$$\psi(x) = \sqrt{\frac{2}{a}} \sin\left(\frac{\pi x}{a}\right)$$

Find $\Delta x \Delta p$ for this state.

SOLUTION
Using $p = i\hbar d/dx$ we have:

$$p\psi(x) = i\hbar\frac{d}{dx}\left[\sqrt{\frac{2}{a}}\sin\left(\frac{\pi x}{a}\right)\right] = \frac{-\hbar\pi}{a}\sqrt{\frac{2}{a}}\cos\left(\frac{\pi x}{a}\right)$$

and:

$$p^2\psi(x) = i\hbar\frac{d}{dx}\left(\frac{-\hbar\pi}{a}\sqrt{\frac{2}{a}}\cos\left(\frac{\pi x}{a}\right)\right) = \frac{-\hbar\pi}{a}\sqrt{\frac{2}{a}}\sin\left(\frac{\pi x}{a}\right)$$

$$\Rightarrow \langle p\rangle = \int \psi^*(x)p\psi(x)dx$$

We found in the example above that $\langle p\rangle = 0$ for this state.

$$\begin{aligned}
\langle p^2\rangle &= \int \psi^*(x)p^2\psi(x)dx \\
&= \int_0^a \sqrt{\frac{2}{a}}\sin\left(\frac{\pi x}{a}\right)\frac{\hbar^2\pi^2}{a^2}\sqrt{\frac{2}{a}}\sin\left(\frac{\pi x}{a}\right)dx \\
&= \frac{\hbar^2\pi^2}{a^2}\int_0^a\left(\frac{2}{a}\right)\left[\sin\left(\frac{\pi x}{a}\right)\right]^2 dx \\
&= \frac{2\hbar^2\pi^2}{a^3}\int_0^a \frac{1-\cos\left(\frac{2\pi x}{a}\right)}{2}dx \\
&= \frac{\hbar^2\pi^2}{a^3}x\bigg|_0^a = \frac{\hbar^2\pi^2}{a^2}
\end{aligned}$$

$$\Delta p = \sqrt{\langle p^2\rangle - \langle p\rangle^2} = \sqrt{\frac{\hbar^2\pi^2}{a^2}} = \frac{\hbar\pi}{a}$$

$$\begin{aligned}
\langle x\rangle &= \int \psi^*(x)x\psi(x)dx = \int_0^a \sqrt{\frac{2}{a}}\sin\left(\frac{\pi x}{a}\right)x\sqrt{\frac{2}{a}}\sin\left(\frac{\pi x}{a}\right)dx \\
&= \frac{2}{a}\int_0^a x\left(\sin\left(\frac{\pi x}{a}\right)\right)^2 dx \\
&= \frac{2}{a}\frac{a^2}{4} = \frac{a}{2}
\end{aligned}$$

$$\langle x^2 \rangle = \int \psi^*(x) x^2 \psi(x) dx = \int_0^a \sqrt{\frac{2}{a}} x^2 \left(\sin\left(\frac{\pi x}{a}\right)\right)^2 dx = \frac{a^2}{6}\left(2 - \frac{3}{\pi^2}\right)$$

$$\Delta x = \sqrt{\langle x^2 \rangle - \langle x \rangle^2} = \sqrt{\frac{a^2(\pi^2 - 6)}{12\pi^2}} = \frac{a}{\pi}\sqrt{\frac{(\pi^2 - 6)}{12}}$$

Notice that $\sqrt{(\pi^2 - 6)/12} = 0.57$, and so $\Delta x \Delta p > \hbar/2$

The Conservation of Probability

The probability density is defined as:

$$\rho(x, t) = \psi^*(x, t)\psi(x, t)$$

We know that $\psi(x, t)$ satisfies the Schrödinger equation. It is also true that $\psi^*(x, t)$ satisfies the Schrödinger equation. Let's take the lime derivative of the product:

$$\frac{\partial \rho}{\partial t} = \frac{\partial}{\partial t}\left[\psi^*(x, t)\psi(x, t)\right] = \psi^* \frac{\partial \psi}{\partial t} + \frac{\partial \psi^*}{\partial t}$$

Recalling the Schrödinger equation:

$$i\hbar \frac{\partial \psi}{\partial t} = \frac{\hbar^2}{2m}\frac{\partial^2 \psi}{\partial x^2} + V\psi$$

$$\Rightarrow \psi^*(x, t)\frac{d\psi(x, t)}{dt}$$

$$= \psi^*(x, t)\left(\frac{-\hbar}{2im}\right)\frac{d^2\psi(x, t)}{dx^2} + \frac{1}{i\hbar}V(x, t)[\psi^*(x, t)\psi(x, t)] \qquad (*)$$

Taking the complex conjugate of the Schrödinger equation, we find the equation satisfied by $\psi^*(x, t)$:

$$-i\hbar \frac{\partial \psi^*}{\partial t} = \frac{\hbar^2}{2m}\frac{\partial^2 \psi^*}{\partial x^2} + V\psi^*$$

$$\Rightarrow \psi \frac{\partial \psi^*}{\partial t}$$

$$= \psi\left(\frac{\hbar}{2im}\right)\frac{\partial^2 \psi^*}{\partial x^2} + \frac{1}{-i\hbar}V\psi\psi^* \qquad (**)$$

Now we add $(*) + (**)$ On the left side, we obtain:

$$\psi^*(x,t)\frac{d\psi}{dt} = \psi(x,t)\frac{d\psi^*}{dt} = \frac{d}{dt}[\psi^*(x,t)\psi(x,t)] = \frac{dp}{dt}$$

On the right side, we get:

$$\psi^*(x,t)\left(\frac{-\hbar}{2im}\right)\frac{d^2\psi(x,t)}{dx^2} + \frac{1}{i\hbar}V(x,t)[\psi^*(x,t)\psi(x,t)]$$
$$+\,\psi(x,t)\left(\frac{\hbar}{2im}\right)\frac{d^2\psi^*(x,t)}{dx^2} + \frac{1}{-i\hbar}V(x,t)[\psi^*(x,t)\psi(x,t)]$$
$$= \left(\frac{\hbar}{2im}\right)\left[\psi^*(x,t)\frac{d^2\psi(x,t)}{dx^2} - \psi(x,t)\frac{d^2\psi^*(x,t)}{dx^2}\right]$$

We use this result as the basis for a new function called the probability current $j(x,t)$, which is defined as:

$$j(x,t) = \left(\frac{\hbar}{2mi}\right)\left[\psi^*(x,t)\frac{d^2\psi(x,t)}{dx^2} - \psi(x,t)\frac{d^2\psi^*(x,t)}{dx^2}\right]$$

We see that the result on the right side is $-\partial j/\partial x$. Setting this equal to the result of the left-hand side, we obtain the continuity equation for probability:

$$\frac{\partial p}{\partial t} + \frac{\partial j}{\partial x} = 0$$

In three dimensions, this generalizes to:

$$J(r,t) = \frac{\hbar}{2mi}[\psi^*(\nabla\psi) - \psi(\nabla\psi^*)]$$
$$\frac{\partial\rho(x,t)}{\partial t} + \nabla\cdot J = 0$$

e.g. **EXAMPLE 2.23**

Suppose that a particle of mass m in a one-dimensional box is described by the wavefunction:

$$\psi(x,t) = \sqrt{\frac{1}{a}}\sin\left(\frac{\pi x}{a}\right)e^{-iE_1t/\hbar} + \sqrt{\frac{1}{a}}\sin\left(\frac{2\pi x}{a}\right)e^{-iE_2t/\hbar}$$

Find the probability current for this wavefunction.

SOLUTION
By definition

$$j(x,t) = \left(\frac{\hbar}{2im}\right)\left[\psi^*(x,t)\frac{d^2\psi(x,t)}{dx^2} - \psi(x,t)\frac{d^2\psi^*(x,t)}{dx^2}\right]$$

Now

$$\psi(x,t) = \sqrt{\frac{1}{a}}\sin\left(\frac{\pi x}{a}\right)e^{-iE_1t/\hbar} + \sqrt{\frac{1}{a}}\sin\left(\frac{2\pi x}{a}\right)e^{-iE_2t/\hbar}$$

and:

$$\frac{d\psi(x,t)}{dx} = \sqrt{\frac{1}{a}}\left(\frac{\pi}{a}\right)\cos\left(\frac{\pi x}{a}\right)e^{-iE_1t/\hbar} + \sqrt{\frac{1}{a}}\left(\frac{2\pi}{a}\right)\cos\left(\frac{2\pi z}{a}\right)e^{-iE_2t/\hbar}$$

$$\frac{d\psi^*(x,t)}{dx} = \sqrt{\frac{1}{a}}\left(\frac{\pi}{a}\right)\cos\left(\frac{\pi x}{a}\right)e^{+iE_1t/\hbar} + \sqrt{\frac{1}{a}}\left(\frac{2\pi}{a}\right)\cos\left(\frac{2\pi z}{a}\right)e^{iE_2t/\hbar}$$

Algebra shows that:

$$\begin{aligned}\psi^*(x,t)\frac{d\psi(x,t)}{dx} &= \frac{\pi}{a^2}\sin\left(\frac{\pi x}{a}\right)\cos\left(\frac{\pi x}{a}\right)\\ &+ \frac{2\pi}{a^2}\sin\left(\frac{2\pi x}{a}\right)\cos\left(\frac{2\pi x}{a}\right)\\ &+ \frac{2\pi}{a^2}\sin\left(\frac{\pi x}{a}\right)\cos\left(\frac{2\pi x}{a}\right)e^{-iE_1t/\hbar}\\ &+ \frac{\pi}{a^2}\sin\left(\frac{2\pi x}{a}\right)\cos\left(\frac{\pi x}{a}\right)e^{-iE_1t/\hbar}\end{aligned}$$

We also find that:

$$\begin{aligned}\psi(x,t)\frac{d\psi^*(x,t)}{dx} &= \frac{\pi}{a^2}\sin\left(\frac{\pi x}{a}\right)\cos\left(\frac{\pi x}{a}\right)\\ &+ \frac{2\pi}{a^2}\sin\left(\frac{2\pi x}{a}\right)\cos\left(\frac{2\pi x}{a}\right)\\ &+ \frac{2\pi}{a^2}\sin\left(\frac{\pi x}{a}\right)\cos\left(\frac{2\pi x}{a}\right)e^{iE_1t\hbar}\\ &+ \frac{\pi}{a^2}\sin\left(\frac{2\pi x}{a}\right)\cos\left(\frac{\pi x}{a}\right)e^{-iE_1t/\hbar}\end{aligned}$$

$$
\begin{aligned}
\Rightarrow \psi^*(x,t)\frac{d\psi(x,t)}{dt} &= \psi^*(x,t)\frac{d\psi(x,t)}{dx} \\
&\quad + \frac{2\pi}{a^2}\sin\left(\frac{\pi x}{a}\right)\cos\left(\frac{2\pi x}{a}\right)e^{-i(E_1-E_2)t/\hbar} \\
&\quad - e^{-i(E_1-E_2)t/\hbar} \\
&= \frac{2\pi}{a^2}\sin\left(\frac{\pi x}{a}\right)\cos\left(\frac{2\pi x}{a}\right)(2i)\sin[(E_1-E_2)t/\hbar] \\
&\quad - \frac{\pi}{a^2}\sin\frac{2\pi x}{a}\cos\left(\frac{\pi x}{a}\right)(2i)\sin[(E_1-E_2)t/\hbar] \\
&= (2i)\sin[(E_1-E_2)t/\hbar]\frac{\pi}{a^2}\left[2\sin\left(\frac{\pi x}{a}\right)\cos\left(\frac{2\pi x}{a}\right)\right. \\
&\quad \left. - \sin\left(\frac{2\pi x}{a}\right)\cos\left(\frac{\pi x}{a}\right)\right]
\end{aligned}
$$

Using

$$
j(x,t) = \left(\frac{\hbar}{2im}\right)\left[\psi^*(x,t)\frac{d\psi(x,t)}{dx} - \psi(x,t)\frac{d\psi^*(x,t)}{dx}\right]
$$

we find that for this wavefunction:

$$
\begin{aligned}
j(x,t) &= \left(\frac{\hbar}{2im}\right)(2i)\sin[(E_1-E_2)t/\hbar]\frac{\pi}{a^2}\left[2\sin\left(\frac{\pi x}{a}\right)\cos\left(\frac{2\pi x}{a}\right)\right. \\
&\quad \left. - \sin\left(\frac{2\pi x}{a}\right)\cos\left(\frac{\pi x}{a}\right)\right] \\
&= \left(\frac{\hbar}{m}\right)\sin[(E_1-E_2)t/\hbar]\frac{\pi}{a^2}\left[2\sin\left(\frac{\pi x}{a}\right)\cos\left(\frac{2\pi x}{a}\right)\right. \\
&\quad \left. - \sin\left(\frac{2\pi x}{a}\right)\cos\left(\frac{\pi x}{a}\right)\right]
\end{aligned}
$$

Quiz

1. (a) Set $\psi(x,t) = \phi(x) f(t)$ and then apply the Schrödinger equation to show that this leads to a solution of the $f(t) = e^{-iEt/\hbar}$

   (b) For the equation

$$\frac{d^2\psi}{dx^2} + k^2\psi = 0$$

   verify that the solution is $\psi(x) = A\sin(kx) + B\cos(kx)$.

2. Let

$$\psi(x) = C\frac{1+ix}{1+ix^2}$$

   defined for $-\infty < X < \infty$

   (a) Show that

$$|\psi(x)|^2 = C^2\frac{1+x^2}{1+x^4}$$

   (b) Normalize ψ and show that

$$C = \frac{1}{\sqrt{\sqrt{2}\,\pi}}$$

   (c) Show that the probability that the particle is found between $0 \le x \le 1$ is 0.52

3. Let a wavefunction be given by:

$$\psi(x,t) = \begin{cases} C\frac{1}{x}e^{i\omega t} & \text{for } 1 \le x \le 2 \\ 0 & \text{otherwise} \end{cases}$$

   (a) Show that if $\psi(x,t)$ is normalized then $C = \sqrt{2}$.

   (b) Show that the probability of finding the particle between $\frac{3}{2} \le x \le 2$ is $\frac{1}{3}$.

4. For the wavefunction:

$$\psi(x) = \sqrt{\frac{54}{\pi}}\frac{1}{x^2+9}$$

   defined for $-\infty < x < \infty$, show that $\langle x \rangle = 0$ and $\langle x^2 \rangle = 9$.

5. Suppose that a wavefunction is given by:

$$\psi(x) = A(x^5 - ax^3)$$

Find Δx and Δp for this wavefunction.

6. Determine whether or not the operator X is Hermitian. What about iX? If not, why not?

7. Let

$$\psi(x,t) = [Ae^{ipx\hbar} + Be^{-ipx\hbar}]e^{-ip^2t/2m\hbar}$$

Find the probability current and check to see if the continuity equation is satisfied.

8. A particle in an infinite square well of dimensions $0 \le x \le a$ is in a state described by:

$$\psi(x,t) = i\sqrt{\frac{3}{2}}\sqrt{\frac{2}{a}}\sin\left(\frac{\pi x}{a}\right)e^{-iE_1t/\hbar} + \frac{1}{\sqrt{2a}}\sin\left(\frac{3\pi x}{a}\right)e^{-iE_3t/\hbar}$$

(a) Is $\psi(x,t)$ normalized?

(b) What values can be measured for the energy and with what probabilities?

(c) Find $\langle x \rangle$ and $\langle p \rangle$ for this state. Do these quantities depend on time? If so, is it true that $md\langle x \rangle/dt = \langle p \rangle$?

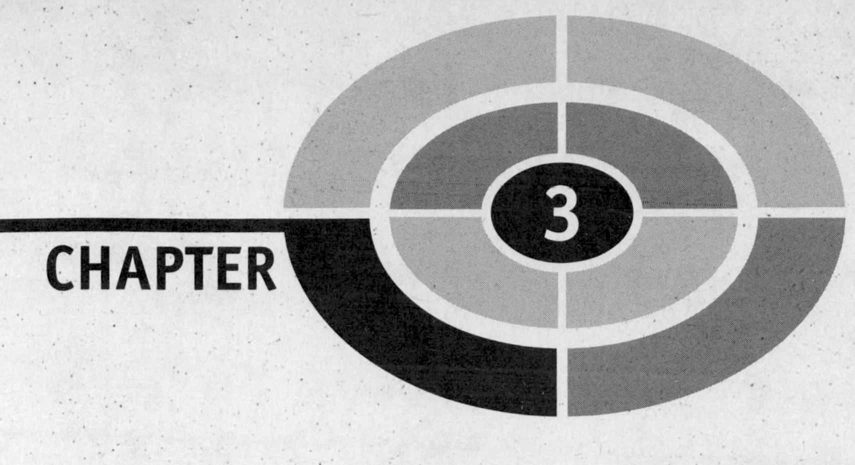

The Time Independent Schrödinger Equation

For a particle of mass m in a potential V, the time-independent Schrödinger equation is written as:

$$-\frac{\hbar^2}{2m}\frac{d^2\psi}{dx} + V\psi = E\psi$$

where E is the energy. The Hamiltonian operator is defined to be:

$$H = -\frac{\hbar^2}{2m}\frac{d^2}{dx^2} + V$$

Therefore the time independent Schrödinger equation has the form of an eigenvalue equation:

$$H\psi = E\psi$$

where the eigenvalues of H are the possible energies E of the system and the eigenfunctions of H are wavefunctions. We proceed to solve this equation under different conditions.

The Free Particle

When $V = 0$ we obtain the solution for a free particle. The Schrödinger equation becomes:

$$-\frac{\hbar^2}{2m}\frac{d^2\psi}{dx} = E\psi$$

Moving all terms to the left side:

$$-\frac{\hbar^2}{2m}\frac{d^2\psi}{dx} - E\psi = 0$$

Now multiply through by $-2m/\hbar^2$:

$$\frac{d^2\psi}{dx^2} + \frac{2mE}{\hbar^2}\psi = 0$$

Now define the wavenumber k using the relationship:

$$k^2 = \frac{2mE}{\hbar^2}$$

We use the wavenumber to define the energy and momentum of the free particle.

Definition: Kinetic Energy and Momentum of a Free Particle

Defined in terms of wavenumber the kinetic energy of a free particle is:

$$E = \frac{\hbar^2 k^2}{2m}$$

The momentum of a free particle is:

$$p = \hbar k$$

The wavenumber k can assume any value. Written in terms of wavenumber, the equation becomes:

$$\frac{d^2\psi}{dx^2} + k^2\psi = 0$$

There are two possible solutions to this equation:

$$\psi_1(x) = Ae^{ikx},\ \psi_2(x) = Ae^{-ikx}$$

As an example, we try the first solution. The first derivative gives:

$$\psi_1(x)' = ikAe^{ikx}$$

For the second derivative, we find:

$$\psi_1(x)'' = (ik)ikAe^{ikx} = -k^2Ae^{ikx} = -k^2\psi_1$$

Therefore it is clear that:

$$\frac{d^2\psi}{dx^2} + k^2\psi = -k^2\psi + k^2\psi = 0$$

and the equation is satisfied. You can try this for the other solution and get the same result. We have found two solutions that solve the equation and correspond to the same energy E. This situation is known as degeneracy.

Definition: Degeneracy

Degeneracy results when two (or more) different solutions to an eigenvalue equation correspond to the *same* eigenvalue.

For the current example, this means that a measurement of the energy that finds:

$$E = \frac{\pi^2k^2}{2m}$$

cannot determine if the particle is in state ψ_1 or state ψ_2. In order to distinguish between the two states, an additional measurement is necessary. This requirement can be accomodated with the momentum operator. Recalling that the momentum operator is represented by:

$$p = -i\hbar\frac{d}{dx}$$

We apply the momentum operator to the two wavefunctions that solve the free particle equation and find:

$$p\psi_1(x) = -i\hbar\frac{d}{dx}(Ae^{ikx}) = -i\hbar(ik)Ae^{ikx} = (\hbar k)Ae^{ikx} = (\hbar k)\psi_1(x)$$

$\Rightarrow Ae^{ikx}$ is an eigenfunction of the momentum operator with eigenvalue $\hbar k$. This wavefunction describes a particle moving to the *right* along the x-axis.

The same procedure applied to $\psi_2(x)$ shows that Ae^{-ikx} is an eigenfunction of the momentum operator with eigenvalue $-\hbar k$. This wavefiinction describes a particle moving to the *left* along the x-axis.

So we see that these eigenfunctions are *simultaneous eigenfunctions of energy and momentum.*

In the last chapter, it was emphasized that normalization is a critical property that a wavefunction must have. The free particle solution cannot be normalized. In either case:

$$\int_{-\infty}^{\infty} |\psi(x)|^2 \, dx = A^2 \int_{-\infty}^{\infty} dx \to \infty$$

So far, we can make the following observations about the free particle solution:

1. The free particle solutions, like for the infinite square well, are given in terms of Ae^{ikx}. This is not surprising because inside the infinite square well, the potential is zero.
2. The boundary conditions of the square well put limits on the values that k could assume. In this case there are no boundary conditions—and therefore k can assume any value.
3. The wavefunction is not normalizable.

In deriving the free particle wavefunction, we solved the time-independent Schrödinger equation. This is an implicit way of stating that we are assuming separable solutions. Since we have found such solutions are not normalizable, this means that a separable solution, which represent stationary states of definite energy, are not possible for a free particle. The way to get around this problem is to take advantage of the linearity of the Schrödinger equation and form a superposition of solutions together (by adding over values of k) to form a new solution. We can always write the exponential solutions in terms of sin and cosine functions, so let's consider a wavefunction of the form:

$$\psi_k(x) = \cos(kx)$$

and see what happens as we add up various values of k. Here are plots of $\cos(2x)$, $\cos(4x)$, and $\cos(6x)$ (Figure 3-1 and Figure 3-2):

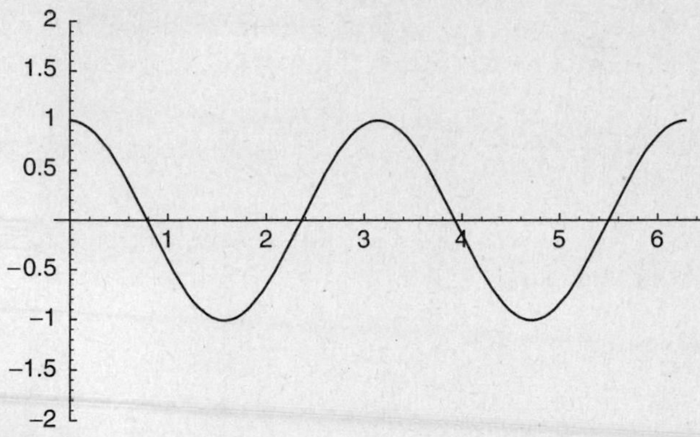
2
1.5
1
0.5
−0.5
−1
−1.5
−2
1 2 3 4 5 6

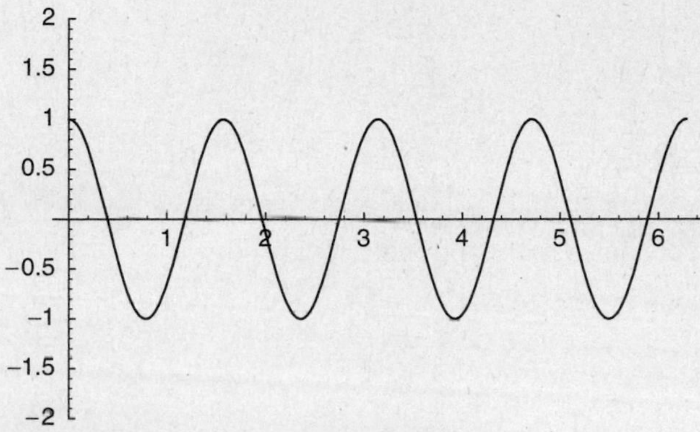
2
1.5
1
0.5
−0.5
−1
−1.5
−2
1 2 3 4 5 6

Fig. 3-1

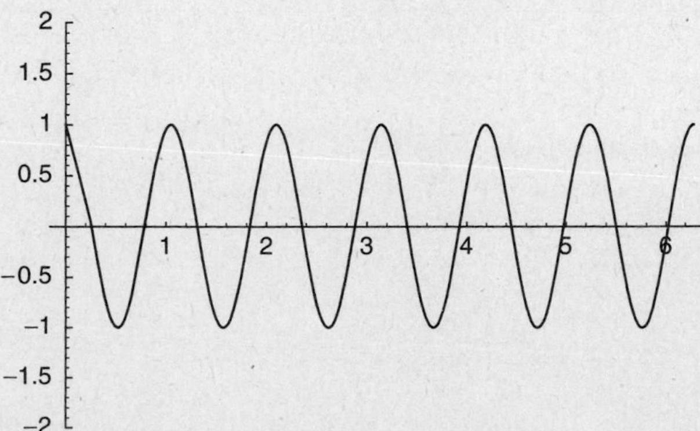
2
1.5
1
0.5
−0.5
−1
−1.5
−2
1 2 3 4 5 6

Fig. 3-2

If we add these waves together, the waves add in some places and cancel in others, and we start to get a localized waveform (Figure 3-3):

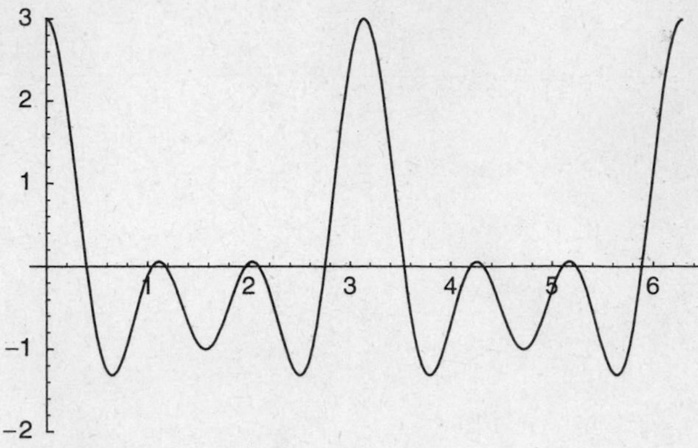

Fig. 3-3

Now look what happens when we add waves with many values of k (Figure 3-4). The wavepacket becomes more and more localized:

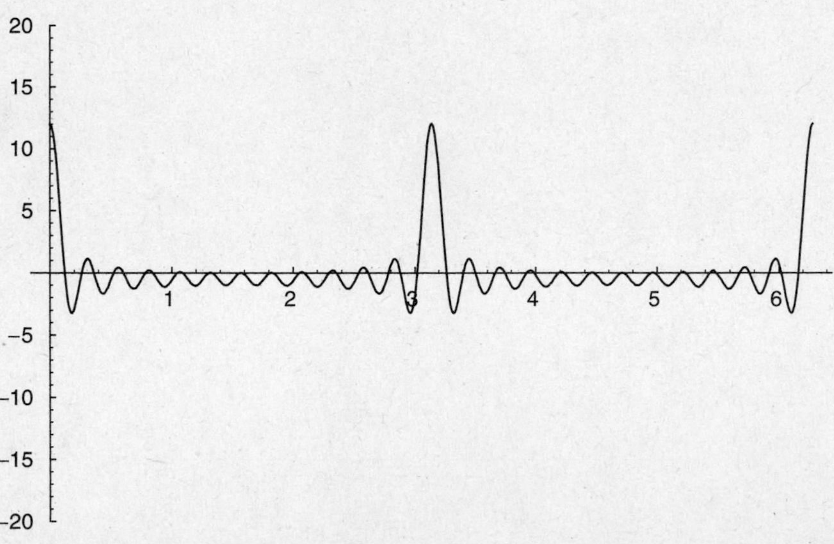

Fig. 3-4

For the free particle, this type of addition process of waves, gives us a localized *wavepacket*.

EXAMPLE 3.1

Let $\psi_1(x) = \cos(2x)$ and $\psi_2(x) = \cos(10x)$. Describe and plot the wavepacket formed by the superposition of these two waves.

SOLUTION

To form the wavepacket, we use the trig identity:

$$cos(A) - cos(B) = -2 \sin \frac{A+B}{2} \sin \frac{A-B}{2}$$

$$\psi_1(x) + \psi_2(x) = \cos(10x) - \cos(2x) = -2 \sin(6x) \sin(4x)$$

The plot of $\psi_2(x)$ is (figure 3-5):

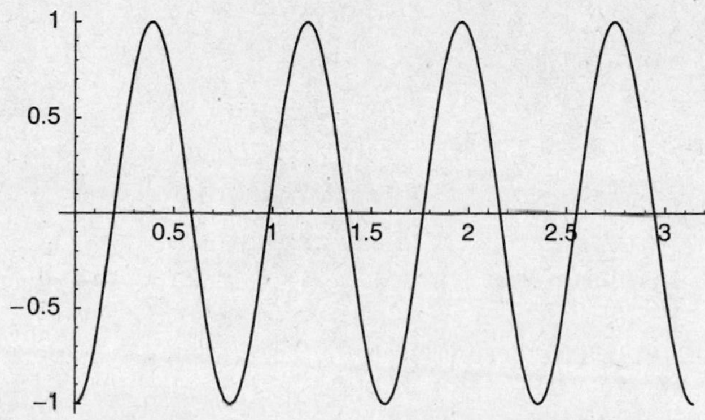

Fig. 3-5

The plot of $\psi_1(x)$ is (Figure 3-6):

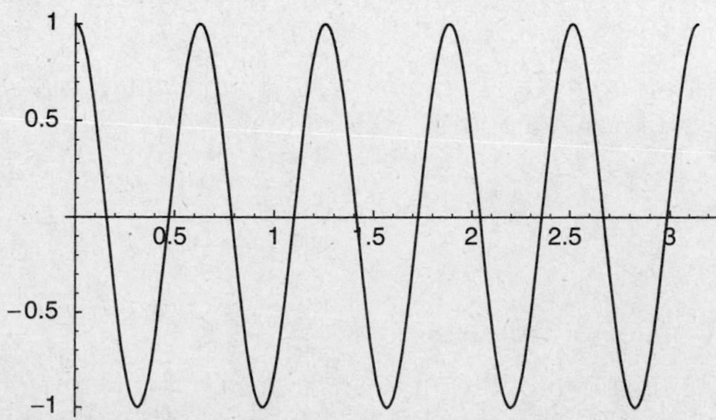

Fig. 3-6

Finally we plot $\psi_1(x)+\psi_2(x)$. Notice how it shows hints of attaining a localized waveform (Figure 3-7):

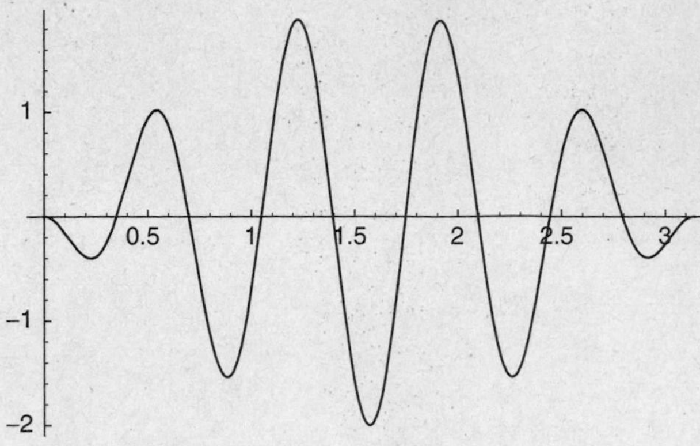

Fig. 3-7

The way to construct a wavepacket by adding up over a large number of values of k is to integrate. In other words, wavepackets are constructed by using the Fourier transform.

Following the procedure described in the last chapter, we append on the time dependence of the wavefunction by writing:

$$\psi(x,t) = Ae^{ikx}e^{-iEt/\hbar}$$

Using $E = \frac{\hbar^2 k^2}{2m}$, we can write:

$$\psi(x,t) = A\exp\left[ik(x - \frac{\hbar k}{2m}t)\right]$$

The wave packet is formed by multiplying by a momentum space wavefunction $\phi(k)$ and adding up over values of k, which means we integrate:

$$\psi(x,t) = \frac{1}{\sqrt{2\pi}}\int_{-\infty}^{\infty}\phi(k)\exp\left[ik\left(x - \frac{\hbar k}{2m}t\right)\right]dk$$

Definition: Dispersion Relation

The dispersion relation is a functional relationship that defines angular Frequency ω in terms of wavenumber k, i.e., $\omega = \omega(k)$. The dispersion relation can be used to find the *phase velocity* and *group velocity* for a wavepacket.

Phase velocity is given by $v_{ph} = \frac{\omega}{k}$, while group velocity $v_g = \frac{d\omega}{dk}$. In the case of the free particle, we have:

$$E = \frac{\hbar^2 k^2}{2m}$$

The Einstein-Planck relations tell us that $E = \hbar\omega$, setting these equal:

$$E = \frac{\hbar^2 k^2}{2m} = \hbar\omega$$

Solving for ω gives:

$$\omega(k) = \frac{\hbar k^2}{2m}$$

Therefore the phase velocity and group velocity are:

$$v_{ph} = \frac{\omega}{k} = \frac{\hbar k}{2m}, \; v_g = \frac{d\omega}{dk} = \frac{\hbar k}{m},$$

Notice that the phase velocity is $\frac{1}{2}$ that of what the velocity of a classical particle would be, while the group velocity gives the correct classical velocity. In general these expressions will be more complicated. We have used the energy relation for a single free particle solution to the Schrödinger equation.

Let's now review the situation where the particle is limited to a fixed region of space and see how this forces a quantization of k (and therefore the energy). Another way to obtain a normalizable solution to the free particle wave function is to impose periodic boundary conditions. If the particle is constrained between $\pm L/2$, then normalization becomes:

$$\int_{-L/2}^{L/2} |\psi(x)|^2 \, dx = A^2 \int_{-L/2}^{L/2} dx$$

This leads to a solution for A:

$$\int_{-L/2}^{L/2} |\psi(x)|^2 \, dx = 1 \Rightarrow A = \frac{1}{\sqrt{L}}$$

The normalized wavefunction is:

$$\psi_1(x) = \frac{1}{\sqrt{L}} e^{ikx}$$

Periodic boundary conditions while requiring that the particle is confined between $\pm L/2$ are manifested by requiring that:

$$\psi_1(x) = \psi_1(x + L)$$

This condition leads to:

$$\frac{1}{\sqrt{L}}e^{ikx} = \frac{1}{\sqrt{L}}e^{ik(x+L)}$$

Which means that:

$$1 = e^{ikx} = \cos(kL) + i\,\sin(kL)$$

The left side is purely real and the right side has an imaginary term, so that imaginary piece must be zero. This can be true if:

$$kL = 2n\pi$$

where $n = 0, \pm 1, \pm 2, \pm 3 \ldots$. The wavenumber k which originally was found to be able to assume any value, is now restricted to the discrete cases defined by:

$$k = 2n\pi/L$$

Bound States and 1-D Scattering

We now consider some examples of scattering and bound states. Scattering involves a beam of particles usually *incident* upon a changing potential from negative x. These types of problems usually involve piecewise constant potentials that may represent barriers or wells of different heights.

Definition: A Wavefunction Bound to an Attractive Potential

If a system is bound to an attractive potential, the wavefunction is localized in space and must vanish as $x \to \pm\infty$. Restricting a wavefunction in this way limits the energy to a discrete spectrum. In this situation the total energy is less than zero.

If the energy is greater than zero, this represents a scattering problem. In a scattering problem we follow these steps:

1. Divide up the problem into different regions, with each region representing a different potential. Here is a plot showing a piecewise constant potential. Each time the potential changes, we define a new region. In the plot that follows (Figure 3-8) there are four regions.

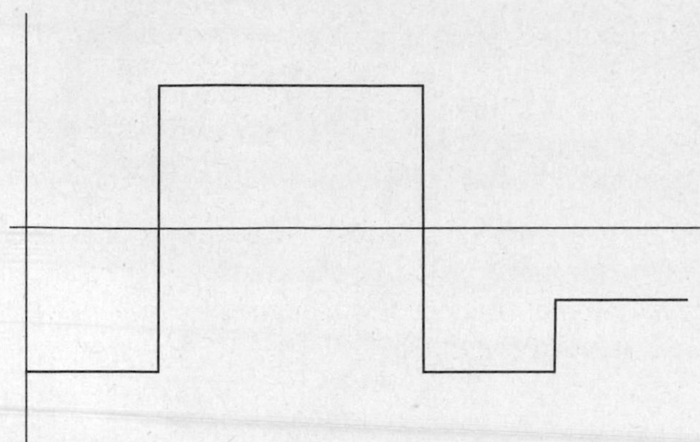

Fig. 3-8

2. Solve the Schrödinger equation in each region. If the particles are in an unbound region ($E > V$), the wavefunction will be in the form of free particle states:

$$\psi = Ae^{ikx} + Be^{-ikx}$$

Remember that the term Ae^{ikx} represents particles moving from the left to the right (from negative x to positive x), while Be^{-ikx} represents particles moving from right to left. In a region where the particles encounter a barrier ($E < V$) the wavefunction will be in the form:

$$\psi = Ae^{\rho x} + Be^{-\rho x}$$

3. The first goal in a scattering problem is to determine the constants A, B, and C. This is done using two facts:
   (a) The wavefunction is continuous at a boundary. If we label the boundary point by a then

   $$\psi_{left}(a) = \psi_{right}(a).$$

   (b) The first derivative of the wavefunction is continuous at a boundary. Therefore we can use:

   $$\frac{d}{dx}\psi_{left}(a) = \frac{d}{dx}\psi_{right}(a).$$

We will see below that there is one exception to rule (b), the derivative of the wavefunction is not continuous across a boundary when that boundary is a Dirac delta function.

4. An important goal in scattering problems is to solve for the *reflection* and *transmission* coefficients. This final piece of the problem tells you whether particles make it past a barrier or not.

Definition: Transmission and Reflection Coefficients

A scattering problem at a potential barrier will have an incident particle beam, reflected particle beam, and transmitted particle beam. The transmission coefficient T and reflection coefficient R are defined by:

$$T = \left| \frac{J_{trans}}{J_{inc}} \right| and R = \left| \frac{J_{ref}}{J_{inc}} \right|$$

where the current density J is given by:

$$J = \frac{\hbar}{2mi} \left(\psi^* \frac{d\psi}{dx} - \psi \frac{d\psi^*}{dx} \right)$$

Continuing our summary of scattering:

5. When particles with $E < V$ encounter a potential barrier represented by V, there will be penetration of the particles into the classically forbidden region. This is represented by a wavefunction that decays exponentially in that region. If particles are moving from left to right, the wavefunction is of the form:

$$\psi = Ae^{-\rho x}$$

While particles penetrate into the region, there is 100% reflection. So we expect $R = 1$, $T = 0$.

6. The reflection and transmission coefficients give us the fraction of particles that are reflected or transmitted, so $R + T = 1$.
7. When the energy of the particles is greater than the potential on both sides of a barrier $(E > V)$, the wavefunction will be:

$$\psi_L = Ae^{ik_i x} + Be^{ik_j x}$$

on the left of the barrier while on the other side of the discontinuity, it will be:

$$\psi_R = Ce^{ik_i x} + De^{-ik_j x}$$

The particular constraints of the problem may force us to set one or more of A, D, C, D to zero. The constants k_1 and k_2 are found by solving the Schrödinger equation in each region:

$$\frac{d^2\psi}{dx^2} + \frac{2m}{\hbar^2}(E - V_i)\psi = 0$$

where E is the energy of the incident particles and V_i is the potential in region i (and we are assuming that $E > V_i$). The constants k_1 and k_2 are set to:

$$k_i^2 = \frac{2m}{\hbar^2}(E - V_1)$$

As E increases we find that $k_2 \to k_1$ and $R \to 0$, $T \to 1$. On the other hand, if E is smaller and approaches V_2 then $R \to 1$ and $T \to 0$.

EXAMPLE 3.2

e.g.

A beam of particles coming from $x = -\infty$ meets a potential barrier described by $V(x) = V$, where V is a positive constant, at $x = 0$. Consider the incident beam of particles to have energy $0 < E < V$. Find the wavefunction for $x < 0$ and for $x > 0$, and describe the transmission and reflection coefficients for this potential.

SOLUTION

A plot of the potential is shown below (Figure 3-9). The energy of the incoming particles is indicated qualitatively by the dashed line, while the solid line represents $V(x) = V$ for $x > 0$:

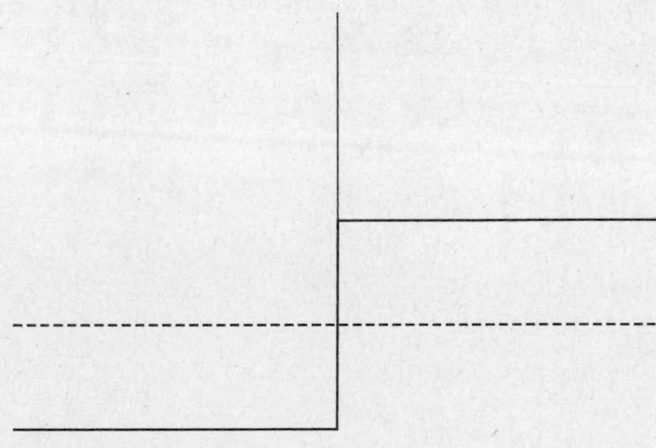

Fig. 3-9

To solve a problem like this, we divide the problem into two regions. We label $x < 0$ as Region I. In this region, the potential is zero and so the beam of particles is described by the free particle wavefunction. A particle coming from the left and moving to the right, which we label as the incident wavefunction, is described by:

$$\psi_{inc}(x) = Ae^{ikx}$$

When the beam encounters a potential barrier, some particles will be reflected back. So the total wavefunction in Region I is given by a sum of incident + reflected wavefunctions. The reflected wavefunction, which moves to the left, is $\sim e^{-ikx}$. We write the total wavefunction in Region I as:

$$\psi = \psi_{inc}(x) + \psi_{ref}(x) = Ae^{ikx} + Be^{-ikx}$$

For $x > 0$, the Schrödinger equation becomes:

$$-\frac{\hbar^2}{2m}\frac{d^2\psi}{dx^2} + V\psi = E\psi$$

Moving all terms to the left and dividing through by $-\frac{\hbar^2}{2m}$ we obtain:

$$\frac{d^2\psi}{dx^2} + \frac{2m}{\hbar^2}(V - E)\psi = 0$$

Since $0 < E < V$ the term:

$$\frac{2m}{\hbar^2}(V - E) > 0$$

Calling this term β^2, the equation becomes:

$$\frac{d^2\psi}{dx^2} + \beta^2\psi = 0$$

There are two solutions to this familiar equation:

$$e^{\beta x} \quad and \quad e^{-\beta x}$$

We reject the first solution because it blows up as $x \to \infty$ and is physically unacceptable. Therefore in Region II, we find that:

$$\psi_{trans}(x) = ce^{-\beta x}$$

So, to the left of the barrier, the wavefunction is oscillatory, while to the right it exponentially decays. We can represent this qualitatively by the following plot (Figure 3-10):

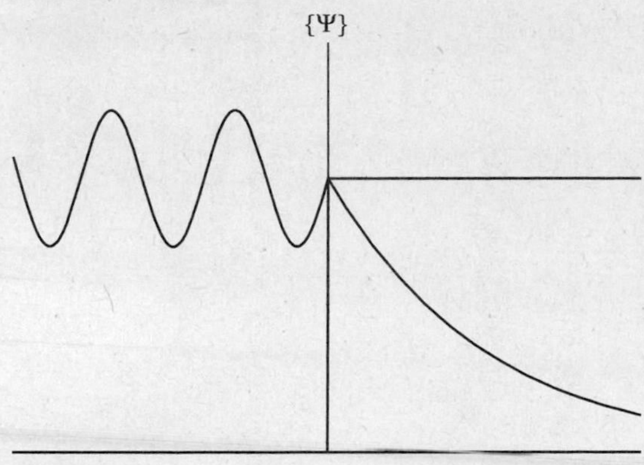

Fig. 3-10

Definition: Conditions the wavefunction must obey at a discontinuity

The wavefunction must obey two conditions:

1. ψ must be continuous
2. The first derivative $d\psi/dx$ must be continuous

The boundary in this problem is at $x = 0$. To the left of the boundary, the wavefunction is:

$$\psi = Ae^{ikx} + Be^{-ikx}$$

Setting $x = 0$ this gives $\psi(0) = A + B$. For $x > 0$,

$$\psi = ce^{-\rho x} \Rightarrow \psi(0) = C$$

Matching up the wavefunction at $x = 0$ gives the condition:

$$A + B = C$$

For $x < 0$, $d\psi/dx$ is:

$$d\psi/dx = ik(Ae^{ikx} - Be^{-ikx})$$

While for $x > 0$ we have:

$$d\psi/dx = \beta Ce^{-\rho x}$$

Letting $x = 0$ and setting both sides equal to each other, we obtain:

$$ik(A - B) = \beta C$$

Using $A + B = C$, we have:

$$ik(A - B) = \beta(A + B)$$

Solving allows us to write:

$$B = \frac{ik + \beta}{ik - \beta} A \quad or \frac{B}{A} = \frac{ik + \beta}{ik - \beta} = \frac{ik\left(1 + \frac{\beta}{ik}\right)}{ik\left(1 - \frac{\beta}{ik}\right)} = \frac{1 - i\frac{\beta}{k}}{1 + i\frac{\beta}{k}}$$

Again using $A + B = C$, we find:

$$C = A + B = A + \frac{ik + \beta}{ik - \beta} A = A(1\frac{ik + \beta}{ik - \beta}) = A\left(\frac{ik + \beta}{ik - \beta} + \frac{ik + \beta}{ik - \beta}\right) = A\frac{i2k}{ik - \beta}$$

$$\Rightarrow \frac{C}{A} = \frac{i2k}{ik - \beta}$$

Now we calculate the probability current density for each wavefunction. This is given by:

$$J = \frac{\hbar}{2mi}\left(\psi^* \frac{d\psi}{dx} - \psi \frac{d\psi^*}{dx}\right)$$

For, $\psi_{inc}(x) = Ae^{ikx}$, we find:

$$\psi^*_{inc}(x) = A^* e^{-ikx}, \frac{d}{dx}\psi_{inc}(x) = ikAe^{ikx}, \frac{d}{dx}\psi^*_{inc}(x) = -ikA^* e^{-ikx}$$

Therefore:

$$\psi^* \frac{d\psi}{dx} = (A^* e^{-ikx})(ikAe^{ikx}) = ikA^2$$

$$\psi \frac{d\psi^*}{dx} = (Ae^{ikx})(-ikA^* e^{ikx}) = -ikA^2$$

$$\Rightarrow \psi^* \frac{d\psi}{dx} = -\psi \frac{d\psi^*}{dx} = ikA^2 - (-ikAe^2) = i2kA^2$$

So we obtain:

$$J_{inc} = \frac{\hbar}{2mi}\left(\psi^* \frac{d\psi}{dx} - \psi \frac{d\psi^*}{dx}\right) = \frac{\hbar}{2mi}(i2kA^2) = \frac{\hbar}{m}(kA^2)$$

A similar procedure shows that:

$$J_{ref} = \frac{\hbar}{m}(kB^2)$$

Now we calculate the current density for the transmitted current. The wavefunction for $x > 0$ is:

$$\psi = ce^{-\beta x}, \Rightarrow \psi^* = c^* e^{-\beta x}$$

$$d\psi/dx = -\beta C e^{-\beta x}, d\psi^*/dx = -\beta C^* e^{-\beta x}$$

$$J_{trans} = \frac{\hbar}{2mi}(\psi^* \frac{d\psi}{dx} - \psi \frac{d\psi^*}{dx}) = \frac{\hbar}{2mi}[C^* e^{-\beta x}(-\beta C e^{-\beta x}) - C e^{-\beta x}(-\beta C^* e^{-\beta x})]$$

$$= \frac{\hbar}{2mi}[-\beta C^2 e^{-2\beta x} + \beta C^2 e^{-2\beta x}] = 0$$

The reflection coefficient is found to be:

$$R = \left|\frac{J_{ref}}{J_{inc}}\right| = \left|\frac{B^2}{A^2}\right| = \left|\frac{ik - \beta}{ik + \beta}\right| = \sqrt{\left(\frac{ik - \beta}{ik + \beta}\right)\left(\frac{-ik - \beta}{-ik + \beta}\right)} = \sqrt{\frac{k^2 + \beta^2}{k^2 + \beta^2}} = 1$$

$$T = \left|\frac{J_{trans}}{J_{inc}}\right| = 0$$

As expected, there is particle penetration into the barrier, but this tells us that no particles make it past the barrier—there is 100% reflection.

EXAMPLE 3.3

In this example we consider the same potential barrier, but this time we take $E > V$.

SOLUTION

This time the energy of the incoming beam of particles is greater than that of the potential barrier. A schematic plot of this situation is shown in Figure 3-11:

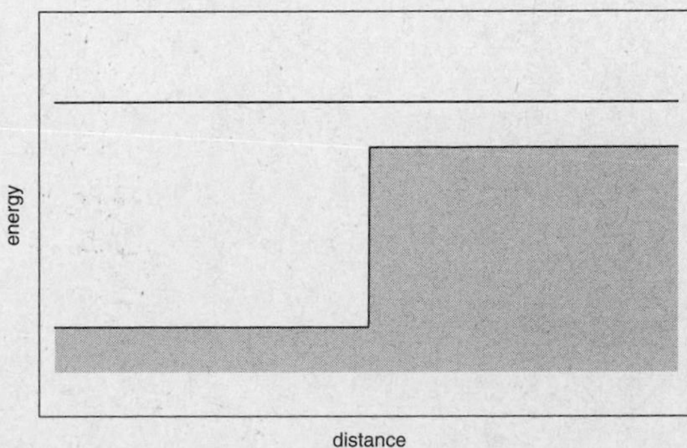

Fig. 3-11

We call the region left of the barrier ($x < 0$) Region I while Region II is $x > 0$. To the left of the barrier in Region I, the situation is exactly the same as it was in the previous example. So the wavefunctions are:

$$\psi_{inc}(x) = Ae^{ik_1x}, \psi_{ref}(x) = Be^{ik_1x}$$

This time, since $E > V$, the Schrödinger equation has the form:

$$\frac{d^2\psi}{dx^2} + \frac{2m}{\hbar^2}(V - E)\psi = 0$$

$E > V$, and so $V - E < 0$ and we have:

$$\frac{d^2\psi}{dx^2} + k_2^2\psi = 0$$

where

$$k_2^2 = \frac{2m}{\hbar^2}(E - V)$$

represents the wavenumber in Region II. The solution to the Schrödinger equation in Region II is:

$$\psi_{trans}(x) = Ce^{ik_2x}, De^{-ik_2x}$$

However, the term De^{-ik_2x} represents a beam of particles moving to the left towards $-x$ and from the direction of $x = +\infty$. The problem states that the incoming beam comes from the left, therefore there are no particles coming from the right in Region II. Therefore $D = 0$ and we take the wavefunction to be:

$$\psi_{trans}(x) = Ce^{ik_2x}$$

The procedure to determine the current densities in Region I is exactly the same as it was in the last problem. Therefore:

$$J_{inc} = \frac{\hbar}{m}(k_1A^2)$$

$$J_{ref} = \frac{\hbar}{m}(k_1B^2)$$

In Region II, this time the calculation of the current density is identical to that used for J_{inc} and J_{ref}, however we substitute $k \rightarrow \alpha$ giving:

$$J_{trans} = \frac{\hbar}{m}(k_2C^2)$$

Continuity of the wavefunction at $x = 0$ leads to the same condition that was found in the previous example:

$$A + B = C$$
$$\Rightarrow 1 + \frac{B}{A} = \frac{C}{A}$$

In Region I, the derivative at $x = 0$ is also the same as it was in the previous example:

$$\psi(x = 0) - ik_1(A - B)$$

In Region II, this time

$$\psi_{trans}(x) = Ce^{ik_2x}, \Rightarrow \psi_{trans} = ik_2Ce^{-ik_2x}$$

And so, the matching condition of the derivatives at $x = 0$ becomes:

$$ik_1(A - B) = ik_2C, \Rightarrow C = \frac{k_1(A - B)}{k_2}$$

From $A + B = C$, we substitute $B = C - A$ and obtain:

$$C = \frac{k_1(A - (C - A))}{k_2} \Rightarrow C(1 + \frac{k_1}{k_2}) = 2\frac{k_1}{k_2}A$$

or:

$$\frac{C}{A} = \frac{2k_1}{k_1 + k_2}$$

This leads to:

$$T = \frac{4k_1^2}{(k_1 + k_2)^2}$$
$$R = (2k_1 - k_2)/2$$

It can be verified that:

$$T + R = 1$$

Recalling that in Region I, we have a free particle so:

$$k_1^2 = \frac{2mE}{\hbar^2}$$

In Region II, we found that:

$$k_1^2 = \frac{2m}{\hbar^2}(E - V)$$

Therefore, the ratio between the wave numbers in the two regions is given by:

$$\frac{k_1^2}{k_2^2} = \frac{\frac{2mE}{\hbar^2}}{\frac{2m}{\hbar^2}(E - V)} = \frac{E}{E - V} = 1 - \frac{E}{V}$$

EXAMPLE 3.4

Incident particles of energy E coming from the direction of $x = -\infty$ are incident on a square potential barrier V, where $E < V$ (Figure 3-12):

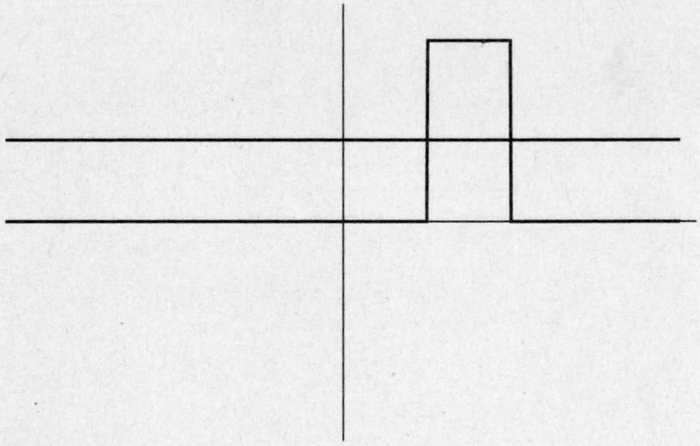

Fig. 3-12

The potential is V for $0 < x < a$ and is 0 otherwise. Find the transmission coefficient.

SOLUTION

We divide this problem into three regions:

Region I: $-\infty < x < 0$

Region II: $0 < x < a$

Region III: $a < x < \infty$

With the definitions:

$$k^2 = \frac{2mE}{\hbar^2}, \beta^2 = \frac{2m(V-E)}{\hbar^2}$$

the wavefunctions for each of the three regions are:

$$\phi_I(x) = Ae^{ikx} + Be^{-ikx}$$
$$\phi_{II}(x) = Ce^{\beta x} + De^{-\beta x}$$
$$\phi_{III}(x) = Ee^{ikx} + Fe^{-ikx}$$

The incident particles are a coming from $x = -\infty$, so $F = 0$. Matching the wavefunctions at $x = 0$ gives:

$$A + B = C + D$$

Matching at $x = a$ gives:

$$Ce^{\beta a} + De^{-\beta a} = Ee^{ika}$$

Continuity of the first derivative at $x = 0$ gives:

$$ik(A - B) = \beta(C - D)$$

and at $x = a$:

$$C\beta e^{\beta a} + D\beta e^{-\beta a} = ikEe^{ika}$$

Combining the matching condition of the wavefunction and the first derivative at $x = 0$ yields:

$$A = \frac{ik+\beta}{2ik}C + \frac{ik-\beta}{2ik}D$$

Doing the same at $x = a$ gives:

$$C = [\frac{ik+\beta}{2\beta}e^{(ik-\beta)a}]E, D = [\frac{\beta-ik}{2\phi}e^{(ik-\beta)a}]E$$

Combining these two results, we find:

$$A = \frac{ik+\beta}{2ik}C + \frac{ik+\beta}{2ik}D = \frac{ik+\beta}{2ik}[\frac{ik+\beta}{2ik}e^{(ik-\beta)a}]E + \frac{ik+\beta}{2ik}[\frac{\beta+ik}{2\beta}e^{(ik-\beta)a}]E$$

$$= [\frac{(ik+\beta)^2}{4ik\beta}e^{(ik-\beta)a} - \frac{(ik+\beta)^2}{4ik\beta}e^{(ik-\beta)a}]E$$

$$\Rightarrow$$

$$\frac{A}{E} = [\frac{(ik+\beta)^2}{4ik\beta}e^{(ik-\beta)a} - \frac{(ik+\beta)^2}{4ik\beta}e^{(ik-\beta)a}]$$

Recalling that:

$$\cosh(x) = \frac{e^x + e^{-x}}{2}, \sinh(x) = \frac{e^x - e^{-x}}{2}$$

We will rewrite the ratio A/E. First we proceed as follows:

$$\begin{aligned}
\frac{A}{E} &= [\frac{(ik+\beta)^2}{4ik\beta} e^{(ik-\beta)a} - \frac{(ik-\beta)^2}{4ik\beta} e^{(ik-\beta)a}] \\
&= [\frac{(ik+\beta)^2}{4ik\beta} e^{ika} e^{-\beta a} - \frac{(ik+\beta)^2}{4ik\beta} e^{ika} e^{\beta a}] \\
&= e^{ika}[\frac{(ik+\beta)^2}{4ik\beta} e^{-\beta a} - \frac{(ik+\beta)^2}{4ik\beta} e^{\beta a}] \\
&= \frac{e^{ika}}{4ik\beta}[(ik+\beta)^2 e^{-\beta a} - (ik+\beta)^2 e^{\beta a}]
\end{aligned}$$

Now we expand the terms:

$$\begin{aligned}
(ik+\beta)^2 &= -k^2 + 2ik\beta + \beta^2 \\
(ik-\beta)^2 &= -k^2 - 2ik\beta + \beta^2
\end{aligned}$$

Putting this into the above result gives:

$$\frac{A}{E} = \frac{e^{ika}}{4ik\beta}[(-k^2 + 2ik\beta + \beta^2)e^{-\beta a} - (-k^2 + 2ik\beta + \beta^2)e^{\beta a}]$$

The transmission coefficient will be calculated using the modulus squared of this quantity. For a complex number z, we recall that $|z|^2 = zz^*$. The complex conjugate of this expression is:

$$\left(\frac{A}{E}\right)^* = \frac{e^{-ika}}{-4ik\beta}[(-k^2 + 2ik\beta + \beta^2)e^{-\beta a} - (-k^2 + 2ik\beta + \beta^2)e^{\beta a}]$$

Now,

$$\left(\frac{e^{ika}}{4ik\beta}\right)\frac{e^{-ika}}{-4ik\beta} = \frac{1}{16k^2\beta^2}$$

Now we make the following definition:

$$\gamma = (-k^2 + 2ik\beta + \beta^2), \Rightarrow \gamma^* = (-k^2 + 2ik\beta + \beta^2)$$

$$\frac{A}{E}\left(\frac{A}{E}\right)^* = \frac{1}{16k^2\beta^2}[\gamma e^{-\beta a} - \gamma^* e^{\beta a}][\gamma^* e^{-\beta a} - \gamma e^{\beta a}]$$

$$= \frac{1}{16k^2\beta^2}\left[|\gamma|^2 (e^{-2\beta a} + e^{2\beta a}) - \gamma^2 - (\gamma^*)^2\right]$$

$$= (-k^2 + 2ik\beta + \beta^2)(-k^2 - 2ik\beta + \beta^2)$$

$$= k^4 + 2ik^3\beta - k^2\beta^2 - 2ik^3\beta + 4k^2\beta^2 + 2ik\beta^3 - \beta^2k^2 - 2ik\beta^3 + \beta^4$$

$$= k^4 - 2k^2\beta^2 + \beta^4 = (k^2 - \beta^2)^2$$

$$(-k^2 + 2ik\beta + \beta^2)(-k^2 + 2ik\beta + \beta^2)$$

$$k^4 - 2ik^3\beta - \beta^2k^2 - 2ik^3\beta - 4k^2\beta^2 + 2ik\beta^3 - \beta^2k^2 + 2ik\beta^3 + \beta^4$$

$$\Rightarrow \gamma^2 + (\gamma^*)^2 = 2k^4 - 12\beta^2k^2 + 2\beta^4 = 2(k^2 - \beta^2)^2 - 8\beta^2k^2$$

Putting these results together we obtain:

$$\frac{A}{E}\left(\frac{A}{E}\right)^* = \frac{1}{16k^2\beta^2}\left[|\gamma|^2 (e^{-2\beta a} + e^{2\beta a}) - \gamma^2 - (\gamma^*)^2\right]$$

$$= \frac{1}{16k^2\beta^2}\left[(k^2 - \beta^2)^2(e^{-2\beta a} + e^{2\beta a}) - 2(k^2 - \beta^2)^2 + 8\beta^2k^2\right]$$

$$= \frac{1}{16k^2\beta^2}\left[2(k^2 - \beta^2)^2\frac{(e^{-2\beta a} + e^{2\beta a})}{2} - 2(k^2 - \beta^2)^2 + 8\beta^2k^2\right]$$

$$= \frac{1}{16k^2\beta^2}\left[2(k^2 - \beta^2)^2\cosh(2\beta a) - 2(k^2 - \beta^2)^2 + 8\beta^2k^2\right]$$

$$= \frac{(k^2 - \beta^2)^2}{8k^2\beta^2}\left[\cosh(2\beta a) - 1 + \frac{4\beta^2k^2}{(k^2 - \beta^2)^2}\right]$$

Now we use the hyperbolic trig identities:

$$\cosh(2a) = \cosh(a)^2 + \sinh(a)^2, \text{ and } 1 = \cosh(a)^2 - \sinh(a)^2$$

$$\Rightarrow \cosh(2a) - 1 = \cosh(a)^2 + \sinh(a)^2 - (\cosh(a)^2 - \sinh(a)^2) = 2\sinh(a)^2$$

The ratio then becomes:

$$\frac{A}{E}\left(\frac{A}{E}\right)^* = \frac{(k^2 - \beta^2)}{8k^2\beta^2}\left[2\sinh(\beta a)^2 + \frac{4\beta^2k^2}{(k^2 - \beta^2)^2}\right]$$

$$= \frac{(k^2 - \beta^2)^2}{4k^2\beta^2} \sinh(\beta a)^2 + \frac{1}{2}$$

Finally, the transmission coefficient is:

$$T = \left|\frac{E}{A}\right|^2 = \frac{1}{\frac{(k^2-\beta^2)^2}{4k^2\beta^2} \sinh(\beta a)^2 + \frac{1}{2}}$$

This differs from the predictions of classical physics. According to classical physics, a particle cannot make it past a barrier when $E < V$. However according to this result there is a non-zero probability that the particle can make it past the barrier. This quantum effect is called *tunneling*.

Parity

The parity operator P acts on a wavefunction in the following way:

$$P\psi(x) = \psi(-x)$$

Now consider the eigenvalue equation for P:

$$P\psi(x) = \lambda\psi(x)$$

Application of P a second time in the first equation gives:

$$P^2\psi(x) = P\psi(-x) = \psi(x)$$

We apply this to the eigenvalue equation:

$$P^2\psi(x) = P(\lambda\psi(x)) = \lambda^2\psi(x)$$

Equating these two results we have:

$$\psi(x) = \lambda^2\psi(x), \Rightarrow \lambda = \pm 1$$

Are the eigenvalues of parity? An even function for which $\psi(-x) = \psi(x)$ is seen to have the eigenvalue $+1$ and can be said to have even parity. An odd function for which $\psi(-x) = -\psi(x)$ corresponds to the eigenvalue -1 and is said to have odd parity. The even and odd parts of any wavefunction can be constructed using:

$$\psi_e(x) = \frac{1}{\sqrt{2}}\left[\psi(x) + \psi(-x)\right]$$

$$\psi_o(x) = \frac{1}{\sqrt{2}}\left[\psi(x) - \psi(-x)\right]$$

When the potential is symmetric, so that $V(x) = V(-x)$, the Hamiltonian H commutes with the parity operator. This means that if $\psi(x)$ is an eigenfunction of H, so is $P\ \psi(x)$.

EXAMPLE 3.5

Find the possible energies for the square well defined by:

$$V(x) = \psi(x) = \begin{cases} \text{-V} & for -a/2 < x < a/2 \\ 0 & Otherwise \end{cases}$$

SOLUTION

A qualitative plot of the potential is shown in Figure 3-13:

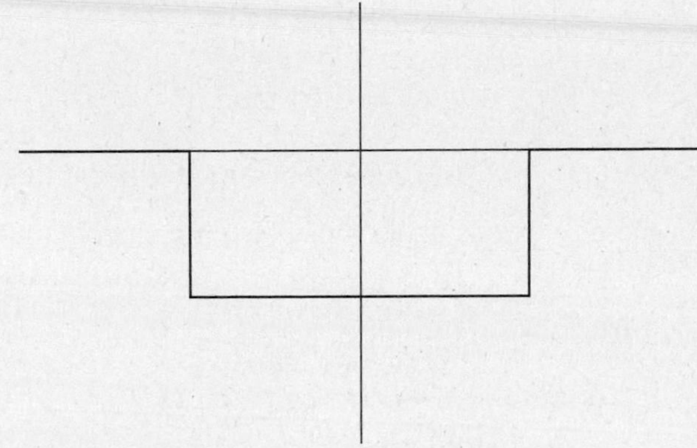

Fig. 3-13

Following the procedure used in the previous problem, we define three regions:

Region I: $-\infty < x < 0$

Region II: $0 < x < a$

Region III: $a < x < \infty$

We again define:

$$k^2 = \frac{2mE}{\hbar^2}, \beta^2 = \frac{2m(V-E)}{\hbar^2}$$

with the wavefunctions:

$$\phi_I(x) = Ae^{\beta x} + Be^{-\beta x}$$

$$\phi_{II}(x) = Ce^{ikx} + De^{-ikx}$$

$$\phi_{III}(x) = Ee^{\beta x} + Fe^{-\beta x}$$

The condition that $\phi \to 0$ as $x \to \pm\infty$ requires us to set $B = 0$ and $E = 0$.

$$\phi_I(x) = Ae^{\beta x}$$
$$\phi_{II}(x) = Ce^{ikx} + De^{-ikx}$$
$$\phi_{III}(x) = Fe^{-\beta x}$$

In Region II, notice that the well is centered about the origin. Therefore the solutions will be either *even* functions or *odd* functions. Even solutions are given in terms of *cos* functions while odd solutions are given in terms of *sin* functions. We proceed with the even solutions. The odd case is similar:

$$\phi_I(x) = Ae^{\beta x}$$
$$\phi_{II}(x) = C\cos(kx)$$
$$\phi_{III}(x) = Fe^{-\beta x}$$

The derivatives of these functions under those conditions are:

$$\phi_I'(x) = \beta Ae^{\beta x}$$
$$\phi_{II}'(x) = -kC\sin(kx)$$
$$\phi_{III}'(x) = -\beta Fe^{-\beta x}$$

The wavefunction must match at the boundaries of the well. We can match it at $x = a/2$, and so:

$$Ae^{\beta a/2} = C\cos(ka/2)$$

The first derivative must also be continuous at this boundary, so:

$$-\beta Ae^{-\beta a/2} = -kC\sin(ka/2)$$

Dividing this equation by the first one gives:

$$k\tan(k - a/2) = \beta$$

Recalling that

$$k^2 = \frac{2mE}{\hbar^2}, \quad \beta^2 = \frac{2m(V-E)}{\hbar^2}$$

The relation above is a transcendental equation that can be used to find the allowed energies. This can be done numerically or graphically. We rewrite the equation slightly:

$$\tan(ka/2) = \frac{\beta}{k}$$

The places where these curves intersect give the allowed energies. These eigenvalues are discrete, the number of them that are found depends on the parameter:

$$\lambda = \frac{2mVa^2}{\hbar^2}$$

If λ is large, there will be several allowed energies, while if λ is small, there might be two or even just one bound state energy. A plot in Figure 3-14 shows an example:

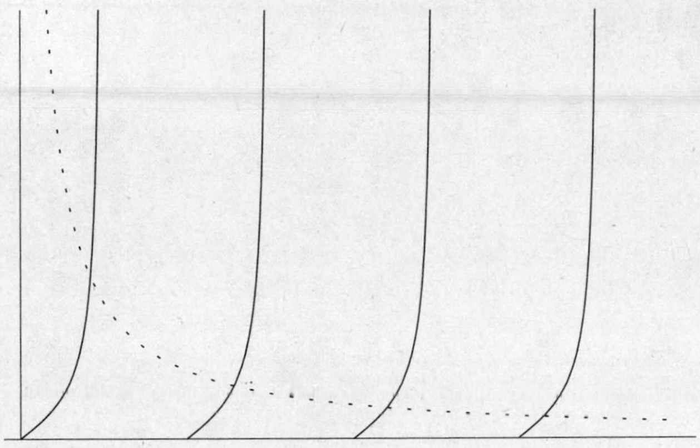

Fig. 3-14

The procedure for the odd solutions is similar, except you will arrive at a *cos* function instead of the tangent.

EXAMPLE 3.6 e.g.

Suppose that:

$$V(x) = -V\delta(x)$$

where $V > 0$.

(a) Let $E < 0$ and find the bound state wavefunction and the energy

(b) Let an incident beam of particles with $E > 0$ approach from the direction of $x = -\infty$ and find the reflection and transmission coefficients.

SOLUTION

(a) If we let $\beta^2 = \frac{-2mE}{\hbar^2}$, the Schrödinger equation is:

$$0 = \frac{d^2\psi}{dx^2} + \frac{2m}{\hbar^2}(V\delta(x) + E)\psi = \frac{d^2\psi}{dx^2} + \frac{2m}{\hbar^2}V\delta(x) - \beta^2\psi$$

We consider two regions. Region I, for $-\infty < x < 0$, has:

$$\psi_I(x) = Ae^{\beta x} + Be^{-\beta x}$$

We take functions of the form $Ae^{\beta x}$ because we are seeking a bound state. The requirement that $\psi \to 0$ as $x \to -\infty$ for a bound state forces us to set $B = 0$. In Region II,

$$\psi_{II}(x) = Ce^{\beta x} + De^{-\beta x}$$

The requirement that $\psi \to 0$ as $x \to \infty$ forces us to set $C = 0$. So the wavefunctions are:

$$\psi_I(x) = Ae^{\beta x}$$
$$\psi_{II}(x) = De^{-\beta x}$$

Even with a delta function potential, continuity of the wavefunction is required. Continuity of Ψ at $x = 0$ tells us that:

$$A = D$$

The derivative of the wavefunction is *not* continuous. We can find the discontinuity by integrating the term $\frac{2m}{\hbar^2}V\delta(x)$. We can integrate about the delta function from $\pm\epsilon$ where ϵ is a small parameter, and then let $\epsilon \to 0$ to find out how the derivative behaves. Let's recall the Schrödinger equation:

$$\frac{d^2\psi}{dx^2} + \frac{2m}{\hbar^2}V\delta(x) - \beta^2\psi = 0$$

Examining the last term, we integrate over $\pm\epsilon$ and use the fact that $\Psi(0) = A$:

$$\int_{-\epsilon}^{\epsilon} \beta^2\psi dx \approx \int_{-\epsilon}^{\epsilon} \beta^2 A dx = \beta^2 A \int_{-\epsilon}^{\epsilon} dx = 2\beta^2 A\epsilon$$

Letting $\varepsilon \to 0$, we see that this term vanishes. This leaves two terms we need to calculate:

$$\int_{-\epsilon}^{\epsilon} \frac{d^2\psi}{dx^2}dx + \int_{-\epsilon}^{\epsilon} \frac{2m}{\hbar^2}V\delta(x)\psi(x)dx = 0$$

The first term is:

$$\int_{-\epsilon}^{\epsilon} \frac{d^2\psi}{dx^2}dx = \frac{d\psi}{dx}\Big|_{x=+\epsilon} - \frac{d\psi}{dx}\Big|_{x=-\epsilon}$$

$x = +\varepsilon$ corresponds to the wavefunction in Region II. Using

$$\psi_{II}(x) = Ae^{-\beta x}$$

We have:

$$\psi_{II}(x)' - \psi_I(x)' = -\beta Ae^{-\beta x} - \beta Ae^{\beta x} = -\beta A(e^{-\beta x} + e^{\beta x})$$

And so:

$$\int_{-\epsilon}^{\epsilon} \frac{d^2\psi}{dx^2}dx = \left.\frac{d\psi}{dx}\right|_{x=+\epsilon} - \left.\frac{d\psi}{dx}\right|_{x=-\epsilon} = -\beta A(e^{-\beta\epsilon} + e^{-\beta\epsilon}) = -2\beta Ae^{-\beta\epsilon}$$

As $\varepsilon \to 0$ this becomes:

$$\left.\frac{d\psi}{dx}\right|_{x=+\epsilon} - \left.\frac{d\psi}{dx}\right|_{x=-\epsilon} = -2\beta A$$

Integrating $\frac{2m}{\hbar^2}V\delta(x)\psi(x)$ we use the sampling property of the Delta function:

$$\int_{-\epsilon}^{\epsilon} \frac{2m}{\hbar^2}V\delta(x)\psi(x)dx = \frac{2m}{\hbar^2}V\psi(0) = \frac{2m}{\hbar^2}VA$$

where we used $\psi_{II}(x) = Ae^{-\beta x}$ and the continuity of the wavefunction at $x = 0$ which required us to set $D = A$.

$$-2\beta A = -\frac{2m}{\hbar^2}VA$$

$$\Rightarrow \beta = \frac{m}{\hbar^2}V$$

Using $\beta^2 = -2\frac{mE}{\hbar^2}$, we find the energy to be:

$$E = -\frac{mV^2}{2\hbar^2}$$

The wavefunction, which we found to be:

$$\psi_I(x) = Ae^{\beta x}$$

$$\psi_{II}(x) = Ae^{-\beta x}$$

can be written compactly as:

$$\psi(x) = Ae^{-\beta|x|}$$

Here is a plot of the wavefunction (Figure 3-15):

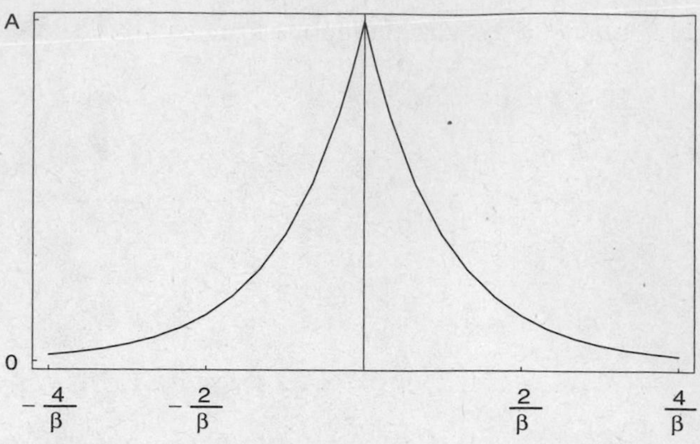

Fig. 3-15

A is found by normalization. In fact we visited this wavefunction in Chapter 2. There we took

$$\psi(x) = Ae^{-\frac{|x|}{2a}}$$

and found the normalization constant to be given by:

$$A = \frac{1}{\sqrt{2a}}$$

In the present case, with $\psi_{II}(x) = Ae^{-\beta|x|}$, we have:

$$A = \sqrt{\beta}, \Rightarrow \psi(x) = \sqrt{\beta}e^{-\beta|x|}$$

(b) For $E > 0$, the wavefunctions are, for particles incident from $x = -\infty$:

$$\psi_I(x) = Ae^{ikx} + Be^{-ikx}$$
$$\psi_{II}(x) = Ce^{ikx}$$

Continuity at $x = 0$ gives:

$$A + B = C$$

The derivatives are:

$$\psi_I(x)' = ik(Ae^{ikx} - Be^{-ikx})$$
$$\psi_{II}(x)' = ikCe^{ikx}$$

The first derivative is again discontinuous in this case. Proceeding as in part (a):

$$\int_{-\epsilon}^{\epsilon} \frac{d^2\psi}{dx^2}dx = \left.\frac{d\psi}{dx}\right|_{x=+\epsilon} - \left.\frac{d\psi}{dx}\right|_{x=-\epsilon} = ikCe^{ik\epsilon} - ik(Ae^{-ik\epsilon} - Be^{ik\epsilon})$$

Letting $\varepsilon \to 0$ we obtain:

$$ik(C - A + B)$$

Since $A+B = C$, we can take $\psi(0)$ to be either term. We will set $\psi(0) = C$. Then:

$$\int_{-\epsilon}^{\epsilon} \frac{d^2\psi}{dx^2}dx + \int_{-\epsilon}^{\epsilon} \frac{2m}{\hbar^2} V\delta(x)\psi(x)dx = 0$$

Gives us in this case:

$$ik(C - A + B) + \frac{2mVC}{\hbar^2} = 0$$

The transmission coefficient is found from C/A:

$$-\left(\frac{2mV}{\hbar^2} - i2k\right)C = -i2kA$$

$$\Rightarrow T = |C/A|^2 = \frac{k^2}{k^2 + \frac{m^2V^2}{\hbar^2}} = \frac{1}{1 + \frac{m^2V^2}{k^2\hbar^2}}$$

Using $T + R = 1$, the reflection coefficient is:

$$R = \frac{2k^2 + \frac{m^2V^2}{\hbar^2}}{k^2 + \frac{m^2V^2}{\hbar^2}}$$

Ehrenfest's Theorem

Consider the time derivative of the expectation value of x (we assume that x does not depend directly on time):

$$\frac{d}{dt}\langle x\rangle = \frac{d}{dt}\langle\psi|x|\psi\rangle = \frac{d}{dt}(\langle\langle\psi|)x|\psi\rangle + \langle\psi|x\frac{d}{dt}(|\psi\rangle) = \frac{i}{\hbar}\langle[H, x]\rangle$$

Now we use $H = p^2/2m + V(x)$:

$$\frac{d}{dt}\langle x\rangle = \frac{i}{\hbar}\langle [H, x]\rangle = \frac{i}{\hbar}\left\langle \left[p^2/2m + V(x), x\right]\right\rangle = \frac{i}{h}\left\langle \left[p^2/2m, x\right]\right\rangle + \frac{i}{h}\left\langle \left[V(x), x\right]\right\rangle$$

Now $[V(x), x] = 0$ and $[p^2, x] = -2i\hbar p$, and so:

$$\frac{d}{dt}\langle x\rangle = \left\langle \frac{p}{m}\right\rangle$$

Now evaluating the time rate of change of $\langle p\rangle$, we find that:

$$\frac{d}{dt}\langle p\rangle = \frac{i}{\hbar}\left\langle [H, p]\right\rangle = \frac{i}{\hbar}\left\langle \left[p^2/2m + V(x), p\right]\right\rangle = \frac{i}{\hbar}\left\langle \left[V(x), p\right]\right\rangle = -\left\langle \frac{dV}{dx}\right\rangle$$

These results together give us the Ehrenfest theorem, which states that the laws of classical mechanics embodied in Newton's laws hold for the expectation values of the quantum operators x and p. This establishes a correspondence between classical and quantum dynamics.

Quiz

1. The Schrödinger equation for a free particle can be written as
   (a) $-\frac{\hbar^2}{2m}\frac{d^2\psi}{dx^2} = 0$
   (b) $-\frac{\hbar^2}{2m}\frac{d^2\psi}{dx^2} = E\psi$
   (c) $-\frac{\hbar^2}{2m}\frac{d^2\psi}{dx^2} + V\psi = 0$
2. Degeneracy with respect to an energy eigenstate can be best described by saying
   (a) Two or more different particle states have the same energy.
   (b) An energy eigenstate has at least two different energies.
   (c) Two or more particle states have zero momentum.
3. A *dispersion relation* is a functional relationship that
   (a) relates energy and momentum.
   (b) relates frequency and wave number.
   (c) defines group velocity in terms of the particles state.

4. The transmission coefficient is the magnitude of
   (a) ρ/J
   (b) J_{trans}
   (c) J_{trans}/J_{inc}
5. The transmission and reflection coefficients obey which of the following
   (a) $T + R = 1$
   (b) $T - R = 1$
   (c) $\frac{T-R}{R} = 1$

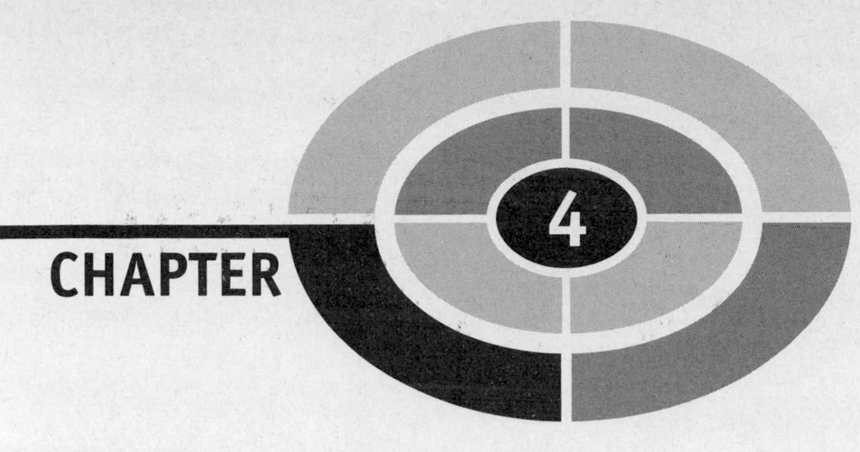

An Introduction to State Space

Hilbert space is the mathematical foundation used for quantum mechanics. This formalism is based on the basic ideas of vector analysis, with functions taking the role of vectors. In this chapter we will highlight a few of the key ideas used in quantum mechanics.

Basic Definitions

Let us quickly review elementary vector analysis and the concepts that will be carried over to define the notion of a Hilbert space. A vector A in three-dimensional euclidean space (using Cartesian coordinates) is defined as:

$$\vec{A} = A_x\hat{x} + A_y\hat{y} + A_x\hat{z}$$

The numbers (A_x, A_y, A_z) are the components of the vector with respect to the basis $(\hat{x}, \hat{y}, \hat{z})$. We could, if desired, represent the vector in another coordinate system (such as spherical coordinates). This would entail representing the vector with respect to a different basis $(\hat{r}, \hat{\theta}, \hat{\Phi})$. In that case the vector would have different components (A_r, A_θ, A_Φ). Let's think about some of the basic operations that can be carried out with vectors.

Given a second vector $\vec{B}$, we can add or subtract the two vectors, producing a new vector $\vec{C}$:

$$\vec{C} = \vec{A} \pm \vec{B} = (A_x \pm B_x)\hat{x} + (A_y \pm B_y)\hat{y} + (A_z \pm B_z)\hat{z}$$

We can multiply a vector by a scalar to produce a new vector:

$$a\vec{A} = aA_x\hat{x} + aA_y\hat{y} + aA_z\hat{z}$$

There exists an inner product between two vectors, which is a *number*:

$$\vec{A} \cdot \vec{B} = A_x B_x + A_y B_y + A_z B_z$$

This allows us to define the length of a vector, which is given by:

$$\sqrt{\vec{A} \cdot \vec{A}} = \sqrt{A_x^2 + A_y^2 + A_z^2}$$

Any vector can be "expanded" in terms of the basis vectors $(\hat{x}, \hat{y}, \hat{z})$. A fancy way of saying this is that the basis vectors "span" the space. Furthermore, the basis vectors are *orthnormal*, meaning:

$$\hat{x} \cdot \hat{x} = \hat{y} \cdot \hat{y} = \hat{z} \cdot \hat{z} = 1 \qquad \hat{x} \cdot \hat{y} = \hat{x} \cdot \hat{y} = \hat{y} \cdot \hat{z} = 0$$

In a Hilbert space, we take notions like these and generalize them to a space in which the "vectors" can have an arbitrary number of n components (which can be complex numbers), or the vectors are *functions*, making the number of "dimensions" infinite.

Hilbert Space Definitions

In a Hilbert space, we abstract all of the basic definitions we are familiar with from vector analysis—such as vector addition, a basis, an inner product—and generalize this to a space where the elements of the space are functions instead of vectors. In quantum mechanics we will have occasion to work with discrete vectors in finite dimensions $v = (v_1, v_2, \ldots, v_n)$, as well as infinite dimensional vectors (functions) $\Phi(x)$ (note that the elements of a Hilbert space can be *complex*).

In analogy with the properties of vectors defined earlier, we define some basic operations and properties of a Hilbert space.

1. Let $\Phi(x)$ belong to a given Hilbert space and let α be a complex number. The element $\alpha\phi(x)$ formed by multiplication by α also belongs to the space. In the discrete case this is also true and is indicated by saying that if $v = (v_1, v_2, \ldots, v_n)$ belongs to the space, so does $(\alpha v_1, \alpha v_2, \ldots, \alpha v_n)$.
2. Let $\Phi(x)$ and $\psi(x)$ be two elements of the Hilbert space. Then $\Phi(x) \pm \psi(x)$ also belongs to the space. In the discrete case, if $v = (v_1, v_2, \ldots, v_n)$ and $w = (w_1, w_2, \ldots, w_n)$ belong to the space, then so does $v \pm w = (v_1 \pm w_1, v_2 \pm w_2, \ldots, v_n \pm w_n)$. This property along with property #1 characterizes a Hilbert space as *linear*.
3. There exists an inner product on the space which is a *complex number*. In the infinite dimensional or continuous case, we define this as:

$$(\phi, \psi) = \int_{-\infty}^{\infty} \phi^*(x)\psi(x)\,dx$$

In the event that the interval of definition is restricted to $a \leq x \leq b$, this becomes:

$$(\phi, \psi) = \int_a^b \phi^*(x)\psi(x)\,dx$$

In the discrete case of a dimensional space, the inner product is defined to be:

$$(w, v) = \sum_{i=1}^{n} w_i^* v_i$$

The inner product, being a complex number, is *not* an element of the Hilbert space. We define the "length" or *norm* of a vector by computing the inner product of the vector with itself. Specifically, in the continuous case, we define the norm $||\phi||$ as:

$$\text{norm}(\phi)^2 = \int \phi^*(x)\phi(x)\,dx$$

where the integral is taken over the range of definition for the space. For a discrete vector with n components, the square of the norm $||v||$ is given by:

$$(\text{norm}(v))^2 = (v, v) = \sum_{i=l}^{n} v_i^* v_i$$

Since elements of Hilbert space can be complex, we see that we take the complex conjugate in the computation of the inner product so that the norm will be a real number. We take the positive square root so that $||\Phi|| \geq 0$ and $||v|| \geq 0$.

For a given space, there exist a set of "basis vectors" that span the space. Any function on that space can be expanded as a linear combination of the basis vectors. This also applies in the discrete case. If we label the basis by e_i, we can write:

$$v = \sum_{i=1}^{n} v_i e_i$$

Lastly, there exists a zero vector and identity such that if f belongs to the Hilbert space (continuous or discrete) then:

$$f(0) = 0, \; f \cdot 1 = f$$

Let's consider two examples of a Hilbert space in the continuous case. L_2 is the set of functions $\psi(x)$ defined over all space with finite norm, i.e.:

$$\int_{-\infty}^{\infty} \psi^*(x)\psi(x)\, dx < \infty$$

EXAMPLE 4.1

Show that

$$f(x) = xe^{-x}\left[\theta(x) - \theta(x - 10)\right]$$

where θ is the "Heaviside" or "unit step" function, belongs to the space L_2.

SOLUTION

The unit step function is defined to be 1 for $x \geq 0$ and is zero otherwise. Here is a plot Figure 4-1:

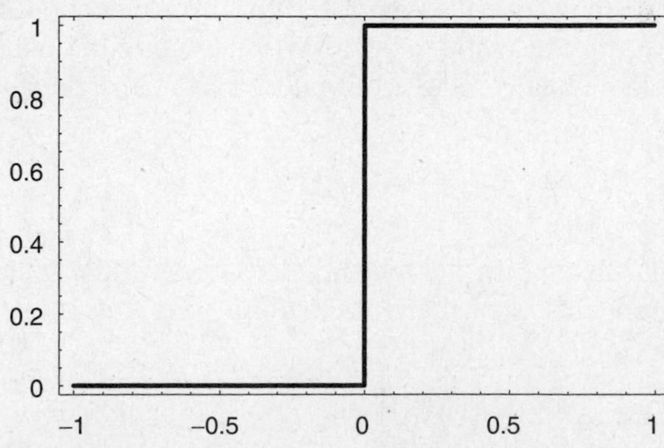

Fig. 4-1

$\theta(x - 10)$ shifts the discontinuity to $x = 10$. Therefore this function is 1 for $x \geq 10$ and is zero otherwise Figure 4-2:

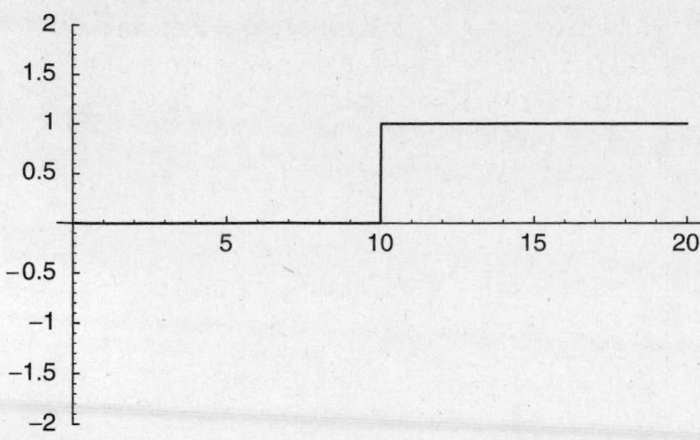

Fig. 4-2

Subtracting this from $\theta(x)$ to get $\theta(x)$–$\theta(x-10)$, we obtain a function that is 1 for $0 \leq x \leq 10$, and is zero otherwise Figure 4-3:

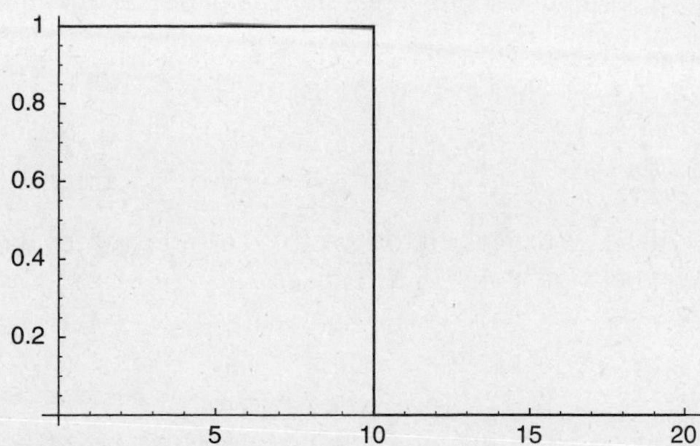

Fig. 4-3

Therefore we see that the function $f(x) = xe^{-x}(\theta(x) - \theta(x-10))$ is only going to be non-zero for $0 \leq x \leq 10$. Here is a plot of the function Figure 4-4 to get an idea of its behavior:

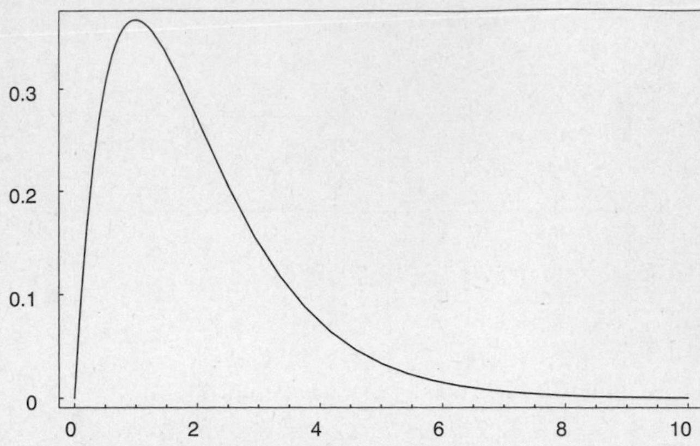

Fig. 4-4

Since this function is non-zero over a finite interval, we expect it to have a finite integral and belong to L_2. We compute $\int f^2(x)\, dx$:

$$\int x^2 e^{-2x} dx = -\frac{1}{4} e^{-2x}(1 + 2x + 2x^2)$$

(This integral can be computed using integration by parts). Evaluating it at the limits $x = 10$ and $x = 0$ we find that:

$$\int_0^{10} xe^{-2x}\, dx = -\frac{1}{4} e^{-20}(221) + \frac{1}{4} = 0.25$$

(e^{-20} is very small, so we can take it to be zero). The norm of the function is finite, and so this integral belongs to the Hilbert space L_2.

---

e.g. **EXAMPLE 4.2**

Let a Hilbert space consist of functions defined over the range $0 < x < 3$. Does the function

$$\varphi(x) = \sin \frac{\pi}{3} x$$

satisfy the requirement that $\int_0^3 \varphi^*(x)\varphi(x)\, dx < \infty$?

✔ **SOLUTION**

A plot of the function follows Figure 4-5:

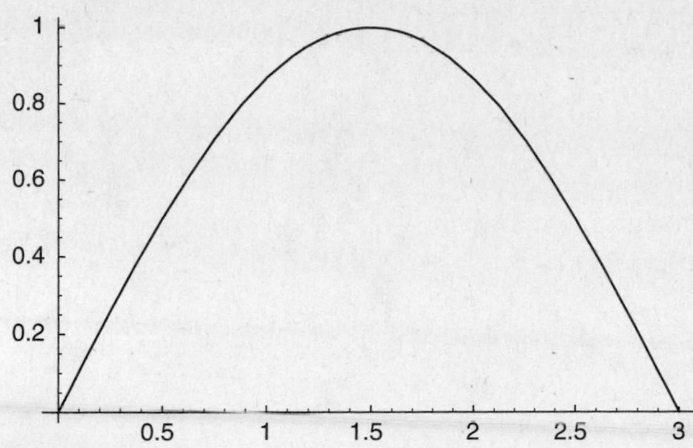

Fig. 4-5

Here we plot the square of the function Figure 4-6:

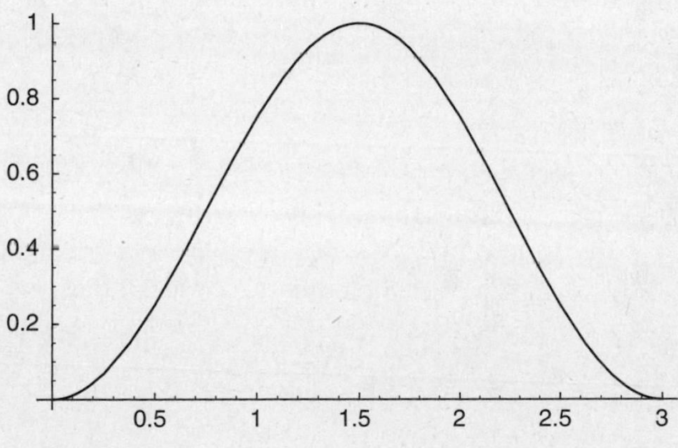

Fig. 4-6

From the plots ii, is apparent that the function does belong to the Hilbert space. But let's compute the integral explicitly.

$$\int_0^3 \varphi^*(x)\varphi(x)\,dx = \int_0^3 \sin\left(\frac{\pi x}{3}\right)^2 dx$$

We use a familiar trig identity to rewrite the integrand:

$$\int_0^3 \sin\left(\frac{\pi x}{3}\right)^2 dx = \frac{1}{2}\int_0^3 dx - \frac{1}{2}\int_0^3 \cos\left(\frac{2\pi x}{3}\right) dx$$

The result of the first integral is immediately apparent:

$$\frac{1}{2}\int_0^3 dx = \frac{1}{2}x|_0^3 = 3/2$$

For the second integral, ignoring the 1/2 out in front, we obtain:

$$\int_0^3 \cos\left(\frac{2\pi x}{3}\right) dx = \frac{3}{2\pi}\sin\frac{2\pi x}{3}|_0^3 = \frac{3}{2\pi}\left[\sin(2\pi) - \sin(0)\right] = 0$$

So the integral is finite:

$$\int_0^3 \sin\left(\frac{\pi x}{3}\right)^2 dx = \frac{3}{2}$$

and this function satisfies the requirement that $\int_0^3 \varphi^*(x)\varphi(x)\, dx \leq \infty$

The task of expanding a function in terms of a given basis is one that is encountered frequently in quantum mechanics. We have already seen this in previous chapters. Let's examine how to find the components of an arbitrary function expanded in some basis.

EXAMPLE 4.3

Suppose that we have an infinite square well of width a. The wavefunctions that are the solutions to the Schrodinger equation are the basis functions for a Hilbert space defined over $0 \leq x \leq a$. We recall that these basis functions are given by:

$$\phi_n(x) = \sqrt{\frac{2}{a}}\sin\frac{n\pi x}{a}$$

We can expand any function in terms of these basis functions using $\varphi(x) = \sum_{i=1}^{\infty} x_i \phi_i(x)$ where the expansion coefficients are found using:

$$a_i = \int_0^a \phi_i^*(x)\varphi(x)\, dx$$

Let $\varphi(x) = \cosh(x)$ where cosh is the hyperbolic cosine function. Find the expansion coefficients necessary to express φ in terms of the basis functions of the square well.

SOLUTION

A plot of $\varphi(x)$ over a finite range $0 \leq x \leq a$, looks like this Figure 4-7:

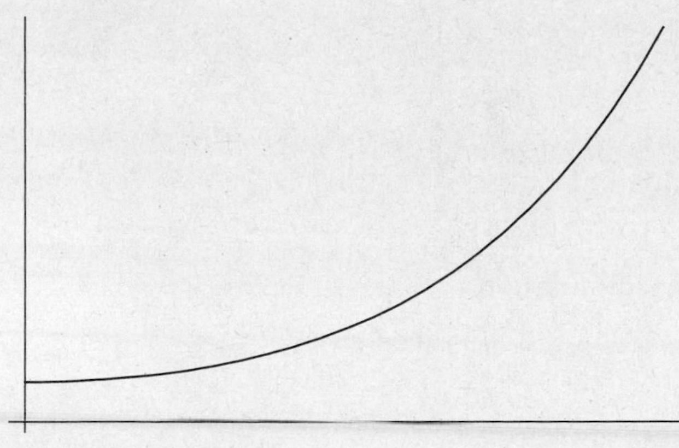

Fig. 4-7

Does this wavefunction have finite norm, as the plot indicates?

$$\int \cosh(x)^2\, dx = \int_0^a \frac{1}{2}\,(x + \cosh[x]\sinh[x])\; dx$$

Since $\cosh(0) = 1$ and $\sinh(0) = 0$, the integral vanishes at the lower limit and:

$$\int \cosh(x)^2\, dx = \frac{1}{2}\,(a + \cosh[a]\sinh[a])$$

So the wavefunction has finite norm. The integral required to determine the coefficients of expansion is given by:

$$a_n = \int_0^a \phi_n^*(x)\cosh(x)\, dx = \int_0^{12} \sqrt{\frac{2}{a}}\sin\left(\frac{n\pi x}{a}\right)\cosh(x)\, dx$$

To compute this integral, we can expand both functions in the integrand in terms of exponentials. First, to simplify notation we let:

$$\beta = \frac{n\pi}{a}$$

Then:

$$\sin(\beta x) = \frac{1}{2i}(e^{i\beta x} - e^{-i\beta x}),\ \cosh(x) = \frac{1}{2}(e^x + e^{-x})$$

Substituting these relations into the integral we obtain:

$$\int_0^a \sqrt{\frac{2}{a}}\sin(\beta x)\cosh(x)\, dx = \int_0^a \sqrt{\frac{2}{a}}\frac{1}{2i}(e^{i\beta x} - e^{-i\beta x})\frac{1}{2}(e^x + e^{-x})\, dx$$

$$= \int_0^a \sqrt{\frac{2}{a}}\frac{1}{4i}(e^{i\beta x}e^x + e^{i\beta x}e^{-x} - e^{-i\beta x}e^x - e^{-x\beta x} + e^{-x})\,dx$$

$$= \int_0^a \sqrt{\frac{2}{a}}\frac{1}{4i}(e^{(i\beta+1)x} + e^{(i\beta-1)x} - e^{(-i\beta+1)x} - e^{(-i\beta-1)x})\,dx$$

This integral can be done term by term using the u-substitution technique. We illustrate this with the first term only. Let $u = (i\beta + 1)x$, then $du = (i\beta + 1)\,dx$, and the integral becomes:

$$= \int_0^a \sqrt{\frac{2}{a}}\frac{1}{4i}(e^{(i\beta+1)x})\,dx = \frac{1}{i\beta+1}\int_0^{a(i\beta+1)} \sqrt{\frac{2}{a}}\frac{1}{4i}e^u\,du$$

$$= \frac{1}{i\beta+1}\sqrt{\frac{2}{a}}\frac{1}{4i}(e^{a(i\beta+1)} - 1)$$

Carrying out a similar procedure for the remaining terms and rewriting the exponentials back in terms of sin and cosh functions, we obtain;

$$a_n = \int_0^a \sqrt{\frac{2}{a}}\sin(\beta x)\cosh(x)dx$$

$$= \sqrt{\frac{2}{a}}\frac{a(n\pi\cos[n\pi]\cosh[a] + \sin[n\pi]\sinh[a])}{a^2 + n^2\pi^2}$$

THE DIRAC DELTA FUNCTION

In this section we briefly touch on a tool from mathematics we will have occasion to use called the Dirac Delta function. The Dirac Delta function is an infinite "spike" located at the origin denoted by $\delta(x)$. If we define:

$$\delta_\epsilon(x) = \begin{cases} \frac{1}{\epsilon} & \text{for } -\frac{\epsilon}{2} < x < \frac{\epsilon}{2} \\ 0 & \text{otherwise} \end{cases}$$

Then we can think of δ(x) in terms of the limit $\delta_\epsilon(x)$. We can see what some of the properties of $\delta(x)$ are by examining how $\delta(x)$ behaves. For example:

$$\int_{-\infty}^{\infty} \delta_\epsilon(x)\,dx = \int_{-\epsilon/2}^{\epsilon/2} \frac{1}{\epsilon}\,dx = \frac{1}{\epsilon}[\epsilon/2 + \epsilon/2] = 1$$

It is also true that:

$$\int_{-\infty}^{\infty} \delta_\epsilon(x)\,dx = 1$$

One important property of the Dirac Delta function is the sampling property:

$$\int_{-\infty}^{\infty} \delta_\epsilon(x) f(x)\, dx = f(0)$$

We can also shift $\delta(x)$ to another point along the real line to $\delta(x-a)$. Then the sampling property becomes:

$$\int_{-\infty}^{\infty} \delta(x-a) f(x)\, dx = f(a)$$

In this way $\delta(x-a)$ can be viewed as representing the presence of a particle at $x = a$. The Dirac Delta can be extended to three dimensions using:

$$\delta(r - r_o) = \delta(x - x_o)\delta(y - y_o)\delta(z - z_o)$$

Some of the basic properties of the Dirac Delta function follow:

1. $\delta(-x) = \delta(x)$

2. $\delta(ax) = \frac{1}{|a|}\delta(x)$

3. $f(x)\delta(x-a) = f(a)\delta(x-a)$

4. $x\delta^{-1}(x) = -\delta(x)$

We can use the sampling property to obtain the Fourier transform of the Dirac Delta:

$$\frac{1}{\sqrt{2\pi}} \int_{-\infty}^{\infty} \delta(x) e^{-ikx}\, dx = \frac{1}{\sqrt{2\pi}}$$

$$\frac{1}{\sqrt{2\pi}} \int_{-\infty}^{\infty} \delta(x-a) e^{-ikx}\, dx = \frac{1}{\sqrt{2\pi}} e^{-ika}$$

EXAMPLE 4.4 e.g.

Find the Fourier transform of $\delta(ax)$.

SOLUTION ✓

The Fourier transform is:

$$\frac{1}{\sqrt{2\pi}} \int_{-\infty}^{\infty} \delta(ax) e^{-ikx}\, dx$$

Let $u = ax$, then $du = a\, dx$ and the integral becomes:

$$\frac{1}{\sqrt{2\pi}} \int_{-\infty}^{\infty} \delta(ax) e^{-ikx}\, dx = \frac{1}{\sqrt{2\pi}} \frac{1}{a} \int_{-\infty}^{\infty} \delta(u) e^{-ik\frac{u}{a}}\, du$$

$$= \frac{1}{a} \frac{1}{\sqrt{2\pi}}$$

If we denote the Fourier transform of $f(x)$ by $F[f(x)]$, this is the same as saying that:

$$F[\delta(ax)] = \frac{1}{a} F[\delta(x)]$$

Since $F[\delta(x)]$ is given by:

$$\frac{1}{\sqrt{2\pi}} \int_{-\infty}^{\infty} \delta(x) e^{-ikx}\, dx = \frac{1}{\sqrt{2\pi}}$$

Quiz

1. Two vectors in 5-dimensions are given by:

$$A = (2, 4i, 0, 1, -7i) \quad B = (1, 0, 1, 9i, 2)$$

Show that $(A, B) = 2 + 23i$ and $(B, A) = 2 - 23i$

2. Determine which of the following functions belong to L_2:
   (a) $f(x) = \operatorname{sech}(x)$
   (b) $f(x) = e^{-x} \cosh(x)$ for $x \geq 0$, zero otherwise
   (c) $f(x) = e^{-x} \sin(\pi x)$ for $x \geq 0$, zero otherwise

3. Verify the properties of the Dirac Delta listed here by considering integrals.

   1. $x\delta(x) = \delta(x)$

   2. $\delta(ax) = \frac{1}{|a|}\delta(x)$

The Mathematical Structure of Quantum Mechanics I

Linear Vector Spaces

The state of a particle in quantum mechanics is represented by an element of an abstract linear vector space. While vectors in ordinary physics are represented by quantities such as $\vec{A}$, $\vec{V}$, $\vec{B}$, in quantum mechanics we represent vectors with the notation:

$$|\phi\rangle\,,\;|V\rangle\,,\;|\psi\rangle$$

These objects are called "kets," and this type of notation is known as *Dirac notation*.

DEFINITION: Linear Vector Space

A linear vector space V is a set of elements $|a\rangle$, $|b\rangle$, $|c\rangle$ called vectors, or kets, for which the following hold:

1. V is *closed* under addition. This means that if two vectors $|a\rangle$, $|b\rangle$ belong to V, then so does their sum $|a\rangle + |b\rangle$
2. A vector $|a\rangle$ can be multiplied by a scalar α to yield a new, well-defined vector $\alpha\,|a\rangle$ that belongs to V
3. Vector addition is commutative: $|a\rangle + |b\rangle = |b\rangle + |a\rangle$
4. Vector addition is associative: $|a\rangle + (|b\rangle + |c\rangle) = (|a\rangle + |b\rangle) + |c\rangle$

If the scalars associated with a given vector space are real numbers, we say we are working with a *real vector space*. On the other hand, if the α can be complex numbers then we say that we are working with a *complex vector space*. The vector spaces used in quantum mechanics are complex.

e.g. **EXAMPLE 5.1**

Consider sets or lists of n complex numbers $(z_1, z_2, \ldots, z_n)$ called "n-tuples." We can define a vector space of n-tuples of complex numbers where we represent a vector by an $n \times 1$ matrix called a *column vector*. For example, consider two vectors $|\psi\rangle$, $|\phi\rangle$ given by:

$$|\psi\rangle = \begin{bmatrix} z_1 \\ z_2 \\ \vdots \\ z_n \end{bmatrix}, \quad |\phi\rangle = \begin{bmatrix} w_1 \\ w_2 \\ \vdots \\ w_n \end{bmatrix}$$

Vector addition in this space is carried out by adding together the individual components of the vectors

$$|\psi\rangle + |\phi\rangle = \begin{bmatrix} z_1 \\ z_2 \\ \vdots \\ z_n \end{bmatrix} + \begin{bmatrix} w_1 \\ w_2 \\ \vdots \\ w_n \end{bmatrix} = \begin{bmatrix} z_1 + w_1 \\ z_2 + w_2 \\ \vdots \\ z_n + w_n \end{bmatrix}$$

We see that the operation of addition has generated a new list of n complex numbers—a new n-tuple—so we have produced a new vector that still belongs to the space. Since the addition of complex numbers is commutative and associative, we see that the vectors in this space automatically satisfy the other properties listed as well.

Scalar multiplication is carried out by multiplying each component of the vector in the following manner

$$\alpha\,|\psi\rangle = \alpha \begin{bmatrix} z_1 \\ z_2 \\ \vdots \\ z_n \end{bmatrix} = \begin{bmatrix} \alpha\, z_1 \\ \alpha\, z_2 \\ \vdots \\ \alpha z_n \end{bmatrix}$$

MORE PROPERTIES OF VECTOR SPACES

In addition to the previous list of properties, a vector space also satisfies the following:

1. There exists a unique element called 0 that satisfies $|a\rangle + 0 = |a\rangle$ for every $|a\rangle$ in V
2. There exists an identity element in V such that $I\,|a\rangle = |a\rangle$ for every vector in V
3. Scalar multiplication is associative:

$$(\alpha\beta)\,|a\rangle = \alpha(\beta\,|a\rangle)$$

4. Scalar multiplication is linear:

$$\alpha(|a\rangle + |b\rangle) = \alpha\,|a\rangle + \alpha\,|b\rangle\,, \quad (\alpha + \beta)\,|a\rangle = \alpha\,|a\rangle + \beta\,|a\rangle$$

5. For each $|a\rangle$ in V, there exists a unique additive inverse $|-a\rangle$ such that $|a\rangle + |-a\rangle = 0$

Returning to the vector space described in Example 5-1, we can define the zero vector as:

$$|0\rangle = \begin{bmatrix} 0 \\ 0 \\ \vdots \\ 0 \end{bmatrix}$$

It's a trivial matter to see that this satisfies $|0\rangle + |\psi\rangle = |\psi\rangle$ for any $|\psi\rangle$. To define the inverse, we simply form the column vector containing $-z_i$. In other words if:

$$|\psi\rangle = \begin{bmatrix} z_1 \\ z_2 \\ \vdots \\ z_n \end{bmatrix}; \text{ then the additive inverse is } |-\psi\rangle = \begin{bmatrix} -z_1 \\ -z_2 \\ \vdots \\ -z_n \end{bmatrix} \text{ since:}$$

$$|\psi\rangle + |-\psi\rangle = \begin{bmatrix} z_1 \\ z_2 \\ \vdots \\ z_n \end{bmatrix} + \begin{bmatrix} -z_1 \\ -z_2 \\ \vdots \\ -z_n \end{bmatrix} = \begin{bmatrix} z_1 - z_1 \\ z_2 - z_2 \\ \vdots \\ z_n - z_n \end{bmatrix} = \begin{bmatrix} 0 \\ 0 \\ \vdots \\ 0 \end{bmatrix}$$

EXAMPLE 5.2 *e.g.*

Let $\vec{A} = A_1\hat{i} + A_2\hat{j} + A_3\hat{k}$ be an ordinary vector in three-dimensional space. Is the set of all vectors with $A_1 = 13$ a vector space?

SOLUTION

If $A_1 = 13 \Rightarrow \vec{A} = 13\hat{i} + A_2\hat{j} + A_3\hat{k}$. It is easy to see that a set of vectors of this form cannot form a vector space. This is because the addition of two vectors does not produce another vector in the space.

Suppose that

$$\vec{B} = 13\hat{i} + B_2\hat{j} + B_3\hat{k}$$

is another vector in this space. If this set constitutes a vector space, then $\vec{A} + \vec{B}$ must be another vector in the space. But

$$\vec{A}+\vec{B} = 13\hat{i}+A_2\hat{j}+A_3\hat{k}+(13\hat{i}+B_2\hat{j}+B_3\hat{k}) = 26\hat{i}+(A_2+B_2)\hat{j}+(A_3+B_3)\hat{k}$$

Since the component associated with the basis vector $\hat{i}$ is $26 \neq 13$, this vector does not belong in the space and axiom (1) is not satisfied. Therefore the set of all vectors with $A_1 = 13$ is not a vector space.

DEFINITION: Dual Vector

In order to carry over the notion of a "dot product" to an abstract vector space we will need to construct a *dual vector*. In the language of kets, the dual vector is called a "bra." Using Dirac notation, the dual of a vector $|\psi\rangle$ is written as $\langle\psi|$.

Returning to the example of complex n-tuples, we write the list of complex numbers out in a row and then take their complex conjugates to obtain the dual vector. In other words:

$$|\psi\rangle = \begin{bmatrix} z_1 \\ z_2 \\ \vdots \\ z_n \end{bmatrix} \quad \Rightarrow \quad \langle\psi| = \left[z_1^*, z_2^*, \ldots, z_n^*\right]$$

The dual space of V is denoted V^*. With a definition of dual vectors in hand, we can generalize the dot product to an "inner product" between two abstract vectors.

THE INNER PRODUCT

The inner product for a vector space V is a map from $V \times V$ to the complex numbers. We can express this more clearly by saying that the inner product is a function on two vectors $|\psi\rangle$, $|\phi\rangle$ that produces a complex number which we represent by $\langle\phi|\psi\rangle$. A vector space that also has an inner product is referred to as an *inner product space*.

To illustrate how we compute the inner product for a given vector space, we once again consider ordinary vectors in three-dimensional space. We want to get away from physically based geometric notions, so we are going to focus on the definition of the dot product that is based on using the components of vectors.

Recall that if we have two vectors:

$$\vec{A} = A_1\hat{i} + A_2\hat{j} + A_3\hat{k}$$
$$\vec{B} = B_1\hat{i} + B_2\hat{j} + B_3\hat{k}$$

The dot product is defined by:

$$\vec{A} \cdot \vec{B} = A_1B_1 + A_2B_2 + A_3B_3 = \sum_{i=1}^{3} A_iB_i$$

We can generalize this procedure to n-tuples of complex numbers in the following way. We calculate the inner product on C^n by using ordinary matrix multiplication. More specifically, given two vectors of complex numbers $|\psi\rangle$, $|\phi\rangle$ such that:

$$|\psi\rangle = \begin{bmatrix} z_1 \\ z_2 \\ \vdots \\ z_n \end{bmatrix}, |\phi\rangle = \begin{bmatrix} w_1 \\ w_2 \\ \vdots \\ w_n \end{bmatrix}$$

the inner product between these two vectors is defined by:

$$\langle\phi|\psi\rangle = (w_1^* w_2^* \ldots w_n^*) \begin{bmatrix} z_1 \\ z_2 \\ \vdots \\ z_n \end{bmatrix} = w_1^*(z_1) + w_2^*(z_2) + \cdots + w_n^*(z_n) = \sum_{i=1}^{n} w_i^* z_i$$

We call $\langle\phi|\psi\rangle$ a *bracket*.

EXAMPLE 5.3

e.g.

Two vectors in a three-dimensional complex vector space are defined by:

$$|A\rangle = \begin{pmatrix} 2 \\ -7i \\ 1 \end{pmatrix}, |B\rangle = \begin{pmatrix} 1+3i \\ 4 \\ 8 \end{pmatrix}$$

Let $a = 6 + 5i$

(a) Compute $a\,|A\rangle$, $a\,|B\rangle$, and $a(|A\rangle + |B\rangle)$. Show that $a(|A\rangle + |B\rangle) = a\,|A\rangle + a\,|B\rangle$.

(b) Find the inner products $\langle A|B\rangle$, $\langle B|A\rangle$.

SOLUTION

(a)

$$a\,|A\rangle = (6+5i)\begin{pmatrix} 2 \\ -7i \\ 1 \end{pmatrix} = \begin{pmatrix} (6+5i)2 \\ (6+5i)(-7i) \\ (6+5i)1 \end{pmatrix} = \begin{pmatrix} 12+10i \\ 35-42i \\ 6+5i \end{pmatrix}$$

$$a\,|B\rangle = (6+5i)\begin{pmatrix} 1+3i \\ 4 \\ 8 \end{pmatrix} = \begin{pmatrix} (6+5i)(1+3i) \\ (6+5i)4 \\ (6+5i)8 \end{pmatrix} = \begin{pmatrix} -9+23i \\ 24+20i \\ 48+40i \end{pmatrix}$$

$$\Rightarrow a\,|A\rangle + a\,|B\rangle = \begin{pmatrix} 12+10i \\ 35-42i \\ 6+5i \end{pmatrix} + \begin{pmatrix} -9+23i \\ 24+20i \\ 48+40i \end{pmatrix}$$

$$= \begin{pmatrix} 12+10i+(-9+23i) \\ 35-42i+(24+20i) \\ 6+5i+(48+40i) \end{pmatrix} = \begin{pmatrix} 3+33i \\ 59-22i \\ 54+45i \end{pmatrix}$$

Now adding the vectors first, we have

$$|A\rangle + |B\rangle = \begin{pmatrix} 2 \\ -7i \\ 1 \end{pmatrix} + \begin{pmatrix} 1+3i \\ 4 \\ 8 \end{pmatrix} = \begin{pmatrix} 2+1+3i \\ -7i+4 \\ 1+8 \end{pmatrix} = \begin{pmatrix} 3+3i \\ 4-7i \\ 9 \end{pmatrix}$$

$$\Rightarrow a(|A\rangle + |B\rangle) = (6+5i)\begin{pmatrix} 3+3i \\ 4-7i \\ 9 \end{pmatrix} = \begin{pmatrix} (6+5i)(3+3i) \\ (6+i)(4-7i) \\ (6+5i)(9) \end{pmatrix}$$

$$= \begin{pmatrix} 18+15i+18i-15 \\ 24+20i-42i+35 \\ 54+45i \end{pmatrix} = \begin{pmatrix} 3+33i \\ 59-22i \\ 54+45i \end{pmatrix} = a\,|A\rangle + a\,|B\rangle$$

(b) First we compute $\langle A|B\rangle$. To form the dual vector of $|A\rangle$, we compute the complex conjugate of its elements, and then transpose the result to form a row vector:

$$|A\rangle^* = \begin{pmatrix} 2 \\ -7i \\ 1 \end{pmatrix}^* = \begin{pmatrix} 2 \\ 7i \\ 1 \end{pmatrix}, \Rightarrow \langle A| = \begin{pmatrix} 2 & 7i & 1 \end{pmatrix}$$

And so the inner product is

$$\langle A|B\rangle = \begin{pmatrix} 2 & 7i & 1 \end{pmatrix}\begin{pmatrix} 1+3i \\ 4 \\ 8 \end{pmatrix} = 2(1+3i) + 7i(4) + 1(8) = 10+34i$$

Now we compute $\langle B|A\rangle$. The complex conjugate of $|B\rangle$ is given by

$$|B\rangle^* = \begin{pmatrix} 1+3i \\ 4 \\ 8 \end{pmatrix}^* = \begin{pmatrix} 1-3i \\ 4 \\ 8 \end{pmatrix}$$

Now we transpose this to get the dual vector

$$\langle B| = (\, 1-3i \quad 4 \quad 8\,)$$

And so the inner product is

$$\langle B|A\rangle = (\, 1-3i \quad 4 \quad 8\,)\begin{pmatrix} 2 \\ -7i \\ 1 \end{pmatrix} = (1-3i)(2)+(4)(-7i)+(8)(1) = 2-6i-28i+8 = 10-34i$$

Notice that $\langle B|A\rangle = \langle A|B\rangle^*$, a result that holds in general for the inner product in a complex vector space. We now list this and other important properties of the inner product.

PROPERTIES OF THE INNER PRODUCT

Let $|\psi\rangle$, $|\phi\rangle$ be two vectors belonging to a complex vector space V and let α and β be complex numbers. Then the following hold for the inner product :

1. $\langle\psi|\phi\rangle = \langle\phi|\psi\rangle^*$
2. $\langle\psi|\left(\alpha\,|\phi\rangle + \beta\,|\omega\rangle\right) = \alpha\,\langle\psi|\phi\rangle + \beta\,\langle\psi|\omega\rangle$
3. $\left(\langle\alpha\psi| + \langle\beta\omega|\right)|\phi\rangle = \alpha^*\,\langle\psi|\phi\rangle + \beta^*\,\langle\psi|\phi\rangle$
4. $\langle\psi|\psi\rangle \geq 0$ with equality if and only if $|\psi\rangle = 0$

If the inner product between two vectors is zero, $\langle\psi|\Phi\rangle = 0$, we say that the vectors are *orthogonal*. We now say some more about property (4), which generalizes the notion of length to give us the "norm" of a vector.

DEFINITION: The Norm of a Vector

The square root of the inner product of a vector with itself is called the *norm*, and is designated by:

$$\|\psi\| = \sqrt{\langle\psi|\psi\rangle}$$

DEFINITION: A Normalized Vector

A vector $|\psi\rangle$ is said to be normalized if:

$$\|\psi\| = \sqrt{\langle\psi|\psi\rangle} = 1$$

EXAMPLE 5.4

Let two vectors be defined by

$$|A\rangle = \begin{pmatrix} 2 \\ -7i \\ 1 \end{pmatrix}, |B\rangle = \begin{pmatrix} 1+3i \\ 4 \\ 8 \end{pmatrix}$$

Find the norm of each vector.

SOLUTION

$$\langle A|A\rangle = (2 \quad 7i \quad 1)\begin{pmatrix} 2 \\ -7i \\ 1 \end{pmatrix} = 2(2) + 7i(-7i) + 1(1) = 4 + 49 + 1 = 54$$

$$\Rightarrow ||A|| = \sqrt{\langle A|A\rangle} = \sqrt{54} = 3\sqrt{6}$$

Now we compute the norm of B:

$$\langle B|B\rangle = (1-3i \quad 4 \quad 8)\begin{pmatrix} 1+3i \\ 4 \\ 8 \end{pmatrix}$$

$$= (1-3i)(1+3i) + 4(4) + 8(8) = 10 + 16 + 64 = 90$$

$$\Rightarrow ||B|| = \sqrt{\langle B|B\rangle} = \sqrt{90} = 3\sqrt{10}$$

EXAMPLE 5.5

Show that the vectors

$$|\psi\rangle = \begin{pmatrix} \frac{1}{\sqrt{2}} \\ \frac{1}{\sqrt{2}} \end{pmatrix}, |\phi\rangle = \begin{pmatrix} \frac{1}{\sqrt{2}} \\ -\frac{1}{\sqrt{2}} \end{pmatrix}$$

are orthogonal. Is $|\psi\rangle$ normalized?

SOLUTION

If the vectors are orthogonal, then $\langle\phi|\psi\rangle = 0$, which we easily verify:

$$\langle\phi|\psi\rangle = \begin{pmatrix} \frac{1}{\sqrt{2}} & -\frac{1}{\sqrt{2}} \end{pmatrix}\begin{pmatrix} \frac{1}{\sqrt{2}} \\ \frac{1}{\sqrt{2}} \end{pmatrix} = \frac{1}{\sqrt{2}}\left(\frac{1}{\sqrt{2}}\right) + \frac{1}{\sqrt{2}}\left(-\frac{1}{\sqrt{2}}\right) = \frac{1}{2} - \frac{1}{2} = 0$$

Now we check to see if $|\psi\rangle$ is normalized:

$$\langle\psi|\psi\rangle = \begin{pmatrix} \frac{1}{\sqrt{2}} & \frac{1}{\sqrt{2}} \end{pmatrix}\begin{pmatrix} \frac{1}{\sqrt{2}} \\ \frac{1}{\sqrt{2}} \end{pmatrix} = \frac{1}{\sqrt{2}}\left(\frac{1}{\sqrt{2}}\right) + \frac{1}{\sqrt{2}}\left(\frac{1}{\sqrt{2}}\right) = \frac{1}{2} + \frac{1}{2} = 1$$

$$\Rightarrow ||\Psi|| = \sqrt{\langle\psi|\psi\rangle} = 1$$

and so the vector is normalized. If a vector is *not* normalized, we can divide by its norm to make it so. In the example where

$$|A\rangle = \begin{pmatrix} 2 \\ -7i \\ 1 \end{pmatrix}$$

we found that $||A|| = \sqrt{\langle A|A\rangle} = \sqrt{54} = 3\sqrt{6}$. We can form a new normalized vector that we will call $\left|\tilde{A}\right\rangle$ by dividing $|A\rangle$ by its norm:

$$\left|\tilde{A}\right\rangle = \frac{|A\rangle}{||A||} = \frac{1}{3\sqrt{6}}\begin{pmatrix} 2 \\ -7i \\ 1 \end{pmatrix} = \begin{pmatrix} \frac{2}{3\sqrt{6}} \\ \frac{-7i}{3\sqrt{6}} \\ \frac{1}{3\sqrt{6}} \end{pmatrix}$$

EXAMPLE 5.6

e.g.

A vector $|u\rangle = \begin{pmatrix} x \\ 3x \\ -2x \end{pmatrix}$ where x is an unknown real number. Find x such that $|u\rangle$ is normalized.

SOLUTION

Noting that x is real and so $x^* = x$,

$$\langle u|u\rangle = (x \quad 3x \quad -2x)\begin{pmatrix} x \\ 3x \\ -2x \end{pmatrix} = x^2 + 9x^2 + 4x^2 = 14x^2$$

$$||u|| = \sqrt{\langle u|u\rangle} = \sqrt{14x}$$

In order for the vector to be normalized, we must have $||u|| = 1$, and so:

$$||u|| = 1 = \sqrt{14}x, \Rightarrow x = \frac{1}{\sqrt{14}}$$

The normalized vector is then:

$$|u\rangle = \begin{pmatrix} \frac{1}{\sqrt{14}} \\ \frac{3}{\sqrt{14}} \\ \frac{-2}{14} \end{pmatrix}$$

We now prove some important theorems involving inner products.

EXAMPLE 5.7

Show that $2Re\,\langle\psi|\phi\rangle \le \langle\psi|\psi\rangle + \langle\phi|\phi\rangle$

SOLUTION

We use the fact that $\langle f|f\rangle \ge 0$ for any ket $|f\rangle$. We let $|f\rangle = |\psi - \phi\rangle$

$$0 \le \langle\psi - \phi|\psi - \phi\rangle = \langle\psi|\psi\rangle - \langle\psi|\phi\rangle - \langle\phi|\psi\rangle + \langle\phi|\phi\rangle$$

Now recall that $\langle\psi|\phi\rangle = \langle\phi|\psi\rangle^*$, and $z + z^* = 2Re(z)$ for any complex number z. This means that $\langle\psi|\phi\rangle + \langle\phi|\psi\rangle = \langle\psi|\phi\rangle + \langle\psi|\phi\rangle^* = 2Re(\langle\psi|\phi\rangle)$, giving us

$$\langle\psi|\psi\rangle + \langle\phi|\phi\rangle - 2Re(\langle\psi|\phi\rangle) \ge 0$$

$$\Rightarrow \langle\psi|\psi\rangle + \langle\phi|\phi\rangle \ge 2Re(\langle\psi|\phi\rangle)$$

EXAMPLE 5.8

The Cauchy-Schwartz Inequality states that

$$|\langle\phi|\psi\rangle|^2 \le \langle\psi|\psi\rangle\langle\phi|\phi\rangle$$

Prove this result.

SOLUTION

The proof also relies on the fact that $\langle f|f\rangle \ge 0$ for any vector. We can minimize this expression if we let

$$|f\rangle = |\phi\rangle - \frac{\langle\psi|\phi\rangle}{\langle\phi|\phi\rangle}|\psi\rangle$$

And so we have:

$$\langle f|f\rangle = \langle\phi|\phi\rangle - \frac{\langle\psi|\phi\rangle}{\langle\psi|\psi\rangle}\langle\phi|\psi\rangle - \frac{\langle\phi|\psi\rangle}{\langle\psi|\psi\rangle}\langle\psi|\phi\rangle + \frac{\langle\phi|\psi\rangle}{\langle\psi|\psi\rangle}\frac{\langle\psi|\phi\rangle}{\langle\psi|\psi\rangle}\langle\psi|\psi\rangle$$

Now we use the fact that

$$\langle\phi|\psi\rangle\langle\psi|\phi\rangle = |\langle\phi|\psi\rangle|^2$$

to rewrite this as

$$\langle f|f\rangle = \langle\phi|\phi\rangle - 2\frac{|\langle\phi|\psi\rangle|^2}{\langle\psi|\psi\rangle} + \frac{|\langle\phi|\psi\rangle|^2}{\langle\psi|\psi\rangle} = \langle\phi|\phi\rangle - \frac{|\langle\phi|\psi\rangle|^2}{\langle\psi|\psi\rangle}$$

Using $\langle f \,|f\,\rangle \geq 0$, we can move the second term to the other side:

$$\langle \phi \,|\phi \,\rangle \geq \frac{|\langle \phi \,|\psi \,\rangle|^2}{\langle \psi \,|\psi \,\rangle}$$

Now we multiply both sides by $\langle \psi \,|\psi \,\rangle$, allowing us to arrive at the final result:

$$|\langle \phi \,|\psi \,\rangle|^2 \leq \langle \psi \,|\psi \,\rangle \langle \phi \,|\phi \,\rangle$$

EXAMPLE 5.9

The Triangle Inequality states that

$$\sqrt{\langle \psi + \phi \,|\psi + \phi \,\rangle} \leq \sqrt{\langle \psi \,|\psi \,\rangle} + \sqrt{\langle \phi \,|\phi \,\rangle}$$

Prove this result.

SOLUTION 

For any complex number z, it is true that $|Re(z)| \leq |z|$. Since the inner product is a complex number this tells us that $Re(|\langle \psi \,|\phi \,\rangle|) \leq |\langle \psi \,|\phi \,\rangle|$.

To derive the result, we use this fact together with the Schwartz inequality. First, we expand the inner product $\langle \psi + \phi \,|\psi + \phi \,\rangle$:

$$\langle \psi + \phi \,|\psi + \phi \,\rangle = \langle \psi \,|\psi \,\rangle + \langle \psi \,|\phi \,\rangle + \langle \phi \,|\psi \,\rangle + \langle \phi \,|\phi \,\rangle$$

Once again, we note that $\langle \psi \,|\phi \,\rangle + \langle \phi \,|\psi \,\rangle = \langle \psi \,|\phi \,\rangle + \langle \psi \,|\phi \,\rangle^* = 2Re(\langle \psi \,|\phi \,\rangle)$, and so this can be written as

$$\langle \psi \,|\psi \,\rangle + \langle \phi \,|\phi \,\rangle + 2Re(\langle \phi \,|\psi \,\rangle)$$

At this point we can use $|Re(z)| \leq |z|$ to write

$$\begin{aligned} &\langle \psi \,|\psi \,\rangle + \langle \phi \,|\phi \,\rangle + 2Re(\langle \phi \,|\psi \,\rangle) \\ &\leq \langle \psi \,|\psi \,\rangle + \langle \phi \,|\phi \,\rangle + 2\,|\langle \phi \,|\psi \,\rangle| \end{aligned}$$

From Cauchy-Schwarz, we have

$$\begin{aligned} &\langle \psi \,|\psi \,\rangle + \langle \phi \,|\phi \,\rangle + 2\,|\langle \phi \,|\psi \,\rangle| \leq \langle \psi \,|\psi \,\rangle + \langle \phi \,|\phi \,\rangle + 2\sqrt{\langle \psi \,|\psi \,\rangle}\sqrt{\langle \phi \,|\phi \,\rangle} \\ &= (\sqrt{\langle \psi \,|\psi \,\rangle} + \sqrt{\langle \phi \,|\phi \,\rangle})^2 \end{aligned}$$

Putting our results together allows us to conclude that

$$\sqrt{\langle \psi + \phi \,|\psi + \phi \,\rangle} \leq \sqrt{\langle \psi \,|\psi \,\rangle} + \sqrt{\langle \phi \,|\phi \,\rangle}$$

DEFINITION: Orthonormal Vectors

If there are two vectors $\{|u\rangle\,|v\rangle\}$ such that $||u|| = ||v|| = 1$ and $\langle u\,|v\,\rangle = 0$ we say that $|u\rangle$ and $|v\rangle$ are *orthonormal*. If we have an entire set of vectors $|u_i\rangle$ that are orthonormal, we can write this compactly as

$$\langle u_i\,|u_j\,\rangle = \delta_{ij}$$

where δ_{ij} is the *Kronecker delta* function. This is defined by

$$\delta_{ij} = \begin{cases} 1 & \text{for } i = j \\ 0 & \text{otherwise} \end{cases}$$

Basis Vectors

We call a set of vectors $\{|\phi_1\rangle\,, |\phi_2\rangle\,, \ldots, |\phi_n\rangle\}$ a *basis* if the set satisfies three criteria:

1. The set $\{|\phi_1\rangle\,, |\phi_2\rangle\,, \ldots, |\phi_n\rangle\}$ *spans* the vector space V, meaning that every vector $|\psi\rangle$ in V can be written as a unique linear combination of the $\{|\phi_i\rangle\}$.
2. The set $\{|\phi_1\rangle\,, |\phi_2\rangle\,, \ldots, |\phi_n\rangle\}$ is linearly independent
3. The closure relation is satisfied.

We now describe each of these items in turn.

DEFINITION: Spanning the Space

If the set $\{|\phi_1\rangle\,, |\phi_2\rangle\,, \ldots, |\phi_n\rangle\}$ spans the vector space V, we can write an arbitrary vector $|\psi\rangle$ from this space as a linear combination:

$$|\psi\rangle = c_1\,|\phi_1\rangle + c_2\,|\phi_2\rangle + \ldots + c_n\,|\phi_n\rangle$$

Next we will show how to calculate the expansion coefficients c_i, which are in general complex numbers.

DEFINITION: Linearly Independent

A collection of vectors $\{|\phi_1\rangle\,, |\phi_2\rangle\,, \ldots, |\phi_n\rangle\}$ are *linearly independent* if the equation

$$a_1\,|\phi_1\rangle + a_2\,|\phi_2\rangle + \ldots + a_n\,|\phi_n\rangle = 0$$

implies that $a_1 = a_2 = \cdots = a_n = 0$. If this condition is not met we say that the set is *linearly dependent*. If a set of vectors is linearly dependent this means that one of the vectors can be expressed as a linear combination of the others. We demonstrate this with an example.

EXAMPLE 5.10

Show that the following vectors are linearly dependent:

$$|a\rangle = \begin{pmatrix} 1 \\ 2 \\ 1 \end{pmatrix} |b\rangle = \begin{pmatrix} 0 \\ 1 \\ 0 \end{pmatrix} |c\rangle = \begin{pmatrix} -1 \\ 0 \\ -1 \end{pmatrix}$$

SOLUTION

Notice that:

$$2\,|b\rangle - |c\rangle \quad = \quad 2\begin{pmatrix} 0 \\ 1 \\ 0 \end{pmatrix} - \begin{pmatrix} -1 \\ 0 \\ -1 \end{pmatrix} = \begin{pmatrix} 0 \\ 2 \\ 0 \end{pmatrix} + \begin{pmatrix} 0 \\ 1 \\ 0 \end{pmatrix} = \begin{pmatrix} 1 \\ 2 \\ 1 \end{pmatrix} = |a\rangle$$

Since $|a\rangle$ can be expressed as a linear combination of the other two vectors in the set, the set is linearly dependent.

EXAMPLE 5.11

Is the following set of vectors linearly independent?

$$|a\rangle = \begin{pmatrix} 2 \\ 0 \\ 0 \end{pmatrix}, |b\rangle = \begin{pmatrix} -1 \\ 0 \\ -1 \end{pmatrix}, |c\rangle = \begin{pmatrix} 0 \\ 0 \\ -4 \end{pmatrix}$$

SOLUTION

Let a_1, a_2, a_3 be three unknown constants. To check linear independence, we write

$$a_1\,|a\rangle + a_2\,|b\rangle + a_3\,|c\rangle = 0$$

With the given column vectors, we obtain

$$a_1\begin{pmatrix} 2 \\ 0 \\ 0 \end{pmatrix} + a_2\begin{pmatrix} 0 \\ -1 \\ 0 \end{pmatrix} + a_3\begin{pmatrix} 0 \\ 0 \\ -4 \end{pmatrix} = \begin{pmatrix} 2a_1 \\ 0 \\ 0 \end{pmatrix} + \begin{pmatrix} 0 \\ -a_2 \\ 0 \end{pmatrix} + \begin{pmatrix} 0 \\ 0 \\ -4a_3 \end{pmatrix} = \begin{pmatrix} 2a_1 \\ -a_2 \\ -4a_3 \end{pmatrix} = 0$$

This equation can only be true if $a_1 = a_2 = a_3 = 0$. Therefore the set of vectors is linearly independent.

DIMENSION OF A SPACE

For a vector space V, the number of vectors in a basis determines the dimension of the space. Therefore if a basis of V consists of n linearly independent vectors that span the space, the dimension of V is n. The number of basis vectors can be finite, but infinite dimensional vector spaces exist as well.

DEFINITION: The Closure Relation

An orthonormal set $\{|\phi_1\rangle, |\phi_2\rangle, \ldots, |\phi_n\rangle\}$ constitutes a basis if and only if the set satisfies the closure relation

$$\sum_{i=1}^{n} |\phi_i\rangle \langle\phi_i| = 1$$

Expanding a Vector in Terms of a Basis

Given a set of basis vectors $\{|\phi_1\rangle, |\phi_2\rangle, \ldots, |\phi_n\rangle\}$ for a vector space V, we can expand any vector $|\psi\rangle \in V$ in terms of this basis as

$$|\psi\rangle = c_1 |\phi_1\rangle + c_2 |\phi_2\rangle + \ldots + c_n |\phi_n\rangle$$

where the c_i are complex numbers found from

$$c_i = \langle\phi_i|\psi\rangle$$

Orthonormal Sets and the Gram-Schmidt Procedure

In quantum mechanics it is desirable to have a set of basis vectors that is also orthonormal. If we have a set of basis vectors $\{|u_i\rangle\}$ that is not orthonormal, we can use the Gram-Schmidt procedure to build an orthonormal basis set. Here we illustrate the procedure for three-dimensions. If we call the orthonormal set $\{|v_i\rangle\}$, we begin by first defining:

$$|w_1\rangle = |u_1\rangle$$

We then construct successive vectors by subtracting off the projection of $|w_1\rangle$ on $|u_1\rangle$

$$|w_2\rangle = |u_2\rangle - \frac{\langle w_1|u_2\rangle}{\langle w_1|w_1\rangle} |w_1\rangle$$

$$|w_3\rangle = |u_3\rangle - \frac{\langle w_1|u_3\rangle}{\langle w_1|w_1\rangle} |w_1\rangle - \frac{\langle w_2|u_3\rangle}{\langle w_2|w_2\rangle} |w_2\rangle$$

To obtain the orthonormal set of basis vectors, we then normalize each of these quantities:

$$|v_1\rangle = \frac{|w_1\rangle}{\sqrt{\langle w_1 | w_1 \rangle}}, \quad |v_2\rangle = \frac{|w_2\rangle}{\sqrt{\langle w_2 | w_2 \rangle}}, \quad |v_3\rangle = \frac{|w_3\rangle}{\sqrt{\langle w_3 | w_3 \rangle}}$$

Dirac Algebra with Bras and Kets

In this section we focus on representing vectors entirely as kets and learn the basics of manipulating bras and kets. We will assume the existence of an orthonormal basis $|\phi_i\rangle$ where:

$$\langle \phi_i | \phi_j \rangle = \delta_{ij}$$

First, let's learn how to represent an arbitrary ket as a bra.

DEFINITION: Representing a Ket as a Bra

To obtain the bra corresponding to a given ket use the complex conjugation operation:

$$(\alpha |\psi\rangle)^* = \alpha^* \langle \psi|$$

For convenience we can write a ket $|\psi\rangle$ multiplied by a scalar α in the following way:

$$|\alpha\psi\rangle = \alpha |\psi\rangle$$

We can also use this notation for bras, but if we pull the scalar outside of the bra, we take its complex conjugate:

$$\langle \alpha\psi| = \alpha^* \langle \psi|$$

EXAMPLE 5.12 e.g.

Suppose that $|u_1\rangle, |u_2\rangle, |u_3\rangle$ is an orthonormal basis. In this basis let,

$$|\psi\rangle = 2i |u_1\rangle - 3 |u_2\rangle + i |u_3\rangle$$

$$|\phi\rangle = 3 |u_1\rangle - 2 |u_2\rangle + 4 |u_3\rangle$$

(a) Find $\langle \psi|$ and $\langle \phi|$.

(b) Compute the inner product $\langle \phi | \psi \rangle$ and show that $\langle \phi | \psi \rangle = \langle \psi | \phi \rangle^*$.

(c) Let $a = 3 + 3i$ and compute $|a\psi\rangle$.

(d) Find $|\psi + \phi\rangle, |\psi - \phi\rangle$.

SOLUTION

(a) We find the bra corresponding to each ket by changing the base kets to bras and taking the complex conjugate of each coefficient. Therefore

$$\langle\psi| = (2i)^* \langle u_1| - 3\langle u_2| + (i)^* \langle u_3| = -2i\langle u_1| - 3\langle u_2| - i\langle u_3|$$

$$\langle\phi| = 3\langle u_1| - 2\langle u_2| + 4\langle u_3|$$

(b) To compute the inner product, we rely on the fact that the basis is orthonormal, i.e. $\langle u_i | u_j \rangle = \delta_{ij}$. And so we obtain

$$\begin{aligned}
\langle\phi|\psi\rangle &= (3\langle u_1| - 2\langle u_2| + 4\langle u_3|)(2i|u_1\rangle - 3|u_2\rangle + i|u_3\rangle) \\
&= (3)(2i)\langle u_1|u_1\rangle + (3)(-3)\langle u_1|u_2\rangle + (3)(i)\langle u_1|u_3\rangle \\
&\quad + (-2)(2i)\langle u_2|u_1\rangle + (-2)(-3)\langle u_2|u_2\rangle + (-2)(i)\langle u_2|u_3\rangle \\
&\quad + (4)(2i)\langle u_3|u_1\rangle + (4)(-3)\langle u_3|u_2\rangle + (4)(i)\langle u_3|u_3\rangle \\
&= 6i\langle u_1|u_1\rangle + 6\langle u_2|u_2\rangle + 4i\langle u_3|u_3\rangle \\
&= 6 + 10i
\end{aligned}$$

Now the inner product $\langle\psi|\phi\rangle$ is

$$\begin{aligned}
\langle\psi|\phi\rangle &= (-2i\langle u_1| - 3\langle u_2| - i\langle u_3|)(3|u_1\rangle - 2|u_2\rangle + 4|u_3\rangle) \\
&= -6i\langle u_1|u_1\rangle + 6\langle u_2|u_2\rangle - 4i\langle u_3|u_3\rangle \\
&= 6 - 10i \\
&\Rightarrow \langle\phi|\psi\rangle = \langle\psi|\phi\rangle^*
\end{aligned}$$

(c) To compute $|a\psi\rangle$, we multiply each coefficient in the expansion by a:

$$\begin{aligned}
|\psi\rangle &= (3+3i)(2i|u_1\rangle - 3|u_2\rangle + i|u_3\rangle) \\
&= (3+3i)2i|u_1\rangle - (3+3i)3|u_2\rangle + (3+3i)i|u_3\rangle \\
&= (-6+6i)|u_1\rangle - (9+9i)|u_2\rangle + (-3+3i)|u_3\rangle
\end{aligned}$$

(d) To compute $|\psi \pm \phi\rangle$, we add/subtract components:

$$\begin{aligned}
|\psi\rangle &= 2i|u_1\rangle - 3|u_2\rangle + i|u_3\rangle \\
|\phi\rangle &= 3|u_1\rangle - 2|u_2\rangle + 4|u_3\rangle \\
\Rightarrow |\psi + \phi\rangle &= (3+2i)|u_1\rangle - 6|u_2\rangle + (4+i)|u_3\rangle \\
|\psi - \phi\rangle &= (-3+2i)|u_1\rangle - |u_2\rangle + (-4+i)|u_3\rangle
\end{aligned}$$

EXAMPLE 5.13

e.g.

Let

$$|\psi\rangle = 3i\,|\phi_1\rangle + 2\,|\phi_2\rangle - 4i\,|\phi_3\rangle$$

where the $|\phi_i\rangle$ are an orthonormal basis. Normalize $|\psi\rangle$.

SOLUTION

The first step is to write down the bra corresponding to $|\psi\rangle$. Remember we need to complex conjugate each expansion coefficient:

$$\langle\psi| = -3i\,\langle\phi_1| + 2\,\langle\phi_2| + 4i\,\langle\phi_3|$$

Now we can compute the norm of the vector:

$$\begin{aligned}\langle\psi\,|\psi\rangle &= (-3i\,\langle\phi_1| + 2\,\langle\phi_2| + 4i\,\langle\phi_3|)(3i\,|\phi_1\rangle + 2\,|\phi_2\rangle - 4i\,|\phi_3\rangle) \\ &= (-3i)(3i)\,\langle\phi_1\,|\phi_1\rangle + 4\,\langle\phi_2\,|\phi_2\rangle + (4i)(-4i)\,\langle\phi_3\,|\phi_3\rangle \\ &= 9 + 4 + 16 = 29\end{aligned}$$

where we have used the fact that the basis is orthonormal. The norm is the square root of this quantity:

$$\|\,|\psi\rangle\| = \sqrt{\langle\psi\,|\psi\rangle} = \sqrt{29}$$

And so the normalized vector is found by dividing $|\psi\rangle$ by the norm to give

$$\left|\tilde{\psi}\right\rangle = \frac{3i}{\sqrt{29}}\,|\phi_1\rangle + \frac{2}{\sqrt{29}}\,|\phi_2\rangle - \frac{4i}{\sqrt{29}}\,|\phi_3\rangle$$

Finding the Expansion Coefficients in the Representation of Bras and Kets

If a ket is written in terms of a set of basis vectors, we find a given component of the ket by taking the inner product with that basis vector. Let's make this statement clear with an example. In an earlier section we considered:

$$|\psi\rangle = 2i\,|u_1\rangle - 3\,|u_2\rangle + i\,|u_3\rangle$$

Remember, we are assuming the basis set is orthonormal and so $\left\langle u_i\,\middle|u_j\right\rangle = \delta_{ij}$. So, for example, we can find the third component of the ket this way:

$$\begin{aligned}\langle u_3\,|\psi\rangle &= 2i\,\langle u_3\,|u_1\rangle - 3\,\langle u_3\,|u_2\rangle + i\,\langle u_3\,|u_3\rangle \\ &= i\,\langle u_3\,|u_3\rangle = i\end{aligned}$$

More generally, we can also index the components and write the ket as:

$$|\psi\rangle = c_1 |u_1\rangle + c_2 |u_2\rangle + \cdots + c_n |u_n\rangle = \sum_{i=1}^{n} c_i |u_i\rangle$$

where $c_i = \langle u_i |\psi\rangle$. The numbers c_i are the components of the ket $|\Psi\rangle$ in the basis $|u_i\rangle$. These numbers are arranged into a column vector, this is how we got the column vectors of complex numbers we worked with earlier in the chapter. In n dimensions, we represent a ket as:

$$|\Psi >\rightarrow \begin{pmatrix} \langle u_1|\psi\rangle \\ \langle u_2|\psi\rangle \\ \vdots \\ \langle u_n|\psi\rangle \end{pmatrix} = \begin{pmatrix} c_1 \\ c_2 \\ \vdots \\ c_n \end{pmatrix}$$

Note that a given vector space has many different bases, and in a different basis, the ket will be represented by a different set of components.

EXAMPLE 5.14

Let $|u_1\rangle, |u_2\rangle, |u_3\rangle$ be an orthonormal basis for a three-dimensional vector space. Suppose that

$$|\psi\rangle = 2i\,|u_1\rangle - 3\,|u_2\rangle + i\,|u_3\rangle$$

Write the column vector representing this vector in the given basis. Then write down the row vector that represents $\langle\psi|$ in this basis.

SOLUTION

The components of the column vector representing $|\psi\rangle$ are found by taking the inner product of the vector with each of the basis vectors. We have

$$\begin{aligned}\langle u_1 |\psi\rangle &= \langle u_1| (2i\,|u_1\rangle - 3\,|u_2\rangle + i\,|u_3\rangle) \\ &= 2i\,\langle u_1 |u_1\rangle - 3\,\langle u_1 |u_2\rangle + i\,\langle u_1 |u_3\rangle\end{aligned}$$

Since the basis is orthonormal, $\langle u_1 |u_1\rangle = 1$ and the other terms are zero. Therefore

$$\langle u_1 |\psi\rangle = 2i$$

The inner product with the second basis vector is

$$\begin{aligned}\langle u_2 |\psi\rangle &= \langle u_2| (2i\,|u_1\rangle - 3\,|u_2\rangle + i\,|u_3\rangle) \\ &= 2i\,\langle u_2 |u_1\rangle - 3\,\langle u_2 |u_2\rangle + i\,\langle u_2 |u_3\rangle \\ &= -3\end{aligned}$$

and for the third basis vector:

$$\begin{aligned}\langle u_3 | \psi \rangle &= \langle u_3 | (2i | u_1 \rangle - 3 | u_2 \rangle + i | u_3 \rangle) \\ &= 2i \langle u_3 | u_1 \rangle - 3 \langle u_3 | u_2 \rangle + i \langle u_3 | u_3 \rangle \\ &= i\end{aligned}$$

Therefore the column vector representation of $|\psi\rangle$ is given by

$$|\psi\rangle = \begin{pmatrix} \langle u_1 | \psi \rangle \\ \langle u_2 | \psi \rangle \\ \langle u_3 | \psi \rangle \end{pmatrix} = \begin{pmatrix} 2i \\ -3 \\ i \end{pmatrix}$$

The dual vector or "bra" corresponding to $|\psi\rangle$ is represented by the row vector

$$\langle \psi | = (\langle \psi | u_1 \rangle \ \langle \psi | u_2 \rangle \ \langle \psi | u_3 \rangle) = \left(\langle u_1 | \psi \rangle^* \ \langle u_2 | \psi \rangle^* \ \langle u_3 | \psi \rangle^* \right)$$

and so we have

$$\langle \psi | = \left((2i)^* \ \ (-3)^* \ \ (i)^* \right) = (-2i \ \ -3 \ \ -i)$$

Quiz

1. Let

$$|u\rangle = \begin{pmatrix} 3i \\ 2 \end{pmatrix}, |v\rangle = \begin{pmatrix} 6i \\ 4 \end{pmatrix}$$

   (a) Find the norm of $|u\rangle$ and the norm of $|v\rangle$. Are these vectors normalized?
   (b) If $a = 4 - 2i$, find $a\,|u\rangle$.
   (c) Find $\langle u | v \rangle$, $\langle v | u \rangle$.

2. Consider the set of 2×2 matrices:

$$A = \begin{pmatrix} a & b \\ c & d \end{pmatrix}$$

   where a, b, c, d are arbitrary complex numbers. Does this set of matrices form a vector space? If so, find a basis.

3. Are $\vec{A} = (1, -2, 1)$, $\vec{B} = (0, 2, 3)$ and $\vec{C} = (0, 0, 5)$ linearly independent?

4. Show that $(1, 1, 0)$, $(0, 0, 2)$ and (i, i, i) are linearly dependent.

5. Using the fact that $Re(z) \leq |z|$, prove the triangle inequality for complex numbers, i.e.

$$|z_1 + z_2| \leq |z_1| + |z_2|$$

6. Apply the Gram-Schmidt procedure to the vectors:

$$[(1, 2, 1), (2, 8, 3), (3, -2, -i)]$$

to find an orthonormal basis set.

7. Suppose that

$$|A\rangle = (2 + 3i)\,|u_1\rangle + 4\,|u_2\rangle\,,\ |B\rangle = 7\,|u_1\rangle - 8i\,|u_2\rangle$$

where $\{|u_1\rangle\,,\ |u_2\rangle\}$ is an orthonormal basis.

(a) Normalize the vectors $|A\rangle$ and $|B\rangle$.
(b) Find $|A + B\rangle$.
(c) Show that $|A\rangle\,,\ |B\rangle$ satisfy the triangle inequality.
(d) Show that $|A\rangle\,,\ |B\rangle$ satisfy the Cauchy-Schwarz inequality.

8. Suppose that

$$|A\rangle = (9 - 2i)\,|u_1\rangle + 4i\,|u_2\rangle - |u_3\rangle + i\,|u_4\rangle$$

(a) Find the dual vector $\langle A|$.
(b) Assuming that the $|u_i\rangle$ are an orthonormal basis, determine the representation of $|A\rangle$ as a column vector in this basis.
(c) Find the row vector representing the dual vector.
(d) Compute the inner product $\langle A\,|A\rangle$ and normalize $|A\rangle$.

CHAPTER 6

The Mathematical Structure of Quantum Mechanics II

Physical observables—that is, quantities that can be measured such as position and momentum—are represented within the mathematical structure of quantum mechanics by operators. Mathematically, an operator, which can be represented by a matrix, is a map that takes a vector to another vector. The eigenvalues of the matrix tell us the possible results of a measurement of the quantity the operator represents, while the eigenvectors of the matrix give us a basis we can use to represent states.

LINEAR OPERATORS

An operator is a mathematical rule or instruction that transforms one vector into a new, generally different vector. Operators are frequently denoted by capital letters with a "hat" or carat on top—for example $\hat{A}$, $\hat{B}$, $\hat{C}$. Sometimes in this book we will denote them simply by capital italicized letters, A, B, C. We write the action of an arbitrary operator $\hat{T}$ on a ket as:

$$\hat{T}\,|u\rangle = |v\rangle$$

The operators that are most interesting to us are linear operators. Suppose V is a linear vector space over the complex Field C. An operator $\hat{T} : V \to V$ is a linear operator on V if given complex, scalars α, β in C and vectors $|u\rangle$, $|v\rangle$ in V:

$$\hat{T}(\alpha |u\rangle + \beta |v\rangle) = \alpha\hat{T} |u\rangle + \beta\hat{T} |v\rangle$$

The sum of two linear operators T and S acts on a single vector as:

$$(\hat{T} + \hat{S}) |u\rangle = \hat{T} |u\rangle + \hat{S} |u\rangle$$

The product of two linear operators T and S acts on a vector in the following manner:

$$(\hat{T}\hat{S}) |u\rangle = \hat{T}(\hat{S} |u\rangle)$$

Since an operator transforms a ket into a new ket, if $S |u\rangle = |w\rangle$ then

$$\hat{T}(\hat{S} |u\rangle) = \hat{T} |w\rangle$$

Operators stand to the left of kets and to the right of bras.

$$A |u\rangle , \quad \langle u| A$$

are valid expressions, but

$$\hat{A} \langle u| \, and \, |u\rangle \hat{A}$$

are not. We now consider some simple operators.

DEFINITION: The Identity Operator

The simplest operator of all is the identity operator, which does nothing to a ket:

$$I |u\rangle = |u\rangle$$

DEFINITION: Outer Product

The outer product between a ket and a bra is written as follows:

$$|\psi\rangle \langle\phi|$$

This expression is an operator. We apply it to a ket $|\chi\rangle$ and show that it produces a new ket which is proportional to $|\Psi\rangle$:

$$(|\psi\rangle \langle\phi|) |\chi\rangle = |\psi\rangle \langle\phi |\chi\rangle$$

Now the inner product $\langle\Phi|\chi\rangle$ is just a complex number, which we denote by α. Therefore we have a new ket proportional to $|\Psi\rangle$:

$$|\psi\rangle\langle\phi|\chi\rangle = \alpha|\psi\rangle$$

DEFINITION: The Closure Relation

The closure relation tells us that given a basis $|u_i\rangle$ we can write the identity operator as a summation of outer products of the form $|u_i\rangle\langle u_i|$ To see this, recall that in such a basis the components of a ket $|\psi\rangle$ are given by the inner products $\langle u_i|\psi\rangle$. The decomposition of a ket $|\psi\rangle$is given by:

$$|\psi\rangle = \sum_{i=1}^{n} |u_i\rangle\langle u_i|\psi\rangle$$

Notice that we can write this as:

$$|\psi\rangle = \sum_{i=1}^{n} |u_i\rangle\langle u_i|\psi\rangle = \left(\sum_{i=1}^{n} |u_i\rangle\langle u_i|\right)|\psi\rangle$$

This can only be true if $\sum_{i=1}^{n}|u_i\rangle\langle u_i| = 1$, the identity operator.

The Representation of an Operator

The representation of an operator is formed by considering its action on a given set of basis vectors. In a basis that we label $|u_i\rangle$, the components of an operator $\hat{T}$ are found by forming the following inner product:

$$T_{ij} = \langle u_i|\hat{T}|u_j\rangle$$

When the given vector space is n dimensional, the components of the operator can be arranged into an $n \times n$ matrix, where T_{ij} is the element at row i and column j:

$$\hat{T} \to (T_{ij}) = \begin{pmatrix} T_{11} & T_{12} & \dots & T_{1n} \\ T_{21} & T_{22} & \dots & T_{2n} \\ \vdots & \vdots & \ddots & \vdots \\ T_{n1} & T_{n2} & \dots & T_{nn} \end{pmatrix}$$

$$= \begin{pmatrix} \langle u_1|\hat{T}|u_1\rangle & \langle u_1|\hat{T}|u_2\rangle & \dots & \langle u_1|\hat{T}|u_n\rangle \\ \langle u_2\hat{T}|u_1\rangle & \langle u_2|\hat{T}|u_2\rangle & \dots & \langle u_2|\hat{T}|u_n\rangle \\ \vdots & \vdots & \ddots & \vdots \\ \langle u_n|\hat{T}|u_1\rangle & \langle u_n|\hat{T}|u_2\rangle & \dots & \langle u_n|\hat{T}|u_n\rangle \end{pmatrix}$$

EXAMPLE 6.1

Suppose that in some orthonormal basis $\{|u_1\rangle, |u_2\rangle, |u_3\rangle\}$ an operator A acts as follows:

$$A|u_1\rangle = 2|u_1\rangle$$

$$A|u_2\rangle = 3|u_1\rangle - i|u_3\rangle$$

$$A|u_3\rangle = -|u_2\rangle$$

Write the matrix representation of the operator.

SOLUTION

The matrix representation of the operator is given by:

$$(A_{ij}) = \begin{pmatrix} \langle u_1|A|u_1\rangle & \langle u_1|A|u_2\rangle & \langle u_1|A|u_3\rangle \\ \langle u_2|A|u_1\rangle & \langle u_2|A|u_2\rangle & \langle u_2|A|u_3\rangle \\ \langle u_3|A|u_1\rangle & \langle u_3|A|u_2\rangle & \langle u_3|A|u_3\rangle \end{pmatrix}$$

$$= \begin{pmatrix} \langle u_1|(2|u_1\rangle) & \langle u_1|(3|u_1\rangle - i|u_3\rangle) & \langle u_1|(-|u_2\rangle) \\ \langle u_2|(2|u_1\rangle) & \langle u_2|(3|u_1\rangle - i|u_3\rangle) & \langle u_2|(-|u_2\rangle) \\ \langle u_3|(2|u_1\rangle) & \langle u_3|(3|u_1\rangle - i|u_3\rangle) & \langle u_3|(-|u_2\rangle) \end{pmatrix}$$

$$= \begin{pmatrix} 2\langle u_1|u_1\rangle & 3\langle u_1|u_1\rangle - i\langle u_1|u_3\rangle & -\langle u_1|u_2\rangle \\ 2\langle u_2|u_1\rangle & 3\langle u_2|u_1\rangle - i\langle u_2|u_3\rangle & -\langle u_2|u_2\rangle \\ 2\langle u_3|u_1\rangle & 3\langle u_3|u_1\rangle - i\langle u_3|u_3\rangle & -\langle u_3|u_2\rangle \end{pmatrix}$$

Since the basis is orthonormal, we have $\langle u_i|u_j\rangle = \delta_{ij}$ and so the matrix representation of A in this basis is:

$$(A_{ij}) = \begin{pmatrix} 2\langle u_1|u_1\rangle & 3\langle u_1|u_1\rangle - i\langle u_1|u_3\rangle & -\langle u_1|u_2\rangle \\ 2\langle u_2|u_1\rangle & 3\langle u_2|u_1\rangle - i\langle u_2|u_3\rangle & -\langle u_2|u_2\rangle \\ 2\langle u_3|u_1\rangle & 3\langle u_3|u_1\rangle - i\langle u_3|u_3\rangle & -\langle u_3|u_2\rangle \end{pmatrix}$$

$$= \begin{pmatrix} 2 & 3 & 0 \\ 0 & 0 & -1 \\ 0 & -i & 0 \end{pmatrix}$$

Note: In a different basis, A will be represented by a different matrix.

EXAMPLE 6.2

e.g.

The outer product $|\phi\rangle\langle\psi|$ is an operator, and is therefore can be represented by a matrix. Show this for:

$$|\Phi\rangle = \begin{pmatrix} 2 \\ 3i \\ 4 \end{pmatrix} |\Psi\rangle = \begin{pmatrix} -1 \\ 0 \\ i \end{pmatrix}$$

Given that $\langle\Psi|\Psi\rangle = 2$, the action of this operator on a ket

$$(|\Phi\rangle\langle\Psi|)(3|\Psi\rangle) = 3(\langle\Psi|\Psi\rangle)|\Phi\rangle = 3(2)|\Phi\rangle = \begin{pmatrix} 12 \\ 18i \\ 24 \end{pmatrix}$$

Show this with matrix multiplication.

SOLUTION

✔

First we write:

$$\langle\Psi| = (\,-1 \quad 0 \quad -i\,) \Rightarrow$$

$$|\Phi\rangle\langle\Psi| = \begin{pmatrix} 2 \\ 3i \\ 4 \end{pmatrix} (\,-1 \quad 0 \quad -i\,) = \begin{pmatrix} 2(-1) & 2(0) & 2(-i) \\ 3i(-1) & 3i(0) & 3i(-i) \\ 4(-1) & 4(0) & 4(-i) \end{pmatrix} = \begin{pmatrix} -2 & 0 & -2i \\ -3i & 0 & 3 \\ -4 & 0 & -4i \end{pmatrix}$$

$$3|\Psi\rangle = 3\begin{pmatrix} -1 \\ 0 \\ i \end{pmatrix} = \begin{pmatrix} -3 \\ 0 \\ 3i \end{pmatrix}$$

$$\Rightarrow (|\Phi\rangle\langle\Psi|)3|\Psi\rangle = \begin{pmatrix} -2 & 0 & -2i \\ -3i & 0 & 3 \\ -4 & 0 & -4i \end{pmatrix} \begin{pmatrix} -3 \\ 0 \\ 3i \end{pmatrix} = \begin{pmatrix} -2(-3) + (-2i)(3i) \\ -3i(-3) + 3(3i) \\ -4(-3) + (-4i)(3i) \end{pmatrix} = \begin{pmatrix} 12 \\ 18i \\ 24 \end{pmatrix}$$

EXAMPLE 6.3

e.g.

Consider a two-dimensional space in which a basis is given by

$$|0\rangle = \begin{pmatrix} 1 \\ 0 \end{pmatrix}, |1\rangle = \begin{pmatrix} 0 \\ 1 \end{pmatrix}$$

and an operator $\hat{A}$ is given by:

$$A = \begin{pmatrix} 1 & -2i \\ 2i & 0 \end{pmatrix}$$

Express A in outer product notation.

SOLUTION

First we write A in terms of outer products of $\{|0\rangle, |1\rangle\}$ with unknown coefficients:

$$A = a\,|0\rangle\,\langle 0| + b\,|0\rangle\,\langle 1| + c\,|1\rangle\,\langle 0| + d\,|1\rangle\,\langle 1|$$

The matrix A is

$$\begin{pmatrix} 1 & -2i \\ 2i & 0 \end{pmatrix} = \begin{pmatrix} \langle 0|\,A\,|0\rangle & \langle 0|\,A\,|1\rangle \\ \langle 1|\,A\,|0\rangle & \langle 1|\,A\,|1\rangle \end{pmatrix}$$

This gives us four equations for the unknowns a, b, c, d. We use the orthonormality of the basis to evaluate each term, i.e. $\langle 0\,|0\rangle = \langle 1\,|1\rangle = 1$, $\langle 1\,|0\rangle = \langle 0\,|1\rangle = 0$:

$$\begin{aligned}
\langle 0|\,A\,|0\rangle &= \langle 0|\,(a\,|0\rangle\,\langle 0| + b\,|0\rangle\,\langle 1| + c\,|1\rangle\,\langle 0| + d\,|1\rangle\,\langle 1|)\,|0\rangle \\
&= a\,\langle 0\,|0\rangle\,\langle\,0\,|0\rangle + b\,\langle 0\,|0\rangle\,\langle 1\,|0\rangle + c\,\langle 0\,|1\rangle\,\langle 0\,|0\rangle + d\,\langle 0\,|1\rangle\,\langle 1\,|0\rangle \\
&= a \\
&\Rightarrow a = 1
\end{aligned}$$

$$\begin{aligned}
-2i = \langle 0|\,A\,|1\rangle &= \langle 0|\,(a\,|0\rangle\,\langle 0| + b\,|0\rangle\,\langle 1| + c\,|1\rangle\,\langle 0| + d\,|1\rangle\,\langle 1|)\,|1\rangle \\
&= a\,\langle 0\,|0\rangle\,\langle 0\,|1\rangle + b\,\langle 0\,|0\rangle\,\langle 1\,|1\rangle + c\,\langle 0\,|1\rangle\,\langle 0\,|1\rangle + d\,\langle 0\,|1\rangle\,\langle 1\,|1\rangle \\
&= b \\
&\Rightarrow b = -2i
\end{aligned}$$

The same procedure can be applied to the other two terms

$$\langle 1|A|0\rangle = 2i \text{ and } \langle 1|A|1\rangle = 0$$

yielding $c = 2i$ and $d = 0$. Therefore, we can write A as:

$$A = |0\rangle\,\langle 0| - 2i\,|0\rangle\,\langle 1| + 2i\,|1\rangle\,\langle 0|$$

DEFINITION: The Trace of an Operator

The trace of an operator $\hat{T}$ is the sum of the diagonal elements of its matrix and is denoted $t_r\,(\hat{T})$. If

$$\hat{T} = (T_{ij}) = \begin{pmatrix} T_{11} & T_{12} & \dots & T_{1n} \\ T_{21} & T_{22} & \dots & T_{2n} \\ \vdots & \vdots & \ddots & \vdots \\ T_{n1} & T_{n2} & \dots & T_{nn} \end{pmatrix}$$

$$= \begin{pmatrix} \langle u_1|\hat{T}|u_1\rangle & \langle u_1|\hat{T}|u_2\rangle & \dots & \langle u_1|\hat{T}|u_n\rangle \\ \langle u_2|\hat{T}|u_1\rangle & \langle u_1|\hat{T}|u_2\rangle & \dots & \langle u_2|\hat{T}|u_n\rangle \\ \vdots & \vdots & \ddots & \vdots \\ \langle u_n|\hat{T}|u_1\rangle & \langle u_n|\hat{T}|u_2\rangle & \dots & \langle u_n|\hat{T}|u_n\rangle \end{pmatrix}$$

then $Tr(\hat{T}) = T_{11} + T_{22} + \ldots + T_{nn} = \sum_{i=1}^{n} T_{ii}$. Alternatively, we can write the trace as:

$$Tr(\hat{T}) = \langle u_1|\hat{T}|u_1\rangle + \langle u_2|\hat{T}|u_2\rangle + \ldots + \langle u_n|\hat{T}|u_n\rangle = \sum_{i=1}^{n} \langle u_i|\hat{T}|u_i\rangle.$$

EXAMPLE 6.4

e.g.

The trace is cyclic; in other words,

$$Tr(ABC) = Tr(BCA) = Tr(CAB)$$

Prove this for the case of two operators A and B, i.e. $Tr(AB) = Tr(BA)$,

SOLUTION

✔

We prove this using the closure relation considering some basis $|u_i\rangle$. Recall that the identity operator can be written as $1 = \sum_{i=1}^{n} |u_i\rangle\langle u_i|$ Then we have:

$$\begin{aligned}
Tr(AB) &= \sum_{i=1}^{n} \langle u_i|\, AB\, |u_i\rangle = \sum_{i=1}^{n} \langle u_i|\, A(I)B\, |u_i\rangle \\
&= \sum_{i=1}^{n} \langle u_i|\, A \left(\sum_{j=1}^{n} |u_j\rangle\langle u_j| \right) B|u_i\rangle \\
&= \sum_{i=1}^{n} \sum_{j=1}^{n} \langle u_j|\, B|u_i\rangle\langle u_i|A|u_j\rangle \\
&= \sum_{j=1}^{n} \langle u_j|B \left(\sum_{i=1}^{n} |u_i\rangle\langle u_i| \right) A|u_j\rangle \\
&= \sum_{j=1}^{n} \langle u_j|B(I)A|u_j\rangle \\
&= \sum_{j=1}^{n} \langle u_j|BA|u_j\rangle = Tr(BA)
\end{aligned}$$

EXAMPLE 6.5

Suppose that in some basis an operator A has the following matrix representation:

$$A = \begin{pmatrix} i & 0 & -2i \\ 1 & 4 & 6 \\ 0 & 8 & -1 \end{pmatrix}$$

Find the trace of A.

SOLUTION

The trace can be calculated from the matrix representation of an operator by summing its diagonal elements, so:

$$Tr(A) = i + 4 - 1 = 3 + i$$

EXAMPLE 6.6

An operator

$$T = -|\Phi_1\rangle\langle\Phi_1| + |\Phi_2\rangle\langle\Phi_2| + 2|\Phi_3\rangle\langle\Phi_3| - i|\Phi_1\rangle\langle\Phi_2| + |\Phi_2\rangle\langle\Phi_1|$$

in some basis $|\Phi_1\rangle, |\Phi_2\rangle, |\Phi_3\rangle$. Calculate $Tr(T)$. The basis is orthonormal.

SOLUTION

$$Tr(T) = \sum \langle\Phi_i|T|\Phi_i\rangle = \langle\Phi_1|T|\Phi_1\rangle + \langle\Phi_2|T|\Phi_2\rangle + \langle\Phi_3|T|\Phi_3\rangle$$

Begin by finding the action of T on each of the basis vectors $|\Phi_1\rangle, |\Phi_2\rangle, |\Phi_3\rangle$

$$T|\Phi_1\rangle = (-|\Phi_1\rangle\langle\Phi_1| + |\Phi_2\rangle\langle\Phi_2| + 2|\Phi_3\rangle\langle\Phi_3| - i|\Phi_1\rangle\langle\Phi_2| + |\Phi_2\rangle\langle\Phi_1|)|\Phi_1\rangle$$

$$= -|\Phi_1\rangle\langle\Phi_1|x\Phi_1\rangle + |\Phi_2\rangle\langle\Phi_2|\Phi_1\rangle + 2|\Phi_3\rangle\langle\Phi_3|\Phi_1\rangle$$

$$-i|\Phi_1\rangle\langle\Phi_2||\Phi_1\rangle + |\Phi_2\rangle\langle\Phi_1||\Phi_1\rangle$$

$$= -|\Phi_1\rangle + |\Phi_2\rangle$$

$$\Rightarrow \langle\Phi_1|T|\Phi_1\rangle = \langle\Phi_1|(-|\Phi_1\rangle + |\Phi_2\rangle) = -\langle\Phi_1|\Phi_1\rangle + \langle\Phi_1|\Phi_2\rangle$$

$$= -\langle\Phi_1|\Phi_1\rangle = -1$$

$$T|\Phi_2\rangle = (-|\Phi_1\rangle\langle\Phi_1| + |\Phi_2\rangle\langle\Phi_2| + 2|\Phi_3\rangle\langle\Phi_3| - i|\Phi_1\rangle\langle\Phi_2| + |\Phi_2\rangle\langle\Phi_1|)|\Phi_2\rangle$$

$$= -|\Phi_1\rangle\langle\Phi_1|\Phi_2\rangle + |\Phi_2\rangle\langle\Phi_2|\Phi_2\rangle + 2|\Phi_3\rangle\langle\Phi_3|\Phi_2\rangle$$

$$-i|\Phi_1\rangle\langle\Phi_2|\Phi_2\rangle + |\Phi_2\rangle\langle\Phi_1|\Phi_2\rangle$$

$$= -i|\Phi_1\rangle + |\Phi_2\rangle$$

$$\Rightarrow \langle\Phi_2 T|\Phi_2\rangle = \langle\Phi_2|(-i|\Phi_1\rangle + |\Phi_2\rangle) = -i\langle\Phi_2|\Phi_1\rangle + \langle\Phi_2|\Phi_2\rangle = +1$$

$$\begin{aligned} T\,|\Phi_3\rangle &= (-\,|\Phi_1\rangle\langle\Phi_1| + |\Phi_2\rangle\langle\Phi_2| + 2\,|\Phi_3\rangle\langle\Phi_3| - i\,|\Phi_1\rangle\langle\Phi_2| + |\Phi_2\rangle\langle\Phi_1|)\,|\Phi_3\rangle \\ &= -\,|\Phi_1\rangle\langle\Phi_1\,|\Phi_3\rangle + |\Phi_2\rangle\langle\Phi_2|\,\Phi_3\rangle + 2\,|\Phi_3\rangle\langle\Phi_3\,|\Phi_3\rangle \\ &\quad - i\,|\Phi_1\rangle\langle\Phi_2|\,|\Phi_3\rangle + |\Phi_2\rangle\langle\Phi_1\,|\Phi_3\rangle \\ &= 2\,|\Phi_3\rangle \\ &\Rightarrow \langle\Phi_3|\,T\,|\Phi_3\rangle = 2 \end{aligned}$$

Therefore $Tr(T) = \langle\Phi_1|T|\Phi_1\rangle + \langle\Phi_2|T|\Phi_2\rangle + \langle\Phi_3|T|\Phi_3\rangle = -1+1+2 = 2$

EXAMPLE 6.7

e.g.

Earlier we saw an operator A and a basis $u_j\rangle$ for which:

$$\begin{aligned} A\,|u_1\rangle &= 2\,|u_1\rangle \\ A\,|u_2\rangle &= 3\,|u_1\rangle - i\,|u_3\rangle \\ A\,|u_3\rangle &= -\,|u_2\rangle \end{aligned}$$

We found that the matrix representing A in this basis was given by:

$$(A_{ij}) = \begin{pmatrix} 2 & 3 & 0 \\ 0 & 0 & -1 \\ 0 & -i & 0 \end{pmatrix}$$

Find $Tr(A)$ from $\sum_{i=1}^{n}\langle u_i|A|u_i\rangle$ and show that this is equal to the sum of the diagonal elements of the matrix.

From $(A_{ij}) = \begin{pmatrix} 2 & 3 & 0 \\ 0 & 0 & -1 \\ 0 & -i & 0 \end{pmatrix}$ we see that $Tr(A) = 2+0+0 = 2$

$$\begin{aligned} &= \sum_{j=1}^{n}\langle u_i|\,A\,|u_i\rangle = \langle u_1|\,A\,|u_1\rangle + \langle u_2|\,A\,|u_2\rangle + \langle u_3|\,A\,|u_3\rangle \\ &= 2\,\langle u_1\,|u_1\rangle + \langle u_2|\,(3\,|u_1\rangle - i\,|u_3\rangle) - \langle u_3\,|u_2\rangle \\ &= 2*1 + 3*0 - i*0 - 0 \\ &= 2 \end{aligned}$$

THE ACTION OF AN OPERATOR ON KETS IN MATRIX REPRESENTATION

In matrix representation in n dimensions a ket is an $n \times 1$ column vector. To act an operator, which is represented by an $n \times n$ matrix on the ket, we perform basic matrix multiplication.

EXAMPLE 6.8

$$\text{Let } A = \begin{pmatrix} 1 & 1 & -3 \\ 3 & 2 & 0 \\ -3 & 0 & -2 \end{pmatrix}, |\psi\rangle = \begin{pmatrix} 2 \\ 3 \\ -1 \end{pmatrix}$$

Find $A\,|\Psi\rangle$.

SOLUTION

$$A\,|\Psi\rangle = \begin{pmatrix} 1 & 1 & -3 \\ 3 & 2 & 0 \\ -3 & 0 & -2 \end{pmatrix}\begin{pmatrix} 2 \\ 3 \\ -1 \end{pmatrix} = \begin{pmatrix} 1(2)+1(3)-3(-1) \\ 3(2)+2(3)+0(-1) \\ -3(2)+0(3)-2(-1) \end{pmatrix} = \begin{pmatrix} 8 \\ 12 \\ -4 \end{pmatrix}$$

CALCULATING EXPECTATION VALUES OF OPERATORS

The epectation value of an operator A with respect to a state $|\Psi\rangle$ is given by

$$\left\langle \hat{A} \right\rangle = \langle\Psi|\,\hat{A}\,|\Psi\rangle$$

EXAMPLE 6.9

A particle is in a state

$$|\Psi\rangle = 2i\,|u_1\rangle - |u_2\rangle + 4i\,|u_3\rangle$$

and an operator

$$A = |u_1\rangle\langle u_1| - 2i\,|u_1\rangle\langle u_2| + |u_3\rangle\langle u_3|$$

(the basis is orthonormal). Find $\langle A\rangle$ in this state.

SOLUTION

$$|\Psi\rangle = 2i\,|u_1\rangle - |u_2\rangle + 4i\,|u_3\rangle\,, \Rightarrow \langle\Psi| = -2i\,\langle u_1| - \langle u_2| - 4i\,\langle u_3|$$

$$A\,|\Psi\rangle = A(2i\,|u_1\rangle - |u_2\rangle + 4i\,|u_3\rangle) = 2iA\,|u_1\rangle - A\,|u_2\rangle + 4iA\,|u_3\rangle$$

We consider the action of A on the basis states alone. For the first state we find

$$\begin{aligned} A\left|u_1\right\rangle &= \left(\left|u_1\right\rangle\left\langle u_1\right| - 2i\left|u_1\right\rangle\left\langle u_2\right| + \left|u_3\right\rangle\left\langle u_3\right|\right)\left|u_1\right\rangle \\ &= \left|u_1\right\rangle\left\langle u_1 \middle| u_1\right\rangle - 2i\left|u_1\right\rangle\left\langle u_2 \middle| u_1\right\rangle + \left|u_3\right\rangle\left\langle u_3 \middle| u_1\right\rangle \end{aligned}$$

Now the basis is orthonormal, and so

$$\begin{aligned} \left\langle u_1 \middle| u_1\right\rangle &= 1 \\ \left\langle u_2 \middle| u_1\right\rangle &= 0 \\ \left\langle u_3 \middle| u_1\right\rangle &= 0 \end{aligned}$$

Therefore we find

$$A\left|u_1\right\rangle = \left|u_1\right\rangle$$

A similar procedure applied to the other states shows that

$$\begin{aligned} A\left|u_2\right\rangle &= -2i\left|u_1\right\rangle \\ A\left|u_3\right\rangle &= \left|u_3\right\rangle \end{aligned}$$

and so we have

$$\begin{aligned} A\left|\Psi\right\rangle &= A(2i\left|u_1\right\rangle - \left|u_2\right\rangle + 4i\left|u_3\right\rangle) \\ &= 2iA\left|u_1\right\rangle - A\left|u_2\right\rangle + 4iA\left|u_3\right\rangle \\ &= 2i\left|u_1\right\rangle + 2i\left|u_1\right\rangle + 4i\left|u_3\right\rangle \\ &= 4i(\left|u_1\right\rangle + \left|u_3\right\rangle) \\ \left\langle A\right\rangle &= \left\langle\Psi\right|A\left|\Psi\right\rangle \\ &= (-2i\left\langle u_1\right| - \left\langle u_2\right| - 4i\left\langle u_3\right|)4i(\left|u_1\right\rangle + \left|u_3\right\rangle) \\ &= (-2i)(4i)\left\langle u_1 \middle| u_1\right\rangle + (-4i)(4i)\left\langle u_3 \middle| u_3\right\rangle \\ &= 8 + 16 = 24 \end{aligned}$$

Eigenvalues and Eigenvectors

The eigenvalue problem from linear algebra plays a central role in quantum mechanics. There is a correspondence that works in the following way. To each physical observable, such as energy or momentum, there exists an operator, which can be represented by a matrix; the eigenvalues of the matrix are the possible results of measurement for that operator. As an example, we might know the Hamiltonian for a given physical system. We can form a matrix representing this Hamiltonian; the eigenvalues of that matrix will be the possible energies the system can have. This amazing mathematical system has passed every experimental test for more than 100 years.

Finding the eigenvectors is also important, for they give us a basis for the space and therefore give us a way to represent any state. In this section we briefly review the concepts and techniques from linear algebra that are necessary to calculate eigenvalues and eigenvectors.

DEFINITION: Eigenvalue and Eigenvector

Let T be a linear operator on a vector space V, and λ be a complex number. We say that λ is an eigenvalue of T if:

$$T\left|u\right\rangle = \lambda\left|u\right\rangle$$

for some vector $|u\rangle$ in V. The vector $|u\rangle$ is called an eigenvector of T.

The eigenvalues and eigenvectors for a given operator can be found by examining the operator in some matrix representation. To find the eigenvalues and eigenvectors of a matrix A, we find the *characteristic polynomial* and set it equal to zero. The polynomial is found by considering the determinant of the following quantity:

$$\det(A - \lambda I) = 0$$

where I is the identity matrix. The characteristic polynomial is found from $\det(A - \lambda I)$; solving the equation above gives us the eigenvalues λ. We can then use them to find the eigenvectors for the matrix. First let's quickly review how to compute the determinant of a 2×2 and 3×3 matrix.

DEFINITION: The Determinant of a Matrix

The determinant of a matrix is a number (which can be complex) associated with that matrix. It can be calculated for square matrices only, and was originally discovered in the study of systems of linear equations. The determinant of a matrix A can be represented by $|A|$. Sometimes we will write $\det(A)$. The determinant of a scalar is just the scalar itself. For the 2×2 case, one that shows up frequently in quantum mechanics, the determinant is given by:

$$A = \begin{pmatrix} a & b \\ c & d \end{pmatrix} \Rightarrow |A| = ad - bc$$

For a 3 × 3 matrix $A = \begin{pmatrix} a_{11} & a_{12} & a_{13} \\ a_{21} & a_{22} & a_{23} \\ a_{31} & a_{32} & a_{33} \end{pmatrix}$, the determinant is:

$$\det|A| = a_{11} \det\left|\begin{pmatrix} a_{22} & a_{23} \\ a_{32} & a_{33} \end{pmatrix}\right| - a_{12} \det\left|\begin{pmatrix} a_{21} & a_{23} \\ a_{31} & a_{33} \end{pmatrix}\right| + a_{13} \det\left|\begin{pmatrix} a_{21} & a_{22} \\ a_{31} & a_{32} \end{pmatrix}\right|$$
$$= a_{11}(a_{22}a_{33} - a_{23}a_{32}) - a_{12}(a_{21}a_{33} - a_{23}a_{31}) + a_{13}(a_{21}a_{32} - a_{22}a_{31})$$

EXAMPLE 6.10

e.g.

Find the determinants of:

$$A = \begin{pmatrix} 2 & 8 \\ 1 & 6 \end{pmatrix}, B = \begin{pmatrix} 1 & 0 \\ i & 2i \end{pmatrix}$$

SOLUTION

Applying the formula described above, we find:

$$\det|A| = \det\left|\begin{pmatrix} 2 & 8 \\ 1 & 6 \end{pmatrix}\right| = 2(6) - 1(8) = 12 - 8 = 4$$

$$\det|B| = \det\left|\begin{pmatrix} 1 & 0 \\ i & 2i \end{pmatrix}\right| = 1(2i) - i(0) = 2i$$

EXAMPLE 6.11

e.g.

Let

$$A = \begin{pmatrix} 3 & 1 & 1 \\ 0 & 4 & -1 \\ 2 & -5 & 0 \end{pmatrix} \text{ and } B = \begin{pmatrix} 1 & 0 & 4 \\ i & 7i & 0 \\ 2 & 8 & -1 \end{pmatrix}$$

Find $\det(A)$ and $\det(B)$.

SOLUTION

✔

$$\det(A) = \det\begin{pmatrix} 3 & 1 & 1 \\ 0 & 4 & -1 \\ 2 & -5 & 0 \end{pmatrix}$$
$$= 3\det\begin{pmatrix} 4 & -1 \\ -5 & 0 \end{pmatrix} - \det\begin{pmatrix} 0 & -1 \\ 2 & 0 \end{pmatrix} + \det\begin{pmatrix} 0 & 4 \\ 2 & -5 \end{pmatrix}$$
$$= 3[4(0) - (5)(-1)] - [0(0) - 2(-1)] + [0(-5) - 2(4)]$$
$$= 3(-5) - -2 - 8 = -15 - -10 = -25$$

Repeating the procedure for B:

$$\det(B) = \det\begin{pmatrix} 1 & 0 & 4 \\ i & 7i & 0 \\ 2 & 8 & -1 \end{pmatrix}$$

$$= \det\begin{pmatrix} 7i & 0 \\ 8 & -1 \end{pmatrix} + 4\det\begin{pmatrix} i & 7i \\ 2 & 8 \end{pmatrix}$$

$$= -7i + 4[8i - 14i]$$

$$= -31i$$

Now that we have reviewed how to calculate some basic determinants, we find the eigenvalues for some matrices.

---

e.g. **EXAMPLE 6.12**

Find the characteristic polynomial and eigenvalues for each of the following matrices:

$$A = \begin{pmatrix} 5 & 3 \\ 2 & 10 \end{pmatrix}, B = \begin{pmatrix} 7i & -1 \\ 2 & -6i \end{pmatrix}, C = \begin{pmatrix} 2 & 0 & -1 \\ 0 & 3 & 1 \\ 1 & 0 & 4 \end{pmatrix}$$

✔ SOLUTION

Starting with the matrix A, we have:

$$\det(A - \lambda 1) = \det\left|\begin{pmatrix} 5 & 3 \\ 2 & 10 \end{pmatrix} - \lambda\begin{pmatrix} 1 & 0 \\ 0 & 1 \end{pmatrix}\right| = \det\left|\begin{pmatrix} 5 & 3 \\ 2 & 10 \end{pmatrix} - \begin{pmatrix} \lambda & 0 \\ 0 & \lambda \end{pmatrix}\right|$$

$$= \det\left|\begin{pmatrix} 5-\lambda & 3 \\ 2 & 10-\lambda \end{pmatrix}\right| = (5-\lambda)(10-\lambda) - 6 = \lambda^2 - 15\lambda + 44$$

This is the characteristic polynomial. Setting it equal to zero and solving for λ, we find:

$$\lambda^2 - 15\lambda + 44 = 0 \Rightarrow (\lambda - 11)(\lambda - 4) = 0$$

and the eigenvalues are $\lambda_2 = 11, \lambda_2 = 4$. Following the same procedure for B, we find:

$$\det(B - \lambda 1) = \det\left|\begin{pmatrix} 7i & -1 \\ 2 & 6i \end{pmatrix} - \begin{pmatrix} \lambda & 0 \\ 0 & \lambda \end{pmatrix}\right| = \det\left|\begin{pmatrix} 7i-\lambda & -1 \\ 2 & 6i-\lambda \end{pmatrix}\right| =$$

$$= (7i - \lambda)(6i - \lambda) + 2$$

the characteristic polynomial is $\lambda^2 - 13i\lambda - 40$. Now we set this equal to zero to obtain the eigenvalues:

$$\lambda^2 - 13i\lambda - 40 = 0$$

We proceed to solve this equation using the quadratic formula:

$$\lambda_{1,2} = \frac{13i \pm \sqrt{(13i)^2 - 4(-40)}}{2} = \frac{13i \pm \sqrt{-169 + 160}}{2} = \frac{13i \pm \sqrt{-9}}{2} = \frac{13i \pm 3i}{2}$$

Therefore the eigenvalues of the matrix B are:

$$\lambda_1 = 8i, \lambda_2 = 5i$$

Now we obtain the characteristic polynomial and eigenvalues for C:

$$\det(C - \lambda I) = \det\left|\begin{pmatrix} 2 & 0 & -1 \\ 0 & 3 & 1 \\ 1 & 0 & 4 \end{pmatrix} - \lambda \begin{pmatrix} 1 & 0 & 0 \\ 0 & 1 & 0 \\ 0 & 0 & 1 \end{pmatrix}\right| = \det\begin{pmatrix} 2-\lambda & 0 & -1 \\ 0 & 3-\lambda & 1 \\ 1 & 0 & 4-\lambda \end{pmatrix}$$

$$\begin{aligned} &= (2-\lambda)\det\begin{pmatrix} 3-\lambda & 1 \\ 0 & 4-\lambda \end{pmatrix} - \det\begin{pmatrix} 0 & 3-\lambda \\ 1 & 0 \end{pmatrix} \\ &= (2-\lambda)[(3-\lambda)(4-\lambda)] + (3-\lambda) \\ &= (3-\lambda)[(2-\lambda)(4-\lambda) + 1] \\ &= (3-\lambda)[\lambda^2 - 6\lambda + 9] \end{aligned}$$

This is the characteristic polynomial for C. We do not carry through the multiplication of $(3 - \lambda)$ because the equation is in a form that will let us easily find the eigenvalues. Setting equal to zero,

$$(3-\lambda)[\lambda^2 - 6\lambda + 9] = 0$$

We see immediately that the first eigenvalue is $\lambda_1 = 3$. We find the other two eigenvalues by solving:

$$\lambda^2 - 6\lambda + 9 = 0$$

This factors immediately into:

$$(\lambda - 3)^2 = 0$$

Therefore we find that $\lambda_2 = \lambda_3 = 3$ also. When a matrix or operator has repealed eigenvalues as in this example, we say that eigenvalue is *degenerate*. An eigenvalue that repeats n times is said to be *n-fold degenerate*.

EXAMPLE 6.13

In some orthonormal basis an operator $T = |\Phi_1\rangle\langle\Phi_1| + 2|\Phi_1\rangle\langle\Phi_2| + |\Phi_2\rangle\langle\Phi_1|$. Find the matrix, representing T and find its (normalized) eigenvectors and eigenvalues. This vector space is two-dimensional.

SOLUTION

The matrix representing T can be found by calculating

$$T = \begin{pmatrix} \langle\Phi_1| T |\Phi_1\rangle & \langle\Phi_1| T |\Phi_2\rangle \\ \langle\Phi_2| T |\Phi_1\rangle & \langle\Phi_2| T |\Phi_2\rangle \end{pmatrix}$$

To perform the calculations, the fact that the basis is orthonormal tells us that

$$\langle\Phi_2 |\Phi_1\rangle = \langle\Phi_1 |\Phi_2\rangle = 0$$
$$\langle\Phi_1 |\Phi_1\rangle = \langle\Phi_2 |\Phi_2\rangle = 1$$

Starting with $\langle\Phi_1| T |\Phi_1\rangle$, we have

$$\begin{aligned}\langle\Phi_1| T |\Phi_1\rangle &= \langle\Phi_1| \left(|\Phi_1\rangle \langle\Phi_1| + 2 |\Phi_1\rangle \langle\Phi_2| + |\Phi_2\rangle \langle\Phi_1|\right) |\Phi_1\rangle \\ &= \langle\Phi_1 |\Phi_1\rangle \langle\Phi_1| \Phi_1\rangle + 2 \langle\Phi_1 |\Phi_1\rangle \langle\Phi_2| \Phi_1\rangle + \langle\Phi_1 |\Phi_2\rangle \langle\Phi_1| \Phi_1\rangle \\ &= 1\end{aligned}$$

$$\begin{aligned}\langle\Phi_1| T |\Phi_2\rangle &= \langle\Phi_1| \left(|\Phi_1\rangle \langle\Phi_1| + 2 |\Phi_1\rangle \langle\Phi_2| + |\Phi_2\rangle \langle\Phi_1|\right) |\Phi_2\rangle \\ &= \langle\Phi_1 |\Phi_1\rangle \langle\Phi_1| \Phi_2\rangle + 2 \langle\Phi_1 |\Phi_1\rangle \langle\Phi_2| \Phi_2\rangle + \langle\Phi_1 |\Phi_2\rangle \langle\Phi_1| \Phi_2\rangle \\ &= 2\end{aligned}$$

$$\begin{aligned}\langle\Phi_2| T |\Phi_1\rangle &= \langle\Phi_2| \left(|\Phi_1\rangle \langle\Phi_1| + 2 |\Phi_1\rangle \langle\Phi_2| + |\Phi_2\rangle \langle\Phi_1|\right) |\Phi_1\rangle \\ &= \langle\Phi_2 |\Phi_1\rangle \langle\Phi_1| \Phi_1\rangle + 2 \langle\Phi_2 |\Phi_1\rangle \langle\Phi_2| \Phi_1\rangle + \langle\Phi_2 |\Phi_2\rangle \langle\Phi_1| \Phi_1\rangle \\ &= 1\end{aligned}$$

$$\begin{aligned}\langle\Phi_2| T |\Phi_2\rangle &= \langle\Phi_2| \left(|\Phi_1\rangle \langle\Phi_1| + 2 |\Phi_1\rangle \langle\Phi_2| + |\Phi_2\rangle \langle\Phi_1|\right) |\Phi_2\rangle \\ &= \langle\Phi_2 |\Phi_1\rangle \langle\Phi_1| \Phi_2\rangle + 2 \langle\Phi_2 |\Phi_1\rangle \langle\Phi_2| \Phi_2\rangle + \langle\Phi_2 |\Phi_2\rangle \langle\Phi_1| \Phi_2\rangle \\ &= 0\end{aligned}$$

$\Rightarrow T = \begin{pmatrix} 1 & 2 \\ 1 & 0 \end{pmatrix}$ in the basis $\{|\Phi_1\rangle, |\Phi_2\rangle\}$.

We now solve $\det(T - \lambda I) = 0$ to find the eigenvalues of T:

$$0 = \det(T - \lambda I) = \det\left|\begin{pmatrix} 1 & 2 \\ 1 & 0 \end{pmatrix} - \lambda\begin{pmatrix} 1 & 0 \\ 0 & 1 \end{pmatrix}\right|$$

$$= \det\left|\begin{pmatrix} 1-\lambda & 2 \\ 1 & -\lambda \end{pmatrix}\right| = (1-\lambda)(-\lambda) - (2)$$

$$\Rightarrow \lambda^2 - \lambda - 2 = 0$$

This leads to

$$\lambda_1 = 2 \text{ and } \lambda_2 = -1.$$

Starting with λ_1, we find the eigenvectors, which we label $\{|a_1\rangle, |a_2\rangle\}$;
Let $|a_1\rangle = \begin{pmatrix} a \\ b \end{pmatrix}$ where a, b are undetermined constants that may be complex.

$$T|a_1\rangle = \lambda_1|a_1\rangle \Rightarrow$$

$$\begin{pmatrix} 1 & 2 \\ 1 & 0 \end{pmatrix}\begin{pmatrix} a \\ b \end{pmatrix} = 2\begin{pmatrix} a \\ b \end{pmatrix}$$

$\Rightarrow a + 2b = 2a$, or $a = 2b$. We can then write:

$$|a_1\rangle = \begin{pmatrix} 2b \\ b \end{pmatrix}$$

We now find b from normalization:

$$1 = \langle a_1|a_1\rangle = (\,2b^* \quad b^*\,)\begin{pmatrix} 2b \\ b \end{pmatrix} = 4|b|^2 + |b|^2 = 5|b|^2, \Rightarrow b = \frac{1}{\sqrt{5}}$$

Using $a = 2b$ gives us the normalized eigenvector of T:

$$|a_1\rangle = \frac{1}{\sqrt{5}}\begin{pmatrix} 2 \\ 1 \end{pmatrix}$$

We now find $|a_2\rangle$ using the eigenvalue $\lambda_2 = -1$:

$$T|a_2\rangle = \lambda_2|a_2\rangle \Rightarrow$$

$$\begin{pmatrix} 1 & 2 \\ 1 & 0 \end{pmatrix}\begin{pmatrix} a \\ b \end{pmatrix} = -\begin{pmatrix} a \\ b \end{pmatrix}$$

$$\Rightarrow a + 2b = -a, \text{ or } b = -a.$$

Writing

$$|a_2\rangle = \begin{pmatrix} a \\ -a \end{pmatrix}$$

and normalizing:

$$1 = \langle a_2|a_2\rangle = (a \; -a^*)\begin{pmatrix} a \\ -a \end{pmatrix} = 2a^2, \Rightarrow a = \frac{1}{\sqrt{2}}, b = -a = -\frac{1}{\sqrt{2}}$$

and so $|a_2\rangle = \frac{1}{\sqrt{2}}\begin{pmatrix} 1 \\ -1 \end{pmatrix}$

In summary, in the $\{|\Phi_1\rangle, |\Phi_2\rangle\}$ basis, we have:

$$T = \begin{pmatrix} 1 & 2 \\ 1 & 0 \end{pmatrix}$$

(notice T is not Hermitian; see below). The normalized eigenvectors of T are:

$$|a_1\rangle = \frac{1}{\sqrt{5}}\begin{pmatrix} 2 \\ 1 \end{pmatrix} \text{ with eigenvalue } \lambda_1 = 2$$

$$|a_2\rangle = \frac{1}{\sqrt{2}}\begin{pmatrix} 1 \\ -1 \end{pmatrix} \text{ with eigenvalue } \lambda_1 = -1$$

---

e.g. **EXAMPLE 6.14**

The Hamiltonian for a two-state system is given by:

$$H = \begin{pmatrix} \omega_1 & \omega_2 \\ \omega_2 & \omega_1 \end{pmatrix}$$

A basis for this system is:

$$|0\rangle = \begin{pmatrix} 1 \\ 0 \end{pmatrix}, |1\rangle = \begin{pmatrix} 0 \\ 1 \end{pmatrix}$$

(a) Find the eigenvalues and eigenvectors of H, and express the eigenvectors in terms of $\{|0\rangle, |1\rangle\}$

(b) Find the time evolution of the system described by the Schrödinger equation:

$$H|\Psi\rangle = i\hbar \frac{\partial |\Psi\rangle}{\partial t}, |\Psi(0)\rangle = |0\rangle$$

✔ **SOLUTION**

(a)

$$\det(H - \lambda 1) = \det\begin{pmatrix} \omega_1 - \lambda & \omega_2 \\ \omega_2 & \omega_1 - \lambda \end{pmatrix} = 0$$

$$\Rightarrow \lambda^2 - 2\omega_1\lambda + (\omega_1^2 - \omega_2^2) = 0$$

This leads to the eigenvalues:

$$\lambda_1 = \omega_1 + \omega_2, \lambda_2 = \omega_1 - \omega_2$$

Let $|A\rangle = \begin{pmatrix} a \\ b \end{pmatrix}$ be the eigenvector for $\lambda_1 = \omega_1 + \omega_2$. Then we have:

$$\begin{pmatrix} \omega_1 & \omega_2 \\ \omega_2 & \omega_1 \end{pmatrix}\begin{pmatrix} a \\ b \end{pmatrix} = (\omega_1 + \omega_2)\begin{pmatrix} a \\ b \end{pmatrix}$$

$$\omega_1 a + \omega_2 b = (\omega_1 + \omega_2)a \Rightarrow b = a$$

Normalizing we obtain:

$$1 = \langle A | A \rangle = \begin{pmatrix} a^* & a^* \end{pmatrix} \begin{pmatrix} a \\ a \end{pmatrix} = 2 |a|^2, \Rightarrow a = \frac{1}{\sqrt{2}}$$

Therefore

$$|A\rangle = \frac{1}{\sqrt{2}} \begin{pmatrix} 1 \\ 1 \end{pmatrix}$$

Recall that the basis vectors are

$$|0\rangle = \begin{pmatrix} 1 \\ 0 \end{pmatrix}, |1\rangle = \begin{pmatrix} 0 \\ 1 \end{pmatrix}$$

With a little algebra we can rewrite $|A\rangle$ in terms of $\{|0\rangle, |1\rangle\}$

$$|A\rangle = \frac{1}{\sqrt{2}} \begin{pmatrix} 1 \\ 1 \end{pmatrix} = \frac{1}{\sqrt{2}} \left[\begin{pmatrix} 1 \\ 0 \end{pmatrix} + \begin{pmatrix} 0 \\ 1 \end{pmatrix} \right] = \frac{|0\rangle + |1\rangle}{\sqrt{2}}$$

Now let $|B\rangle = \frac{1}{\sqrt{2}}$ be the eigenvector for $\lambda_2 = \omega_1 - \omega_2$. Then we have

$$\begin{pmatrix} \omega_1 & \omega_2 \\ \omega_2 & \omega_1 \end{pmatrix} \begin{pmatrix} a \\ b \end{pmatrix} = (\omega_1 - \omega_2) \begin{pmatrix} a \\ b \end{pmatrix}$$

This gives us

$$\omega_1 a + \omega_2 b = (\omega_1 - \omega_2) a, \Rightarrow b = -a$$

Again normalizing, we find:

$$1 = \langle B | B \rangle = (a^* - a^*) \begin{pmatrix} a \\ -a \end{pmatrix} = 2 |a|^2, \Rightarrow a = \frac{1}{\sqrt{2}}$$

And so the second eigenvector, again written in terms of the basis states, is

$$|B\rangle = \frac{1}{\sqrt{2}} \begin{pmatrix} 1 \\ -1 \end{pmatrix} = \frac{1}{\sqrt{2}} \left[\begin{pmatrix} 1 \\ 0 \end{pmatrix} - \begin{pmatrix} 0 \\ 1 \end{pmatrix} \right] = \frac{|0\rangle - |1\rangle}{\sqrt{2}}$$

(b) The time evolution of the system is given by

$$H |\Psi\rangle = i\hbar \frac{\partial |\Psi\rangle}{\partial t}$$

The initial condition is given by

$$|\Psi(0)\rangle = |0\rangle$$

At an arbitrary time t, we can write the state as

$$|\Psi(t)\rangle = \begin{pmatrix} \alpha(t) \\ \beta(t) \end{pmatrix}$$

Using the given matrix representation of the Hamiltonian, we have:

$$H|\Psi\rangle = \begin{pmatrix} \omega_1 & \omega_2 \\ \omega_2 & \omega_1 \end{pmatrix} \begin{pmatrix} \alpha(t) \\ \beta(t) \end{pmatrix} = \begin{pmatrix} \omega_1\alpha(t) + \omega_2\beta(t) \\ \omega_2\alpha(t) + \omega_1\beta(t) \end{pmatrix}$$

The other side of the equation is

$$i\hbar \frac{\partial |\Psi\rangle}{\partial t} = i\hbar \begin{pmatrix} \dot{\alpha} \\ \dot{\beta} \end{pmatrix}$$

where the dot indicates a time derivative. Setting both sides equal to each other leads to the following system

$$\begin{pmatrix} \omega_1\alpha(t) + \omega_2\beta(t) \\ \omega_2\alpha(t) + \omega_1\beta(t) \end{pmatrix} = i\hbar \begin{pmatrix} \dot{\alpha} \\ \dot{\beta} \end{pmatrix}$$

which gives these two equations:

$$\omega_1\alpha + \omega_2\beta = i\hbar\dot{\alpha}$$
$$\omega_2\alpha + \omega_1\beta = i\hbar\dot{\beta}$$

Adding these equations, we obtain

$$(\omega_1 + \omega_2)(\alpha + \beta) = i\hbar(\dot{\alpha} + \dot{\beta})$$

It would seem we have a complicated mess. But we can simplify things quite a bit by defining a new function that we call $\gamma = \alpha + \beta$. Then this is a simple differential equation:

$$(\omega_1 + \omega_2)\gamma = i\hbar \frac{d\gamma}{dt}$$

with solution

$$\gamma = C \exp\left(\frac{\omega_1 + \omega_2}{i\hbar} t \right)$$

We now repeat the procedure. This time we subtract, giving

$$(\omega_1 - \omega_2)(\alpha - \beta) = i\hbar(\dot{\alpha} - \dot{\beta})$$

Now let $\delta = \alpha - \beta$. This gives

$$(\omega_1 - \omega_2)\delta = i\hbar \frac{d\delta}{dt}, \Rightarrow$$

$$\delta = D \exp\left(\frac{\omega_1 - \omega_2}{i\hbar} t\right)$$

Now, $\alpha = \frac{\gamma + \delta}{2} \quad \Rightarrow \alpha = \frac{1}{2}[C \exp\,(\frac{\omega_1 + \omega_2}{i\hbar} t) + D \exp\,(\frac{\omega_1 - \omega_2}{i\hbar} t)]$
Recalling that the initial condition is

$$|\Psi(0)\rangle = |0\rangle = \begin{pmatrix} 1 \\ 0 \end{pmatrix}$$

with $|\Psi(t)\rangle = \begin{pmatrix} \alpha(t) \\ \beta(t) \end{pmatrix}$, this implies that $\alpha(0) = 1$. Therefore:

$$\alpha(0) = 1 = \frac{C + D}{2}$$

The initial condition also tells us that $\beta(0) = 0$. Using $\beta = \frac{\gamma - \delta}{2}$, this leads to the equation

$$0 = \frac{C - D}{2}, \quad \Rightarrow C = D$$

Putting this together with the condition $1 = \frac{C + D}{2}$, we obtain $C = D = 1$. Substitution of C, D into the equation for α gives

$$\begin{aligned}
\alpha(t) &= \frac{1}{2}\left[C \exp\left(\frac{\omega_1 + \omega_2}{i\hbar} t\right) + D \exp\left(\frac{\omega_1 - \omega_2}{i\hbar} t\right)\right] \\
&= \frac{1}{2}\left[\exp\left(\frac{\omega_1 + \omega_2}{i\hbar} t\right) + \exp\left(\frac{\omega_1 - \omega_2}{i\hbar} t\right)\right]
\end{aligned}$$

Pulling out the common multiplier $\exp\left(\frac{\omega_1}{i\hbar} t\right)$, we write this as:

$$\begin{aligned}
\alpha &= \frac{1}{2}\left[\exp\left(\frac{\omega_1 + \omega_2}{i\hbar} t\right) + \exp\left(\frac{\omega_1 - \omega_2}{i\hbar} t\right)\right] \\
&= \exp\left(\frac{\omega_1 t}{i\hbar}\right) \frac{1}{2}\left[\exp\left(\frac{\omega_2}{i\hbar} t\right) + \exp\left(-\frac{\omega_2}{i\hbar} t\right)\right] \\
&= \exp\left(-i\frac{\omega_1 t}{\hbar}\right) \frac{1}{2}\left[\exp\left(i\frac{\omega_2}{\hbar} t\right) + \exp\left(-i\frac{\omega_2}{\hbar} t\right)\right] \\
&= \exp\left(-i\frac{\omega_1 t}{\hbar}\right) \cos\left(\frac{\omega_2 t}{\hbar}\right)
\end{aligned}$$

A similar procedure applied to β (work it out for yourself) leads to

$$\beta = ie^{-i\frac{\omega_1 t}{\hbar}} \sin\left(\frac{\omega_2 t}{\hbar}\right),$$

and therefore the state at time t is:

$$|\Psi(t)\rangle = \begin{pmatrix} \alpha \\ \beta \end{pmatrix} = e^{-i\frac{\omega_1}{\hbar}t} \begin{pmatrix} \cos\left(\frac{\omega_2}{\hbar}t\right) \\ -i\sin\left(\frac{\omega_2}{\hbar}t\right) \end{pmatrix}$$

The Hermitian Conjugate of an Operator

We have seen that an operator acts on a ket to produce a new ket, i.e. $T|u\rangle = |v\rangle$. Now let's consider the action of an operator inside an inner product, that is, $\langle w|T|u\rangle$. Using $T|u\rangle = |v\rangle$, we see that this is just a complex number, the inner product of which is:

$$\langle w|T|u\rangle = \langle w|v\rangle$$

Since the inner product is just a complex number, we can form the complex conjugate which is given by relation $\langle w|v\rangle = \langle v|w\rangle^*$. When an operator is present inside the inner product this becomes

$$\langle w|T^\dagger|v\rangle = \langle v|T|w\rangle^*$$

$\hat{T}^\dagger$ (pronounced "T dagger") is called the *Hermitian conjugate* or *adjoint* of the operator $\hat{T}$. The Hermitian adjoint of a product of operators A, B, C is given by reversing their order, and then forming the adjoint of each operator:

$$(AB)^\dagger = B^\dagger A^\dagger, (ABC)^\dagger = C^\dagger B^\dagger A^\dagger \text{etc.}$$

We can form the adjoint of any expression by applying the following rules.

DEFINITION: Forming the Adjoint of a General Expression

1. Replace any constants by their complex conjugates
2. Replace kets by their associated bras, and bras by their associated kets
3. Replace each operator by its adjoint
4. Reverse the order of all factors in the expression

This example summarizes these steps:

$$\lambda \langle u \,|\hat{A}\hat{B}|\, v\rangle \to \lambda^* \langle v \,|\hat{B}^\dagger \hat{A}^\dagger|\, u\rangle$$

The adjoint operation also has the following properties:

$$(\hat{A} + \hat{B}) = \hat{A}^\dagger + \hat{B}^\dagger$$

$$(a\hat{A})^\dagger = a^* \hat{A}^\dagger$$

$$(\hat{A}\hat{B}|\, u\rangle)^\dagger = \langle u \,|\hat{B}^\dagger \hat{A}^\dagger$$

If an operator is presented inside a bra expression, when pulling it out take the Hermitian conjugate of the operator. For scalars, take the complex conjugate:

$$\langle a\hat{A}u| = a^* \langle u \,|\hat{A}^\dagger$$

In quantum mechanics it is often necessary or desirable to work with the matrix representation of an operator. Therefore we need to know how to form the Hermitian conjugate of a matrix as well.

FINDING THE HERMITIAN CONJUGATE OF A MATRIX

We form the Hermitian conjugate of a matrix H by applying these two steps:

1. Form the transpose of the matrix, H^T, by exchanging rows and columns
2. Take the complex conjugate of each element of H^T.

We summarize this by writing $H^\dagger = \left(H^T\right)^*$.
The transpose of a matrix is formed by swapping the rows and columns of the matrix. Consider the following example. Let the matrix A given by:

$$A = \begin{pmatrix} 1 & 6 & 3 \\ 2 & 8 & 7 \\ -1 & 9 & 2i \end{pmatrix}$$

We begin with the top row, taking its components and using them as the first column of a new matrix:

$$\begin{pmatrix} 1 & '' & ' \\ 6 & '' & ' \\ 3 & '' & ' \end{pmatrix}$$

We repeat the procedure for the next row, using the components of the second row to construct the second column:

$$\begin{pmatrix} 1 & 2 & \\ 6 & 8 & \\ 3 & 7 & \end{pmatrix}$$

Now we do the same for the last row, using its elements to form the final column of the new matrix, which we denote A^T.

$$A^T = \begin{pmatrix} 1 & 2 & -1 \\ 6 & 8 & 9 \\ 3 & 7 & 2i \end{pmatrix}$$

The transpose operation shares many properties with the Hermitian conjugate.

PROPERTIES OF THE TRANSPOSE OPERATION

1. $(A + B)^T = A^T + B^T$
2. $\left(A^T\right)^T = A.$
3. $(aA)^T = aA^T$
4. $(AB)^T = B^T A^T$

where A, B are matrices and a is a scalar.

e.g. **EXAMPLE 6.15**

For the given matrices A and B, find $(A + B)^T$ and verify that it is equal to $A^T + B^T$

$$A = \begin{pmatrix} 6 & -1 \\ 3 & 4i \\ 5 & -2 \end{pmatrix} B = \begin{pmatrix} 7 & 2 \\ 8 & 1 \\ 0 & 3 \end{pmatrix}$$

✔ **SOLUTION**

First we form $A + B$, which we do by adding the individual elements of both matrices:

$$A + B = \begin{pmatrix} 6+7 & -1+2 \\ 3+8 & 4i+1 \\ 5+0 & -2+3 \end{pmatrix} \begin{pmatrix} 13 & 1 \\ 11 & 1+4i \\ 5 & 1 \end{pmatrix}$$

Now we compute the transpose:

$$(A+B)^T = \begin{pmatrix} 13 & 11 & 5 \\ 1 & 1+4i & 1 \end{pmatrix}$$

Next, let's write the transpose of each individual matrix:

$$A^T = \begin{pmatrix} 6 & -1 \\ 3 & 4i \\ 5 & -2 \end{pmatrix} = \begin{pmatrix} 6 & 3 & 5 \\ -1 & 4i & -2 \end{pmatrix}$$

$$B^T = \begin{pmatrix} 7 & 2 \\ 8 & 1 \\ 0 & 3 \end{pmatrix}^T = \begin{pmatrix} 7 & 8 & 0 \\ 2 & 1 & 3 \end{pmatrix}$$

Finally, we form their sum:

$$A^T + B^T = \begin{pmatrix} 6+7 & 3+8 & 5+0 \\ -1+2 & 4i+1 & -2+3 \end{pmatrix} = \begin{pmatrix} 13 & 11 & 5 \\ 1 & 1+4i & 1 \end{pmatrix} = (A+B)^T$$

DEFINITION: Complex Conjugate of a Matrix

The complex conjugate of a matrix 4 is computed by taking the complex conjugate of each element. For example, for a 3×3 matrix we have:

$$\begin{pmatrix} a_{11} & a_{12} & a_{13} \\ a_{21} & a_{22} & a_{23} \\ a_{31} & a_{32} & a_{33} \end{pmatrix}^* = \begin{pmatrix} a_{11}^* & a_{12}^* & a_{13}^* \\ a_{21}^* & a_{22}^* & a_{23}^* \\ a_{31}^* & a_{32}^* & a_{33}^* \end{pmatrix}$$

Now we can apply these two operations to form the Hermitian conjugate of a matrix.

EXAMPLE 6.16

e.g.

Find the Hermitian conjugate of the matrix:

$$A = \begin{pmatrix} 0 & i & 2i \\ -i & 0 & 2i \\ 2i & 7i & 0 \end{pmatrix}$$

SOLUTION

First we apply Step 1, and write down the transpose by exchanging rows and columns:

$$A^T = \begin{pmatrix} 0 & i & 2i \\ -i & 0 & 2i \\ 2i & 7i & 0 \end{pmatrix}^T$$

Now apply Step 2, forming the complex conjugate of each element by letting $i \to -i$:

$$A^\dagger = \begin{pmatrix} 0 & -i & 2i \\ i & 0 & 7i \\ 2i & 2i & 0 \end{pmatrix}^* = \begin{pmatrix} 0 & i & -2i \\ -i & 0 & -7i \\ -2i & -2i & 0 \end{pmatrix}$$

The Hermitian conjugate of a column vector is a row of vectors with each component replaced by its complex conjugate, as this next example shows.

EXAMPLE 6.17

Find the Hermitian conjugate of:

$$\Phi = \begin{pmatrix} 2 \\ -3i \\ 6 \end{pmatrix}$$

SOLUTION

Taking the transpose gives a row vector:

$$\Phi^T = \begin{pmatrix} 2 \\ -3i \\ 6 \end{pmatrix}^T = (2 \quad -3i \quad 6)$$

To find the Hermitian conjugate, take the complex conjugate of each element:

$$\Phi^\dagger = (2 \quad -3i \quad 6)^* = (2 \quad 3i \quad 6)$$

EXAMPLE 6.18

Let

$$A = \begin{pmatrix} 1 & 0 & -3i \\ 3 & 5 & 0 \\ 3i & 0 & -2 \end{pmatrix}, |\Psi\rangle = \begin{pmatrix} 2 \\ 3i \\ -1 \end{pmatrix}, |\Phi\rangle = \begin{pmatrix} 0 \\ -1 \\ 1 \end{pmatrix}$$

(a) Find $A|\Psi\rangle$, $A|\Phi\rangle$

(b) Find $|\Psi\rangle^\dagger$, $|\Phi\rangle^\dagger$ and use these to compute $\langle\Psi|\Phi\rangle$ and $\langle\Phi|\Psi\rangle$

SOLUTION

(a)

$$A|\Psi\rangle = \begin{pmatrix} 1 & 0 & -3i \\ 3 & 5 & 0 \\ 3i & 0 & -2 \end{pmatrix}\begin{pmatrix} 2 \\ 3i \\ -1 \end{pmatrix} = \begin{pmatrix} 1(2)+0(3i)+(-3i)(-1) \\ 3(2)+5(3i)+0(-1) \\ 3i(2)+0(3i)+(-2)(-1) \end{pmatrix} = \begin{pmatrix} 2+3i \\ 6+15i \\ 2+6i \end{pmatrix}$$

$$A|\Phi\rangle = \begin{pmatrix} 1 & 0 & -3i \\ 3 & 5 & 0 \\ 3i & 0 & -2 \end{pmatrix}\begin{pmatrix} 0 \\ -1 \\ 1 \end{pmatrix} = \begin{pmatrix} 1(0)+0(-1)+(-3i)(1) \\ 3(0)+5(-1)+0(1) \\ 3i(0)+0(-1)+(-2)(1) \end{pmatrix} = \begin{pmatrix} -3i \\ -5 \\ -2 \end{pmatrix}$$

(b) First we form the Hermitian conjugate of $|\Psi\rangle$. Taking the transpose we convert $|\Psi\rangle$ into a row vector:

$$|\Psi\rangle^T = (2 \quad 3i \quad -1)$$

We then obtain $|\Psi\rangle^\dagger$ by taking the complex conjugate of each element:

$$|\Psi\rangle^\dagger = (2 \quad 3i \quad -1)^* = (2 \quad -3i \quad -1) = \langle\Psi|$$

Now we compute the dot product:

$$\langle\Psi \mid \Phi\rangle = (2 \quad -3i \quad -1)\begin{pmatrix} 0 \\ -1 \\ 1 \end{pmatrix} = 2(0)+(-3i)(-1)+(-1)(1) = -1+3i$$

We now proceed to calculate $\langle\Phi|\Psi\rangle$. Since the components of Φ are real numbers, $\langle\Phi| = |\Phi\rangle^T = (0 \quad -1 \quad 1)$. So we obtain

$$\langle\Phi \mid \Psi\rangle = (0 \quad -1 \quad 1)\begin{pmatrix} 2 \\ 3i \\ -1 \end{pmatrix} = (0)(2)+(-1)(3i)+(1)(-1) = -1-3i$$

Notice that $\langle\Phi|\Psi\rangle = \langle\Psi|\Phi\rangle^*$, as it should.

In the next section we now turn our attention to *Hermitian* operators, which play a central role in quantum mechanics. Let's summarize the properties of the Hermitian conjugate operation.

Summary: Properties of the Hermitian Conjugate Operation

1. $(A + B)^\dagger = A^\dagger + B^\dagger$
2. $(AB)^\dagger = B^\dagger A^\dagger$
3. $(aA)^\dagger = a^* A^\dagger$

e.g. **EXAMPLE 6.19**

Prove that $(AB)^\dagger = B^\dagger A^\dagger$

✔ **SOLUTION**
Consider

$$\langle \Phi \, |\hat{A}\hat{B}| \, \Psi \rangle$$

We know that $\langle \Phi|\hat{A}\hat{B}|\Psi\rangle^* = \langle \psi|(\hat{A}\hat{B})^\dagger|\Phi\rangle$. Now define a new ket:

$$|\chi\rangle = |\hat{B}\Psi\rangle,$$
$$\Rightarrow \langle \chi| = \langle \Psi \hat{B}^\dagger|$$

This allows us to write

$$\langle \Phi \, |\hat{A}\hat{B}| \, \Psi \rangle = \langle \Phi \, |\hat{A}| \, \chi \rangle$$

So we have

$$\langle \Phi \, |\hat{A}\hat{B}| \, \Psi \rangle^* = \langle \Phi \, |\hat{A}| \, \chi \rangle^* = \langle \chi \, |\hat{A}^\dagger| \, \Phi \rangle = \langle \Psi \, |\hat{B}^\dagger \hat{A}^\dagger| \, \Phi \rangle$$

Since $\langle \Phi \, |\hat{A}\hat{B}| \, \Psi \rangle^* = \langle \Psi \, |\hat{B}^\dagger \hat{A}^\dagger| \, \Phi \rangle$, it follows that $(AB)^\dagger = B^\dagger A^\dagger$.

HERMITIAN AND UNITARY OPERATORS

A *Hermitian* operator is one for which $\hat{T}^\dagger = \hat{T}$. For a Hermitian operator,

$$\langle w \, |\hat{T}| \, u \rangle = \langle u \, |\hat{T}| \, w \rangle^*$$

It is often necessary to work with the matrix representation of an operator. We will need to know how to determine if a matrix is Hermitian.

DEFINITION: Hermitian Matrix

A Hermitian matrix A has the following property:

$$A = A^\dagger$$

EXAMPLE 6.20

Show that the following matrix is Hermitian and find $Tr(A)$.

$$A = \begin{pmatrix} 2 & 0 & 0 \\ 0 & -3 & i \\ 0 & -i & 1 \end{pmatrix}$$

SOLUTION

First we compute the transpose of the matrix:

$$A^T = \begin{pmatrix} 2 & 0 & 0 \\ 0 & -3 & i \\ 0 & -i & 1 \end{pmatrix}^T = \begin{pmatrix} 2 & 0 & 0 \\ 0 & -3 & -i \\ 0 & i & 1 \end{pmatrix}$$

Second, we form the complex conjugate:

$$A^\dagger = \left(A^T\right)^* = \begin{pmatrix} 2 & 0 & 0 \\ 0 & -3 & -i \\ 0 & i & 1 \end{pmatrix}^* = \begin{pmatrix} 2 & 0 & 0 \\ 0 & -3 & i \\ 0 & -i & 1 \end{pmatrix}$$

Examining the original matrix A, we see that $A = A^\dagger$. Therefore the matrix is Hermitian. The trace is just the sum of the diagonal elements:

$$Tr\,(A) = 2 - 3 + 1 = 0$$

Note: The diagonal elements and eigenvalues of Hermitian matrices are real numbers.

EXAMPLE 6.21

Show that Hermitian operators have real eigenvalues.

SOLUTION

Let A be a Hermitian operator, and suppose that $|a\rangle$ is an eigenvector of A with eigenvalue λ. Then

$$\langle a|\, A\, |a\rangle = \langle a|\, (A\, |a\rangle) = \langle a|\, \lambda\, |a\rangle = \lambda\, \langle a\, |a\,\rangle$$

Now let the operator A act to the left

$$\langle a|\, A\, |a\rangle = (\langle a|\, A)\, |a\rangle = \left(\langle a|\, \lambda^*\right) |a\rangle = \lambda^*\, \langle a\, |a\,\rangle$$

Now we subtract this equation from the first one. The left side is just zero:

$$\langle a|\, A\, |a\rangle - \langle a|\, A\, |a\rangle = 0$$

On the right side we have

$$\lambda \langle a|\, a\rangle - \lambda^* \langle a|\, a\rangle = \left(\lambda - \lambda^*\right) \langle a|\, a\rangle$$

Since $\langle a|\, a\rangle$ is not zero, we must have:

$$\lambda - \lambda^* = 0, \quad \Rightarrow \lambda = \lambda^*$$

Therefore, the eigenvalues of a Hermitian operator are real.

DEFINITION: Skew-Hermitian Operator

An operator A is skew-Hermitian or *anti-Hermitian* if:

$$A^\dagger = -A$$

Note: The diagonal elements of the matrix representation of an anti-Hermitian operator are pure imaginary.

EXAMPLE 6.22

Show that $B = \begin{pmatrix} i & 0 & -2 \\ 0 & 3i & 8 \\ 2 & -8 & -7i \end{pmatrix}$ is skew-Hermitian

SOLUTION

First let's write out $-B$:

$$-B = \begin{pmatrix} -i & 0 & 2 \\ 0 & -3i & -8 \\ -2 & 8 & 7i \end{pmatrix}$$

Now let's find B† beginning in the usual way:

$$B^T = \begin{pmatrix} i & 0 & 2 \\ 0 & 3i & -8 \\ -2 & 8 & -7i \end{pmatrix}$$

Now we complex conjugate each element:

$$B^\dagger = \begin{pmatrix} -i & 0 & 2 \\ 0 & -3i & -8 \\ -2 & 8 & 7i \end{pmatrix}$$

So we see that $B^\dagger = -B$, and B is skew-Hermitian. We now consider unitary operators/matrices.

DEFINITION: A Matrix or Operator U is Unitary if:

$$UU^{\dagger} = U^{\dagger}U = 1$$

where 1 is the identity matrix. We can also say that $U^{\dagger} = U^{-1}$.
Another important characteristic of unitary matrices is that the rows or columns of the matrix form an orthonormal set. We demonstrate this for the rows of a matrix using an example.

EXAMPLE 6.23

e.g.

Show that $U = \begin{pmatrix} \frac{1}{3} - \frac{2}{3}i & \frac{2}{3}i \\ -\frac{2}{3}i & -\frac{1}{3} - \frac{2}{3}i \end{pmatrix}$ is a unitary matrix and verify that the rows of U form an orthonormal set.

SOLUTION

First we compute the Hermitian conjugate of U:

$$U^{\dagger} = \left(\begin{pmatrix} \frac{1}{3} - \frac{2}{3}i & \frac{2}{3}i \\ -\frac{2}{3}i & -\frac{1}{3} - \frac{2}{3}i \end{pmatrix}^T \right)^* = \begin{pmatrix} \frac{1}{3} - \frac{2}{3}i & -\frac{2}{3}i \\ \frac{2}{3}i & -\frac{1}{3} - \frac{2}{3}i \end{pmatrix}^* = \begin{pmatrix} \frac{1}{3} + \frac{2}{3}i & +\frac{2}{3}i \\ -\frac{2}{3}i & -\frac{1}{3} + \frac{2}{3}i \end{pmatrix}$$

Now we compute $UU^{\dagger}$:

$$\begin{aligned} UU^{\dagger} &= \begin{pmatrix} \frac{1}{3} - \frac{2}{3}i & \frac{2}{3}i \\ -\frac{2}{3}i & -\frac{1}{3} - \frac{2}{3}i \end{pmatrix} \begin{pmatrix} \frac{1}{3} + \frac{2}{3}i & +\frac{2}{3}i \\ -\frac{2}{3}i & -\frac{1}{3} + \frac{2}{3}i \end{pmatrix} \\ &= \begin{pmatrix} \left[\left(\frac{1}{3} - \frac{2}{3}i\right)\left(\frac{1}{3} + \frac{2}{3}i\right) + \left(\frac{2}{3}i\right)\left(-\frac{2}{3}i\right)\right] & \left[\left(\frac{1}{3} - \frac{2}{3}i\right)\left(\frac{2}{3}i\right) + \left(\frac{2}{3}i\right)\left(-\frac{1}{3} + \frac{2}{3}i\right)\right] \\ \left[\left(-\frac{2}{3}i\right)\left(\frac{1}{3} + \frac{2}{3}i\right) + \left(-\frac{1}{3} - \frac{2}{3}i\right)\left(-\frac{2}{3}i\right)\right] & \left[\left(-\frac{2}{3}i\right)\left(\frac{2}{3}i\right) + \left(-\frac{1}{3} - \frac{2}{3}i\right)\left(-\frac{1}{3} + \frac{2}{3}i\right)\right] \end{pmatrix} \\ &= \begin{pmatrix} \left[\frac{5}{9} + \frac{4}{9}\right] & \left[\frac{2}{9}i + \frac{4}{9} - \frac{2}{9}i - \frac{4}{9}\right] \\ \left[-\frac{2}{9}i + \frac{4}{9} + \frac{2}{9}i - \frac{4}{9}\right] & \left[\frac{4}{9} + \frac{5}{9}\right] \end{pmatrix} \\ &= \begin{pmatrix} 1 & 0 \\ 0 & 1 \end{pmatrix} = 1 \Rightarrow U \text{ is unitary} \end{aligned}$$

Now let's verify that the rows of U form an orthonormal set. First we compute the dot product of the first row with itself:

$$\left(\frac{1}{3} - \frac{2}{3}i, \frac{2}{3}i\right) \cdot \left(\frac{1}{3} - \frac{2}{3}i, \frac{2}{3}i\right) = \left(\frac{1}{9} + \frac{4}{9}\right) + \frac{4}{9} = 1$$

Now the first row with the second row:

$$\left(\frac{1}{3} - \frac{2}{3}i, \frac{2}{3}i\right) \cdot \left(-\frac{2}{3}i, -\frac{1}{3} - \frac{2}{3}i\right) = \left(\frac{2}{9}i + \frac{4}{9}\right) + \left(-\frac{2}{9}i - \frac{4}{9}\right) = 0$$

Finally, the second row with itself:

$$\left(-\frac{2}{3}i, -\frac{1}{3} - \frac{2}{3}i\right) \cdot \left(-\frac{2}{3}-, -\frac{1}{3} - \frac{2}{3}i\right) = \frac{4}{9} + \left(\frac{1}{9} + \frac{4}{9}\right) = 1$$

Note: on the Eigenvalues of a Unitary Matrix:

The eigenvalues of a unitary matrix have *unit* magnitude, that is, $|a_n|^2 = 1$, where a_n is an eigenvalue of a unitary matrix.

---

e.g. **EXAMPLE 6.24**

Show that

$$U = \begin{pmatrix} \frac{1}{\sqrt{2}} & \frac{1}{\sqrt{2}}i & 0 \\ -\frac{1}{\sqrt{2}}i & \frac{1}{\sqrt{2}}i & 0 \\ 0 & 0 & i \end{pmatrix}$$

is unitary and that its eigenvalues have unit magnitude.

✔ **SOLUTION**

We compute $U^\dagger$:

$$U^T = \begin{pmatrix} \frac{1}{\sqrt{2}} & \frac{1}{\sqrt{2}} & 0 \\ -\frac{1}{\sqrt{2}}i & \frac{1}{\sqrt{2}}i & 0 \\ 0 & 0 & i \end{pmatrix}^T = \begin{pmatrix} \frac{1}{\sqrt{2}} & -\frac{1}{\sqrt{2}}i & 0 \\ \frac{1}{\sqrt{2}} & \frac{1}{\sqrt{2}}i & 0 \\ 0 & 0 & i \end{pmatrix}$$

Taking the complex conjugate of each element we find:

$$U^\dagger = \begin{pmatrix} \frac{1}{\sqrt{2}} & \frac{1}{\sqrt{2}} & 0 \\ \frac{1}{\sqrt{2}} & \frac{-1}{\sqrt{2}} & 0 \\ 0 & 0 & -i \end{pmatrix}$$

Therefore:

$$UU^\dagger = \begin{pmatrix} \frac{1}{\sqrt{2}} & \frac{1}{\sqrt{2}} & 0 \\ -\frac{1}{\sqrt{2}}i & \frac{1}{\sqrt{2}}i & 0 \\ 0 & 0 & i \end{pmatrix} \begin{pmatrix} \frac{1}{\sqrt{2}} & \frac{i}{\sqrt{2}} & 0 \\ \frac{1}{\sqrt{2}} & \frac{-i}{\sqrt{2}} & 0 \\ 0 & 0 & -i \end{pmatrix}$$

$$= \begin{pmatrix} \left(\frac{1}{\sqrt{2}}\right)\left(\frac{1}{\sqrt{2}}\right) + \left(\frac{1}{\sqrt{2}}\right)\left(\frac{1}{\sqrt{2}}\right) & \left(\frac{1}{\sqrt{2}}\right)\left(\frac{i}{\sqrt{2}}\right) + \left(\frac{-i}{\sqrt{2}}\right)\left(\frac{1}{\sqrt{2}}\right) & 0 \\ \left(\frac{-1}{\sqrt{2}}\right)\left(\frac{1}{\sqrt{2}}\right) + \left(\frac{1}{\sqrt{2}}\right)\left(\frac{i}{\sqrt{2}}\right) & \left(\frac{-1}{\sqrt{2}}\right)\left(\frac{1}{\sqrt{2}}\right) + \left(\frac{1}{\sqrt{2}}\right)\left(\frac{-1}{\sqrt{2}}\right) & 0 \\ 0 & 0 & (i)(-i) \end{pmatrix}$$

$$= \begin{pmatrix} \frac{1}{2}+\frac{1}{2} & \frac{1}{2}-\frac{1}{2} & 0 \\ -\frac{i}{2}+\frac{i}{2} & \frac{1}{2}-\frac{1}{2} & 0 \\ 0 & 0 & 1 \end{pmatrix}$$

$$= \begin{pmatrix} 1 & 0 & 0 \\ 0 & 1 & 0 \\ 0 & 0 & 1 \end{pmatrix} = 1 \Rightarrow U \text{ is unitary}$$

Now let's find the eigenvalues:

$$U - \lambda I = \begin{pmatrix} \frac{1}{\sqrt{2}} & \frac{i}{\sqrt{2}} & 0 \\ -\frac{1}{\sqrt{2}}i & \frac{1}{\sqrt{2}}i & 0 \\ 0 & 0 & i \end{pmatrix} - \lambda \begin{pmatrix} 1 & 0 & 0 \\ 0 & 1 & 0 \\ 0 & 0 & 1 \end{pmatrix}$$

$$= \begin{pmatrix} \frac{1}{\sqrt{2}} - \lambda & \frac{i}{\sqrt{2}} & 0 \\ -\frac{1}{\sqrt{2}} & \frac{i}{\sqrt{2}} - \lambda & 0 \\ 0 & 0 & i - \lambda \end{pmatrix}$$

$$\det[U - \lambda I] = \det \left| \begin{pmatrix} \frac{1}{\sqrt{2}} - \lambda & \frac{i}{\sqrt{2}} & 0 \\ -\frac{1}{\sqrt{2}} & \frac{i}{\sqrt{2}} - \lambda & 0 \\ 0 & 0 & i - \lambda \end{pmatrix} \right|$$

$$= \left(\frac{1}{\sqrt{2}} - \lambda\right) \det \begin{pmatrix} \frac{i}{\sqrt{2}} - \lambda & 0 \\ 0 & i - \lambda \end{pmatrix} - \frac{i}{\sqrt{2}} \det \begin{pmatrix} -\frac{1}{\sqrt{2}} & 0 \\ 0 & i - \lambda \end{pmatrix}$$

$$= \left(\frac{1}{\sqrt{2}} - \lambda\right)\left(\frac{i}{\sqrt{2}} - \lambda\right)(i - \lambda) - \frac{i}{\sqrt{2}}\left(-\frac{1}{\sqrt{2}}\right)(i - \lambda)$$

Setting equal to zero, we find that:

$$\lambda_1 = i \Rightarrow |\lambda_1|^2 = (i)(-i) = 1$$

$$\lambda_2 = \frac{\sqrt{2} - \sqrt{6}}{4} + i\frac{\sqrt{2} + \sqrt{6}}{4} \Rightarrow |\lambda_2|^2$$

$$= \left(\frac{\sqrt{2} - \sqrt{6}}{4} + i\frac{\sqrt{2} + \sqrt{6}}{4}\right)\left(\frac{\sqrt{2} - \sqrt{6}}{4} - i\frac{\sqrt{2} + \sqrt{6}}{4}\right)$$

$$= \left(\frac{\sqrt{2} - \sqrt{6}}{4}\right)\left(\frac{\sqrt{2} - \sqrt{6}}{4}\right) + \left(\frac{\sqrt{2} + \sqrt{6}}{4}\right)\left(\frac{\sqrt{2} + \sqrt{6}}{4}\right)$$

$$= \frac{2 - 2\sqrt{2}\sqrt{6} + 6}{16} + \frac{2 + 2\sqrt{2}\sqrt{6} + 6}{16} = \frac{4 + 12}{16} = 1$$

Finally, the last eigenvalue is:

$$\lambda_3 = \frac{\sqrt{2}+\sqrt{6}}{4} + i\frac{\sqrt{2}-\sqrt{6}}{4}$$

Following a procedure similar to that used for λ_2, we find that $|\lambda_3|^2 = 1$ also. Now let's consider a 2×2 unitary matrix with unit determinant.

EXAMPLE 6.25

For a general 2×2 unitary matrix $U = \begin{pmatrix} a & b \\ c & d \end{pmatrix}$ with $\det(U) = 1$, show that $a^* = d$, $b = -c^*$ and that $|a|^2 + |b|^2 = 1$. Show that such a matrix has only two independent components.

SOLUTION

Since the $\det(U) = 1$, we must have:

$$ad - bc = 1 \Rightarrow bc = ad - 1$$

It is also true that $UU^\dagger = I$. Now $U^\dagger = \begin{pmatrix} a^* & c^* \\ b^* & d^* \end{pmatrix}$. Therefore:

$$UU^\dagger = \begin{pmatrix} a & b \\ c & d \end{pmatrix}\begin{pmatrix} a^* & c^* \\ b^* & d^* \end{pmatrix} = \begin{pmatrix} |a|^2 + |b|^2 & ac^* + bd^* \\ ca^* + db^* & |c|^2 + |d|^2 \end{pmatrix} = \begin{pmatrix} 1 & 0 \\ 0 & 1 \end{pmatrix}$$

and so:

$$|a|^2 + |b|^2 = 1$$
$$ac^* + bd^* = 0$$
$$ca^* + db^* = 0$$
$$|c|^2 + |d|^2 = 1$$

Considering the second of these equations, $ac^* + bd^* = 0$, and recalling that we found $bc = ad - 1$, we multiply through by c to give:

$$\begin{aligned} c\left(ac^* + bd^*\right) &= a\,|c|^2 + (bc)\,d^* = a\,|c|^2 + (ad-1)\,d^* = a\,|c|^2 + a\,|d|^2 - d^* \\ &= a\left(|c|^2 + |d|^2\right) - d^* \\ &= a - d^* \end{aligned}$$

Since this equals zero, we conclude that $a = d^*$. Returning to $bc = ad - 1$,

$$bc = ad - 1 = d^*d - 1 = |d|^2 - 1 = |d|^2 - \left(|c|^2 + |d|^2\right) = -|c|^2 = -cc^*$$

Dividing both sides by c gives $b = -c^*$. Using the two conditions we have just derived, we can write the general form of a 2×2 unitary matrix with unit determinant as:

$$U = \begin{pmatrix} a & b \\ -b^* & a^* \end{pmatrix}$$

Often the action of an operator simply multiplies a given vector by a number. This number is called an *eigenvalue* and for an operator representing a physical observable, it represents a possible result of measurement.

EXAMPLE 6.26

e.g.

An important operator used in quantum computation is the "Hadamard gate," which is represented by the matrix:

$$H = \frac{1}{\sqrt{2}}\begin{pmatrix} 1 & 1 \\ 1 & -1 \end{pmatrix}$$

(a) Is H Hermitian and unitary?

(b) Find the eigenvalues and eigenvectors of this matrix.

SOLUTION

✔

(a)

$$H^\dagger = \frac{1}{\sqrt{2}}\begin{pmatrix} 1 & 1 \\ 1 & -1 \end{pmatrix}, \Rightarrow H \text{ is Hermitian}$$

$$\begin{aligned} HH^\dagger &= \frac{1}{\sqrt{2}}\begin{pmatrix} 1 & 1 \\ 1 & -1 \end{pmatrix}\frac{1}{\sqrt{2}}\begin{pmatrix} 1 & 1 \\ 1 & -1 \end{pmatrix} \\ &= \frac{1}{2}\begin{pmatrix} 1 & 1 \\ 1 & -1 \end{pmatrix}\begin{pmatrix} 1 & 1 \\ 1 & -1 \end{pmatrix} \\ &= \frac{1}{2}\begin{pmatrix} 1\,(1) + 1\,(1) & 1\,(1) + 1\,(-1) \\ 1\,(1) + (-1)\,(1) & 1\,(1) + (-1)\,(-1) \end{pmatrix} \\ &= \frac{1}{2}\begin{pmatrix} 2 & 0 \\ 0 & 2 \end{pmatrix} = \begin{pmatrix} 1 & 0 \\ 0 & 1 \end{pmatrix} = I, \Rightarrow H \text{ is unitary} \end{aligned}$$

(b)

$$0 = \det(H - \lambda I) = \det(\frac{1}{\sqrt{2}}\begin{pmatrix} 1 & 1 \\ 1 & -1 \end{pmatrix} - \lambda\begin{pmatrix} 1 & 0 \\ 0 & 1 \end{pmatrix})$$

$$= \det(\frac{1}{\sqrt{2}}\begin{pmatrix} 1-\lambda & 1 \\ 1 & -1-\lambda \end{pmatrix})$$

$$= \frac{1}{\sqrt{2}}(1-\lambda)(-1-\lambda) - 1$$

$\Rightarrow \frac{1}{\sqrt{2}}\left(\lambda^{2-2}\right) = 0,$ and the eigenvalues are $\lambda_{1,2} = \pm 1$

Let $|\Phi_1\rangle = \begin{pmatrix} a \\ b \end{pmatrix}$ with eigenvalue + 1. Then:

$$\frac{1}{\sqrt{2}}\begin{pmatrix} 1 & 1 \\ 1 & -1 \end{pmatrix}\begin{pmatrix} a \\ b \end{pmatrix} = \begin{pmatrix} a \\ b \end{pmatrix}$$

This leads to:

$$\frac{1}{\sqrt{2}}(a+b) = a, \Rightarrow b = \left(\sqrt{2}-1\right)a$$

Normalizing to find a:

$$|\Phi_1\rangle = \begin{pmatrix} a \\ b \end{pmatrix} = \begin{pmatrix} a \\ \left(\sqrt{2}-1\right)a \end{pmatrix}, \quad \langle\Phi_1| = \begin{pmatrix} a^* & \left(\sqrt{2}-1\right)a^* \end{pmatrix}$$

$$1 = \langle\Phi_1|\Phi_1\rangle = \begin{pmatrix} a^* & \left(\sqrt{2}-1\right)a^* \end{pmatrix}\begin{pmatrix} a \\ \left(\sqrt{2}-1\right)a \end{pmatrix} = a^2 + \left(\sqrt{2}-1\right)^2 a^2 = a^2\left(4-2\sqrt{2}\right)$$

This leads to:

$$a = \frac{1}{\sqrt{4-2\sqrt{2}}}, \quad b = \frac{\sqrt{2}-1}{\sqrt{4-2\sqrt{2}}} = \frac{1}{\sqrt{2\sqrt{2}}}$$

$$|\Phi_1\rangle = \begin{pmatrix} \frac{1}{\sqrt{4-2\sqrt{2}}} \\ \frac{1}{\sqrt{2\sqrt{2}}} \end{pmatrix}$$

We now consider the eigenvalue $\lambda = -1$ for $|\Phi_2\rangle = \begin{pmatrix} c \\ d \end{pmatrix}$

$$\frac{1}{\sqrt{2}}\begin{pmatrix} 1 & 1 \\ 1 & -1 \end{pmatrix}\begin{pmatrix} c \\ c \end{pmatrix} = -\begin{pmatrix} c \\ d \end{pmatrix}$$

$$\Rightarrow \frac{1}{\sqrt{2}}(c+d) - -c, \Rightarrow d = -\left(1+\sqrt{2}\right)c$$

Normalizing, we find:

$$1 = \langle \Phi_2 | \Phi_2 \rangle = \begin{pmatrix} c^* & -\left(1+\sqrt{2}\right) c^* \end{pmatrix} \begin{pmatrix} c \\ -\left(1+\sqrt{2}\right) c \end{pmatrix} = c^2 + \left(1+\sqrt{2}\right)^2 c^2$$

$$= c^2 \left(4 + 2\sqrt{2}\right)$$

$$c = \frac{1}{\sqrt{4+2\sqrt{2}}}, \; b = \frac{-1\left(1+\sqrt{2}\right)}{\sqrt{4+2\sqrt{2}}} = \frac{-1}{\sqrt{2\sqrt{2}}}$$

$$|\Phi_2\rangle = \begin{pmatrix} \frac{1}{\sqrt{4+2\sqrt{2}}} \\ \frac{-1}{\sqrt{2\sqrt{2}}} \end{pmatrix}$$

The Commutator

Let $\hat{A}$ and $\hat{B}$ be two operators. In general, $\hat{A}\hat{B} \neq \hat{B}\hat{A}$. The quantity

$$[\hat{A}, \hat{B}] = \hat{A}\hat{B} - \hat{B}\hat{A}$$

is called the *commutator* of $\hat{A}$ and $\hat{B}$. If $[\hat{A}, \hat{B}] = 0$, we say that the operators $\hat{A}$ and $\hat{B}$ *commute*. Two operators commute if and only if they share a basis of common eigenvectors.

DEFINITION: Complete Set of Commuting Observables

A set of operators $\hat{A}, \hat{B}, \hat{C}, \ldots$ is a complete set of commuting observables (CSCO) if all subpairs of the operators commute

$$[\hat{A}, \hat{B}] = [\hat{A}, \hat{C}] = [\hat{B}, \hat{C}] = \cdots = 0$$

and there exists a basis of common eigenvectors that is unique to within a multiplicative factor.

PROPERTIES OF THE COMMUTATOR

Let A, B, and C be operators. Then:

1. $[A, B] = -[B, A]$
2. $[A + B, C] = [A, C] + [B, C]$
3. $[A, BC] = [A, B]C + B[A, C]$
4. If $\hat{X}$ is the position operator and $\hat{P}$ the momentum operator, then $[\hat{X}, \hat{P}] = i\hbar$, $[\hat{X}, \hat{X}] = [\hat{P}, \hat{P}] = 0$

EXAMPLE 6.27

Prove that $[\hat{X}, \hat{P}] = i\hbar$

SOLUTION

We apply the commutator to a test wavefunction, $\Psi(x)$ and recall that $\hat{X}\psi(x) = x\psi(x)$ and $\hat{P} = -i\hbar\frac{\partial}{\partial x}$

$$\left[\hat{X}, \hat{P}\right]\psi(x) = \left(\hat{X}\hat{P} - \hat{P}\hat{X}\right)\psi(x) = \hat{X}\hat{P}\psi(x) - \hat{P}\hat{X}\psi(x)$$

$$= -i\hbar x\frac{\partial\psi}{\partial x} + i\hbar\frac{\partial}{\partial x}\left(\hat{X}\psi(x)\right)$$

$$= -i\hbar x\frac{\partial\psi}{\partial x} + i\hbar\frac{\partial}{\partial x}(x\psi(x))$$

$$= i\hbar x\frac{\partial\psi}{\partial x} + i\hbar\left\{\frac{\partial x}{\partial x}\psi(x) + x\frac{\partial\psi}{\partial x}\right\}$$

$$= -i\hbar x\frac{\partial\psi}{\partial x} + i\hbar\left\{\psi(x) + x\frac{\partial\psi}{\partial x}\right\}$$

$$= i\hbar\psi(x) - i\hbar x\frac{\partial\psi}{\partial x} + i\hbar x\frac{\partial\psi}{\partial x}$$

$$= i\hbar\psi(x)$$

So we conclude that $\left[\hat{X}, \hat{P}\right]\psi(x) = i\hbar\psi(x), \Rightarrow \left[\hat{X}, \hat{P}\right] = i\hbar$

EXAMPLE 6.28

Show that $[A, BC] = [A, B]C + B[A, C]$

SOLUTION

We have:

$$[A, BC] = A(BC) - (BC)A$$

Notice that $B[A, C] = B(AC - CA) = BAC - BCA$. We have the second term in this expression, but the first, BAC, is missing. So we use $0 = BAC - BAC$ to add in the missing piece and then rearrange terms:

$$[A, BC] = A(BC) - (BC)A$$

$$= A(BC) - (BC)A + BAC - BAC$$

$$= ABC - BAC + BAC - BCA$$

$$= ABC - BAC + B(AC - CA)$$
$$= ABC - BAC + B[A, C]$$
$$= (AB - BA)C + B[A, C]$$
$$= [A, B]C + B[A, C]$$

EXAMPLE 6.29

e.g.

Let A and B be two operators that commute. If A has non-degenerate eigenvalues, show that an eigenvector of A is also an eigenvector of B.

SOLUTION

Since A and B commute, $[A, B] = AB - BA = 0 \Rightarrow AB = BA$. Let $|a\rangle$ be an eigenvector of A such that $A\,|a\rangle = \lambda\,|a\rangle$. Then:

$$A(B\,|a\rangle) = BA\,|a\rangle = B\lambda\,|a\rangle = \lambda\,(B\,|a\rangle)$$

Therefore $B\,|a\rangle$ is also an eigenvector of A with eigenvalue λ. If A is non-degenerate, $|a\rangle$ is unique up to a proportionality factor. This implies that:

$$B\,|a\rangle = \omega\,|a\rangle$$

for some ω. Therefore $|a\rangle$ is also an eigenvector of B.

EXAMPLE 6.30

e.g.

Let $A = \begin{pmatrix} -1 & 2i & 0 \\ 0 & 4 & 0 \\ 1 & 0 & 1 \end{pmatrix}$ and $B = \begin{pmatrix} 0 & 2 & i \\ -i & 2i & 0 \\ 0 & 1 & 4 \end{pmatrix}$

(a) Find $tr(A)$ and $tr(B)$,
(b) Find $\det(A)$ and $\det(B)$.
(c) Find the inverse of A.
(d) Do A and B commute?

SOLUTION

(a) The trace is the sum of the diagonal elements:

$$Tr\,(A) = Tr \begin{pmatrix} -1 & 2i & 0 \\ 0 & 4 & 0 \\ 1 & 0 & 1 \end{pmatrix} = -1 + 4 + 1 = 4$$

$$Tr\,(B) = Tr \begin{pmatrix} 0 & 2 & i \\ -i & 2i & 0 \\ 0 & 1 & 4 \end{pmatrix} = 0 + 2i + 4$$

(b) We begin with the det (A):

$$\det(A) = \det\begin{pmatrix} -1 & 2i & 0 \\ 0 & 4 & 0 \\ 1 & 0 & 1 \end{pmatrix} = -\det\begin{pmatrix} 4 & 0 \\ 0 & 1 \end{pmatrix} - 2i\det\begin{pmatrix} 0 & 0 \\ 1 & 1 \end{pmatrix}$$

$$= -1\,(4) = -4$$

$$\det(B) = \det\begin{pmatrix} 0 & 2 & i \\ -i & 2i & 0 \\ 0 & 1 & 4 \end{pmatrix} = -2\det\begin{pmatrix} -i & 0 \\ 0 & 4 \end{pmatrix} + i\det\begin{pmatrix} -i & 2i \\ 0 & 1 \end{pmatrix}$$

$$= -2\,(-4i) + i\,(-i) = 8i + 1$$

(c) The det $(A) \neq 0, \Rightarrow A$ does have an inverse. We recall that $A^{-1} = \frac{C^T}{\det}(A)$ where C is the matrix of cofactors. First we compute C, recalling that $cij = (-1)^{i+j}\det(M_{ij})$, where M_{ij} is the minor obtained by crossing out row i and column j:

$$M_{11} = \begin{pmatrix} 4 & 0 \\ 0 & 1 \end{pmatrix}, \det(M_{11}) = 4, \Rightarrow c_{11} = (-1)^{1+1}(4) = (-1)^2\,4 = 4$$

$$M_{12} = \begin{pmatrix} 0 & 0 \\ 1 & 1 \end{pmatrix}, \det(M_{12}) = 0, \Rightarrow c_{12} = 0$$

$$M_{13} = \begin{pmatrix} 0 & 4 \\ 1 & 0 \end{pmatrix}, \det(M_{13}) = -4, \Rightarrow c_{13} = (-1)^{1+3}(-4) = (-1)^4 - 4 = -4$$

$$M_{21} = \begin{pmatrix} 2i & 0 \\ 0 & 1 \end{pmatrix}, \det(M_{21}) = 2i, \Rightarrow c_{21} = (-1)^{2+1}\,2i = (-1)^3\,2i = -2i$$

$$M_{22} = \begin{pmatrix} -1 & 0 \\ 1 & 1 \end{pmatrix}, \det(M_{22}) = -1, \Rightarrow c_{22} = (-1)^{2+2}(-1) = -1$$

$$M_{23} = \begin{pmatrix} -1 & 2i \\ 1 & 0 \end{pmatrix}, \det(M_{23}) = -2i, \Rightarrow c_{23} = (-1)^{2+3}(-2i) = (-1)^5(-2i)$$

$$= (-1)(-2i) = 2i$$

$$M_{31} = \begin{pmatrix} 2i & 0 \\ 4 & 0 \end{pmatrix}, \det(M_{31}) = 0, \Rightarrow c_{31} = 0$$

$$M_{32} = \begin{pmatrix} -1 & 0 \\ 0 & 0 \end{pmatrix}, \det(M_{32}) = 0, \Rightarrow c_{32} = 0$$

$$M_{33} = \begin{pmatrix} -1 & 2i \\ 0 & 4 \end{pmatrix}, \det(M_{33}) = -4, \Rightarrow c_{33} = (-1)^{3+3}(-4)$$

$$= (-1)^6(-4) = -4$$

Putting all of this together, C is given by:

$$C = \begin{pmatrix} 4 & 0 & -4 \\ -2i & -1 & 2i \\ 0 & 0 & -4 \end{pmatrix}, C^T = \begin{pmatrix} 4 & -2i & 0 \\ 0 & -1 & 0 \\ -4 & 2i & -4 \end{pmatrix}$$

Dividing by $\det(A) = -4$ gives us the inverse:

$$A^{-1} = \frac{1}{\det(A)} C^T = -\frac{1}{4} \begin{pmatrix} 4 & -2i & 0 \\ 0 & -1 & 0 \\ -4 & 2i & -4 \end{pmatrix} = \begin{pmatrix} -1 & \frac{1}{2}i & 0 \\ 0 & \frac{1}{4} & 0 \\ 1 & -\frac{1}{2}i & 1 \end{pmatrix}$$

Let's check to see that this is the inverse:

$$AA^{-1} = \begin{pmatrix} -1 & 2i & 0 \\ 0 & 4 & 0 \\ 1 & 0 & 1 \end{pmatrix} \begin{pmatrix} -1 & \frac{1}{2}i & 0 \\ 0 & \frac{1}{4} & 0 \\ 1 & -\frac{1}{2}i & 1 \end{pmatrix}$$

$$= \begin{pmatrix} [-1(-1)] & \left[-1\left(\frac{1}{2}i\right) + 2i\left(\frac{1}{4}\right)\right] & [0] \\ [0] & \left[4\left(\frac{1}{4}\right)\right] & [0] \\ [1(-1) + 1(1)] & \left[1\left(\frac{1}{2}i\right) + 1\left(-\frac{1}{2}i\right)\right] & [1] \end{pmatrix} = \begin{pmatrix} 1 & 0 & 0 \\ 0 & 1 & 0 \\ 0 & 0 & 1 \end{pmatrix} = 1$$

(d) $[A, B] = AB - BA$

$$= \begin{pmatrix} -1 & 2i & 0 \\ 0 & 4 & 0 \\ 1 & 0 & 1 \end{pmatrix} \begin{pmatrix} 0 & 2 & i \\ -i & 2i & 0 \\ 0 & 1 & 4 \end{pmatrix} - \begin{pmatrix} 0 & 2 & i \\ -i & 2i & 0 \\ 0 & 1 & 4 \end{pmatrix} \begin{pmatrix} -1 & 2i & 0 \\ 0 & 4 & 0 \\ 1 & 0 & 1 \end{pmatrix}$$

$$= \begin{pmatrix} 2i(-i) & -1(2) + 2i(2i) & -1(i) \\ 4(-i) & 4(2i) & [0] \\ [0] & 1(2) + 1(1) & 1(i) + 1(4) \end{pmatrix}$$

$$- \begin{pmatrix} i(1) & 2(4) & i(1) \\ -i(-1) & -i(2i) + 2i(4) & 0 \\ 4(1) & 1(4) & 4(1) \end{pmatrix}$$

$$= \begin{pmatrix} 2 & -6 & -i \\ -4i & 8i & 0 \\ 0 & 3 & 4i \end{pmatrix} - \begin{pmatrix} i & 8 & i \\ i & 2 - 8i & 0 \\ 4 & 4 & 4 \end{pmatrix} = \begin{pmatrix} 2 - 1 & -14 & -2i \\ -3i & 2 & 0 \\ -4 & -1 & i \end{pmatrix}$$

$[A, B] \neq 0, \Rightarrow A, B$ do not commute.

DEFINITION: The Anticommutator

The *anticommutator* of operators A and B is:

$$\{A, B\} = AB + BA$$

EXAMPLE 6.31

If $\hat{A}$ and $\hat{B}$ are Hermitian operators, show that their anticommutator is Hermitian.

SOLUTION

$$\begin{aligned}\left\{\hat{A}, \hat{B}\right\}^{\dagger} &= \left(\hat{A}\hat{B} + \hat{B}\hat{A}\right)^{\dagger} = \left(\hat{A}\hat{B}\right)^{\dagger} + \left(\hat{B}\hat{A}\right)^{\dagger}\\ &= B^{\dagger}A^{\dagger} + A^{\dagger}B^{\dagger}\\ &= \hat{A}\hat{B} + \hat{B}\hat{A} = \left\{\hat{A}, \hat{B}\right\}\end{aligned}$$

$$\left\{\hat{A}, \hat{B}\right\} = \left\{\hat{A}, \hat{B}\right\}' \Rightarrow \left\{\hat{A}, \hat{B}\right\} \text{ is Hermitian for } \hat{A}, \hat{B} \text{ Hermitian.}$$

Quiz

1. The Pauli matrices are defined by:

$$\sigma_x = \begin{pmatrix} 0 & 1 \\ 1 & 0 \end{pmatrix}, \sigma_y = \begin{pmatrix} 0 & i \\ -i & 0 \end{pmatrix}, \sigma_z = \begin{pmatrix} 1 & 0 \\ 0 & -1 \end{pmatrix}$$

   (a) Find the eigenvalues and eigenvectors of the Pauli matrices.
   (b) Are they Hermitian? Are they unitary?
   (c) Find $\left[\sigma_x, \sigma_y\right]$ and $\left\{\sigma_x, \sigma_y\right\}$.
   (d) Express the Pauli matrices in outer product notation.

2. Prove that the eigenvalues of an *anti-Hermitian* operator are pure imaginary.

3. Prove that the eigenvectors of a Hermitian operator are orthogonal. (*Hint*: Examine the proof that the eigenvalues of a Hermitian operator are real, but this time consider different eigenvectors).

4. (a) Let A and B be two operators. Find BA in terms of $[A, B]$ and $\{A, B\}$
   (b) Prove that $[A + B, C] = [A, C] + [B, C]$.

5. Show that the eigenvalues of:

$$A = \begin{pmatrix} 0 & 1 & 0 \\ -1 & 2 & 0 \\ 0 & 0 & 4 \end{pmatrix}$$

are $\{4, 1, 1\}$. Find a set of normalized eigenvectors of A.

6. Prove that unitary operators preserve inner products, that is, $\langle U\Phi | U\psi\rangle = \langle \Phi | \psi \rangle$.

7. Show that $[A, B]^\dagger = -[A^\dagger, B^\dagger]$.

8. Let $A = \begin{pmatrix} 1 & -i & 0 \\ i & 2 & 0 \\ 0 & 0 & 4 \end{pmatrix}$.

   (a) Is A Hermitian?

   (b) Is A unitary?

   (c) Find $Tr(A)$.

   (d) Show that the eigenvalues of A are $\left\{4, \frac{1}{2}\left(3+\sqrt{5}\right), \frac{1}{2}\left(3-\sqrt{5}\right)\right\}$.

   (e) Find the normalized eigenvectors of A.

The Mathematical Structure of Quantum Mechanics III

In this chapter we finish our discussion of the mathematical theory underlying quantum mechanics. We first consider forming a basis from the eigenvectors of a Hermitian matrix and unitary transformations and then examine a mathematical derivation of the famous uncertainty relation. We close out the chapter with a look at projection operators and tensor product spaces.

Change of Basis and Unitary Transformations

In this section we first consider the question as to whether or not the eigenvectors of a matrix form a basis. We then see how to construct a unitary transformation matrix based upon those eigenvectors. To find out if a set of eigenvectors forms a basis, we rely on two tests.

Definition: Test to See If Eigenvectors Form a Basis

1. Do they satisfy the completeness relation?
2. Are they orthonormal?

The completeness relation tells us that, given a basis $|u_i\rangle$, the following is true:

$$\sum |u_i\rangle \langle u_i| = 1$$

This fact allows us to verify that the eigenvectors of a Hermitian matrix form a basis. As an example, we review the Pauli matrix:

$$\sigma_z = \begin{pmatrix} 1 & 0 \\ 0 & -1 \end{pmatrix}$$

The eigenvalues of the matrix are found in the usual way:

$$\begin{aligned} 0 = \det|\sigma_z - \lambda I| &= \det\left|\begin{pmatrix} 1 & 0 \\ 0 & -1 \end{pmatrix} - \lambda \begin{pmatrix} 1 & 0 \\ 0 & 1 \end{pmatrix}\right| \\ &= \det\left|\begin{pmatrix} 1-\lambda & 0 \\ 0 & -1-\lambda \end{pmatrix}\right| \\ &= (1-\lambda)(-1-\lambda) \\ &= -1 + \lambda^2 \end{aligned}$$

Setting this equal to zero we find $\Rightarrow \lambda_{1,2} = \pm 1$

For $\lambda_1 = +1$, we have:

$$\begin{pmatrix} 1 & 0 \\ 0 & -1 \end{pmatrix}\begin{pmatrix} a \\ b \end{pmatrix} = \begin{pmatrix} a \\ b \end{pmatrix}$$

This gives two equations:

$$\begin{aligned} a &= a \\ -b &= b \end{aligned}$$

The first equation gives no information while the second tells us that $b = 0$. We find a by normalization

$$1 = \begin{pmatrix} a^* & 0 \end{pmatrix}\begin{pmatrix} a \\ 0 \end{pmatrix} = |a|^2, \Rightarrow a = 1$$

So the first basis vector is:

$$|u_1\rangle = \begin{pmatrix} 1 \\ 0 \end{pmatrix}$$

For $\lambda_2 = -1$, we have:

$$\begin{pmatrix} 1 & 0 \\ 0 & -1 \end{pmatrix}\begin{pmatrix} a \\ b \end{pmatrix} = -\begin{pmatrix} a \\ b \end{pmatrix}$$

This gives two equations:

$$a = -a$$

$$-b = -b$$

Following the same logic used for the previous eigenvector, we see that $a = 0$ and write u_2 as:

$$|u_2\rangle = \begin{pmatrix} 0 \\ 1 \end{pmatrix}$$

It is easily determined that the eigenvectors form an orthonormal set:

$$\langle u_1 | u_1 \rangle = (1 \quad 0) \begin{pmatrix} 1 \\ 0 \end{pmatrix} = 1(1) + 0(0) = 1$$

$$\langle u_2 | u_1 \rangle = (0 \quad 1) \begin{pmatrix} 1 \\ 0 \end{pmatrix} = 0(1) + 1(0) = 0$$

$$\langle u_2 | u_2 \rangle = (0 \quad 1) \begin{pmatrix} 0 \\ 1 \end{pmatrix} = 0(0) + 1(1) = 1$$

Now we need to find out if these vectors satisfy the completeness relation:

$$\begin{aligned} |u_1\rangle \langle u_1| + |u_2\rangle \langle u_2| &= \begin{pmatrix} 1 \\ 0 \end{pmatrix} (1 \quad 0) + \begin{pmatrix} 0 \\ 1 \end{pmatrix} (0 \quad 1) \\ &= \begin{pmatrix} 1(1) & 1(0) \\ 0(1) & 0(0) \end{pmatrix} + \begin{pmatrix} 0(0) & 0(1) \\ 1(0) & 1(1) \end{pmatrix} \\ &= \begin{pmatrix} 1 & 0 \\ 0 & 0 \end{pmatrix} + \begin{pmatrix} 0 & 0 \\ 0 & 1 \end{pmatrix} = \begin{pmatrix} 1 & 0 \\ 0 & 1 \end{pmatrix} = 1 \end{aligned}$$

We see that the completeness relation is satisfied, and that these eigenvectors are orthonormal. Therefore we conclude that they form a basis. It is, of course, easy to see that any vector in two dimensions can be written in terms of $\{|u_1\rangle |u_2\rangle\}$. Let $|\psi\rangle$ be such an arbitrary ket;

$$|\psi\rangle = \begin{pmatrix} \alpha \\ \beta \end{pmatrix} = \begin{pmatrix} \alpha \\ 0 \end{pmatrix} + \begin{pmatrix} 0 \\ \beta \end{pmatrix} = \alpha \begin{pmatrix} 1 \\ 0 \end{pmatrix} + \beta \begin{pmatrix} 0 \\ 1 \end{pmatrix} = \alpha |u_1\rangle + \beta |u_2\rangle$$

SIMILARITY TRANSFORMATIONS

We now consider using the eigenvectors of a Hermitian matrix for what is called a *similarity* transformation. This will allow us to put a matrix into a diagonal form. This means that all entries in the matrix will be zero except those that lie on the diagonal. For example, here are the general forms of diagonal 2×2 and 3×3 matrices:

$$A = \begin{pmatrix} a_{11} & 0 \\ 0 & a_{22} \end{pmatrix}, \; B = \begin{pmatrix} b_{11} & 0 & 0 \\ 0 & b_{22} & 0 \\ 0 & 0 & b_{33} \end{pmatrix}$$

The elements along the diagonal are the eigenvalues of the matrix. In order to diagonalize the matrix, we apply a *similarity* transformation. If D is the diagonal form of a matrix C, then:

$$D = S^{-1}CS$$

where S is an invertible matrix that is composed of the eigenvectors of C. The matrices C and D are said to be "similar" and these two matrices actually represent the same operator with respect to two different bases. If the first basis is orthonormal, then the second will be also if the matrix S is unitary. If the matrix C is Hermitian, then we automatically have a basis by finding its eigenvectors.

FINDING A SIMILARITY TRANSFORMATION FOR A MATRIX C

1. Find the eigenvalues and eigenvectors of the matrix C
2. Normalize the eigenvectors of C
3. Form a new matrix S^{-1} by forming the columns of S^{-1} with the eigenvectors of C

In quantum mechanics, we are concerned primarily with Hermitian and Unitary matrices. In that case, the diagonal transformation of a Hermitian matrix H takes the form:

$$D = U^{\dagger}HU$$

This holds because $U^{\dagger} = U^{-1}$ for a unitary matrix. Such a transformation is called a *unitary transformation*. We demonstrate it with an example.

EXAMPLE 7.1

Consider the matrix $R = \begin{pmatrix} \cos\theta & \sin\theta \\ -\sin\theta & \cos\theta \end{pmatrix}$, the 2×2 *rotation matrix*. Find a unitary transformation that diagonalizes R.

SOLUTION

First we find the eigenvalues of R. Solving $\det(R - \lambda I) = 0$, we obtain:

$$R - \lambda I = \begin{pmatrix} \cos\theta & \sin\theta \\ -\sin\theta & \cos\theta \end{pmatrix} - \lambda \begin{pmatrix} 1 & 0 \\ 0 & 1 \end{pmatrix} = \begin{pmatrix} \cos\theta - \lambda & \sin\theta \\ -\sin\theta & \cos\theta - \lambda \end{pmatrix}$$

$\Rightarrow \det(R - \lambda I) = 0 = (\cos\theta - \lambda)^2 + \sin^2\theta = \cos^2\theta - 2\lambda\cos\theta + \lambda^2 + \sin^2\theta$

Rearranging terms, we have:

$$\lambda^2 - 2\lambda\cos\theta + \cos^2\theta + \sin^2\theta = \lambda^2 - 2\lambda\cos\theta + 1 = 0$$

We now use the quadratic formula to obtain the roots of this equation, which we call $\lambda_{1,2}$:

$$\lambda_{1,2} = \frac{2\cos\theta \pm \sqrt{(2\cos\theta)^2 - 4}}{2} = \cos\theta \pm \sqrt{\cos^2\theta - 1}$$
$$= \cos\theta \pm \sqrt{\cos^2\theta - \left(\cos^2\theta + \sin^2\theta\right)}$$
$$= \cos\theta \pm \sqrt{-\sin^2\theta} = \cos\theta \pm i\sin\theta$$

Now we recall Euler's formula for the *cos* and *sin* functions:

$$\cos\theta = \frac{e^{i\theta} + e^{-i\theta}}{2}, \sin\theta = \frac{e^{i\theta} - e^{-i\theta}}{2i} \Rightarrow i\sin\theta = \frac{e^{i\theta} - e^{-i\theta}}{2}$$

And so we obtain the following eigenvalues:

$$\lambda_1 = \cos\theta + i\sin\theta$$
$$= \frac{e^{i\theta} + e^{-i\theta}}{2} + \frac{e^{i\theta} - e^{-i\theta}}{2} = e^{i\theta}$$
$$\lambda_2 = \cos\theta - i\sin\theta = \frac{e^{i\theta} + e^{-i\theta}}{2} - \frac{e^{i\theta} - e^{i\theta}}{2} = e^{-i\theta}$$

Now let's find the eigenvectors that correspond to each eigenvalue. We will label each eigenvector by $\Phi_{1,2}$ and let $\Phi_1 = \begin{pmatrix} a \\ b \end{pmatrix}$, $\Phi_2 = \begin{pmatrix} c \\ d \end{pmatrix}$. Starting with $\lambda_1 = e^{i\theta}$ we have:

$$R\Phi_1 = \lambda_1\Phi_1 \Rightarrow \begin{pmatrix} \cos\theta & \sin\theta \\ -\sin\theta & \cos\theta \end{pmatrix}\begin{pmatrix} a \\ b \end{pmatrix} = e^{i\theta}\begin{pmatrix} a \\ b \end{pmatrix}$$

Carrying out the multiplication on the left side, we have:

$$\begin{pmatrix} \cos\theta & \sin\theta \\ -\sin\theta & \cos\theta \end{pmatrix}\begin{pmatrix} a \\ b \end{pmatrix} = \begin{pmatrix} a\cos\theta + b\sin\theta \\ -a\sin\theta + b\cos\theta \end{pmatrix}$$
$$\Rightarrow \begin{pmatrix} a\cos\theta + b\sin\theta \\ -a\sin\theta + b\cos\theta \end{pmatrix} = e^{i\theta}\begin{pmatrix} a \\ b \end{pmatrix}$$

This matrix relationship gives us two equations to solve, along with the two unknowns a and b:

$$a\cos\theta + b\sin\theta = e^{i\theta}a$$
$$-a\sin\theta + b\cos\theta = e^{i\theta}b$$

Solving for b in terms of a in the second equation, we obtain:

$$b(\cos\theta - e^{i\theta}) = -a\sin\theta$$

Recalling that $e^{i\theta} = \cos\theta + i\sin\theta$, the right side is just:

$$b\,(\cos\theta - \cos\theta - i\sin\theta) = b\,(-i\sin\theta)$$

which from our previous result is equal to $-a\sin\theta$. So in terms of a, b is given by $b = \frac{a}{-i} = ia$. This allows us to write the eigenvector entirely in terms of a:

$$\Phi_1 = \begin{pmatrix} a \\ b \end{pmatrix} = \begin{pmatrix} a \\ ia \end{pmatrix}$$

To find the constant a, we normalize it. Noting $\Phi_1^\dagger = (a^*\ (ia)^*)$, we obtain:

$$1 = \Phi_1^\dagger \cdot \Phi_1 = \begin{pmatrix} a^* & -ia^* \end{pmatrix} \begin{pmatrix} a \\ ia \end{pmatrix} = a^*(a) + (-ia^*)(ia) = |a|^2 + |a|^2 = 2\,|a|^2$$

$\Rightarrow |a|^2 = \frac{1}{2}$, or $a = \frac{1}{\sqrt{2}}$. Therefore $b = ia = i\frac{1}{\sqrt{2}}$

and we obtain $\Phi_1 = \begin{pmatrix} a \\ b \end{pmatrix} = \begin{pmatrix} \frac{1}{\sqrt{2}} \\ \frac{i}{\sqrt{2}} \end{pmatrix}$.

We follow the same procedure to find $\Phi_2 = \begin{pmatrix} c \\ d \end{pmatrix}$ for $\lambda_2 = e^{-i\theta}$:

$$R\Phi_2 = \lambda_2\Phi_2 \Rightarrow \begin{pmatrix} \cos\theta & \sin\theta \\ -\sin\theta & \cos\theta \end{pmatrix} \begin{pmatrix} c \\ d \end{pmatrix} = e^{-i\theta} \begin{pmatrix} c \\ d \end{pmatrix}$$

$\Rightarrow c\cos\theta + d\sin\theta = e^{-i\theta}c$, $-c\sin\theta + d\cos\theta = e^{-i\theta}d$. Focusing on the second equation, we have:

$$-c\sin\theta = d\left(e^{-i\theta} - \cos\theta\right) = d\,(\cos\theta - i\sin\theta - \cos\theta) = -id\sin\theta$$

Solving for d in terms of c, we obtain:

$$d = -ic$$

Inserting this into $\Phi_2 = \begin{pmatrix} c \\ d \end{pmatrix} = \begin{pmatrix} c \\ -ic \end{pmatrix}$ and normalizing using the same procedure as we did before

$$\Phi_2 = \begin{pmatrix} \frac{1}{\sqrt{2}} \\ \frac{-i}{\sqrt{2}} \end{pmatrix}$$

We form the matrix U from these two eigenvectors. The first column of U is the first eigenvector, and the second column of U is the second eigenvector:

$$U = (\Phi_1 \Phi_2) = \begin{pmatrix} \frac{1}{\sqrt{2}} & \frac{1}{\sqrt{2}} \\ \frac{i}{\sqrt{2}} & \frac{-i}{\sqrt{2}} \end{pmatrix} = \frac{1}{\sqrt{2}} \begin{pmatrix} 1 & 1 \\ i & -i \end{pmatrix}$$

$$\Rightarrow U^{\dagger} = \frac{1}{\sqrt{2}} \begin{pmatrix} 1 & -i \\ 1 & i \end{pmatrix}$$

We check to see that $UU^{\dagger} = I$:

$$UU^{\dagger} = \frac{1}{\sqrt{2}} \begin{pmatrix} 1 & 1 \\ i & -i \end{pmatrix} \frac{1}{\sqrt{2}} \begin{pmatrix} 1 & -i \\ 1 & i \end{pmatrix} = \frac{1}{2} \begin{pmatrix} 1 & 1 \\ i & -i \end{pmatrix} \begin{pmatrix} 1 & -i \\ 1 & i \end{pmatrix}$$

$$= \frac{1}{2} \left(\begin{array}{c|c} 1(1) + 1(1) & 1(-i) + 1(i) \\ \hline i(1) + (-i)(1) & i(-i) + (-i)(i) \end{array} \right)$$

$$= \frac{1}{2} \begin{pmatrix} 2 & 0 \\ 0 & 2 \end{pmatrix} = \begin{pmatrix} 1 & 0 \\ 0 & 1 \end{pmatrix} = I$$

Finally we apply the transformation to diagonalize the matrix R:

$$U^{\dagger} R U = \frac{1}{\sqrt{2}} \begin{pmatrix} 1 & -i \\ 1 & i \end{pmatrix} \begin{pmatrix} \cos\theta & \sin\theta \\ -\sin\theta & \cos\theta \end{pmatrix} \frac{1}{\sqrt{2}} \begin{pmatrix} 1 & 1 \\ i & -i \end{pmatrix}$$

$$= \frac{1}{2} \begin{pmatrix} 1 & -i \\ 1 & i \end{pmatrix} \begin{pmatrix} \cos\theta & \sin\theta \\ -\sin\theta & \cos\theta \end{pmatrix} \begin{pmatrix} 1 & 1 \\ i & -i \end{pmatrix}$$

$$= \frac{1}{2} \begin{pmatrix} 1 & -i \\ 1 & i \end{pmatrix} \begin{pmatrix} \cos\theta + i\sin\theta & \cos\theta - i\sin\theta \\ -\sin\theta + i\cos\theta & -\sin\theta - i\cos\theta \end{pmatrix}$$

$$= \frac{1}{2} \begin{pmatrix} 1 & -i \\ 1 & i \end{pmatrix} \begin{pmatrix} e^{i\theta} & e^{-i\theta} \\ ie^{i\theta} & -ie^{-i\theta} \end{pmatrix}$$

$$= \frac{1}{2} \begin{pmatrix} e^{i\theta} + e^{i\theta} & e^{-i\theta} - e^{-i\theta} \\ e^{i\theta} - e^{i\theta} & e^{-i\theta} + e^{-i\theta} \end{pmatrix} = \frac{1}{2} \begin{pmatrix} 2e^{i\theta} & 0 \\ 0 & 2e^{-i\theta} \end{pmatrix} = \begin{pmatrix} e^{i\theta} & 0 \\ 0 & e^{-i\theta} \end{pmatrix}$$

and we see that the diagonal form of the matrix. The elements on the diagonal are the eigenvalues of R.

EXAMPLE 7.2 *e.g.*

Suppose that an operator

$$\hat{A} = 2\left|\Phi_1\right\rangle\left\langle\Phi_1\right| - i\left|\Phi_1\right\rangle\left\langle\Phi_2\right| + i\left|\Phi_2\right\rangle\left\langle\Phi_1\right| + 2\left|\Phi_2\right\rangle\left\langle\Phi_2\right|,$$

where

$$\left|\Phi_1\right\rangle \text{ and } \left|\Phi_2\right\rangle$$

form an orthonormal and complete basis.

(a) Is $\hat{A}$ Hermitian?

(b) Find the eigenvalues and eigenvectors of $\hat{A}$ and show they satisfy the completeness relation.

(c) Find a unitary transformation that diagonalizes $\hat{A}$.

SOLUTION

(a) First we recall the general rule for finding the adjoint of an expression:

$$\lambda \langle u| \hat{A}\hat{B} |v\rangle \rightarrow \lambda^* \langle v| \hat{B}^\dagger \hat{A}^\dagger |u\rangle$$

The Hermitian conjugate operation is linear, so we examine each piece of $\hat{A}$, replacing any scalars by their complex conjugates, turning kets into bras, bras into kets, and then reversing the order of factors. Therefore:

$$\begin{aligned}
(2\,|\Phi_1\rangle\,\langle\Phi_1|)^\dagger &= 2\,|\Phi_1\rangle\,\langle\Phi_1| \\
(-i\,|\Phi_1\rangle\,\langle\Phi_2|)^\dagger &= i\,|\Phi_2\rangle\,\langle\Phi_1| \\
(i\,|\Phi_2\rangle\,\langle\Phi_1|)^\dagger &= -i\,|\Phi_1\rangle\,\langle\Phi_2| \\
(2\,|\Phi_2\rangle\,\langle\Phi_2|)^\dagger &= 2\,|\Phi_2\rangle\,\langle\Phi_2|
\end{aligned}$$

Therefore, we have:

$$\begin{aligned}
\hat{A}^\dagger &= (2\,|\Phi_1\rangle\,\langle\Phi_1| - i\,|\Phi_1\rangle\,\langle\Phi_2| + i\,|\Phi_2\rangle\,\langle\Phi_1| + 2\,|\Phi_2\rangle\,\langle\Phi_2|)^\dagger \\
&= (2\,|\Phi_1\rangle\,\langle\Phi_1|)^\dagger + (-i\,|\Phi_1\rangle\,\langle\Phi_2|)^\dagger + (i\,|\Phi_2\rangle\,\langle\Phi_1|)^\dagger + (2\,|\Phi_2\rangle\,\langle\Phi_2|)^\dagger \\
&= 2\,|\Phi_1\rangle\,\langle\Phi_1| - i\,|\Phi_1\rangle\,\langle\Phi_2| + i\,|\Phi_2\rangle\,\langle\Phi_1| + 2\,|\Phi_2\rangle\,\langle\Phi_2| \\
&= \hat{A}, \Rightarrow \text{the operator is Hermitian.}
\end{aligned}$$

(b) To find the eigenvalues and eigenvectors of $\hat{A}$, we first find the representation of the operator from the given orthonormal basis:

$$\begin{aligned}
\langle\Phi_1|\hat{A}|\Phi_1\rangle &= \langle\Phi_1| \Big(2\,|\Phi_1\rangle\,\langle\Phi_1| - i\,|\Phi_1\rangle\,\langle\Phi_2| + i\,|\Phi_2\rangle\,\langle\Phi_1| + 2\,|\Phi_2\rangle\,\langle\Phi_2|\Big)\,|\Phi_1\rangle \\
&= 2\,\langle\Phi_1|\,\Phi_1\rangle\,\langle\Phi_1\,|\Phi_1\rangle - i\,\langle\Phi_1\,|\Phi_1\rangle\,\langle\Phi_2\,|\Phi_1\rangle \\
&\quad + i\,\langle\Phi_1\,|\Phi_2\rangle\,\langle\Phi_1\,|\Phi_1\rangle + 2\,\langle\Phi_1\,|\Phi_2\rangle\,\langle\Phi_2\,|\Phi_1\rangle \\
&= 2
\end{aligned}$$

$$\begin{aligned}
\langle\Phi_1|\hat{A}|\Phi_2\rangle &= \langle\Phi_1|\Big(2|\Phi_1\rangle\langle\Phi_1| - i|\Phi_1\rangle\langle\Phi_2| + i|\Phi_2\rangle\langle\Phi_1| + 2|\Phi_2\rangle\langle\Phi_2|\Big)|\Phi_2\rangle \\
&= 2\langle\Phi_1|\Phi_1\rangle\langle\Phi_1|\Phi_2\rangle - i\langle\Phi_1|\Phi_1\rangle\langle\Phi_2|\Phi_2\rangle \\
&\quad + i\langle\Phi_1|\Phi_2\rangle\langle\Phi_1|\Phi_2\rangle + 2\langle\Phi_1|\Phi_2\rangle\langle\Phi_2|\Phi_2\rangle \\
&= -i \\
\langle\Phi_2|\hat{A}|\Phi_1\rangle &= \langle\Phi_2|\Big(2|\Phi_1\rangle\langle\Phi_1| - i|\Phi_1\rangle\langle\Phi_2| + i|\Phi_2\rangle\langle\Phi_1| + 2|\Phi_2\rangle\langle\Phi_2|\Big)|\Phi_1\rangle \\
&= 2\langle\Phi_2|\Phi_1\rangle\langle\Phi_1|\Phi_1\rangle - i\langle\Phi_2|\Phi_1\rangle\langle\Phi_2|\Phi_1\rangle + i\langle\Phi_2|\Phi_2\rangle\langle\Phi_1|\Phi_1\rangle \\
&\quad + 2\langle\Phi_2|\Phi_2\rangle\langle\Phi_2|\Phi_1\rangle \\
&= +i \\
\langle\Phi_2|\hat{A}|\Phi_2\rangle &= \langle\Phi_2|\Big(2|\Phi_1\rangle\langle\Phi_1| - i|\Phi_1\rangle\langle\Phi_2| + i|\Phi_2\rangle\langle\Phi_1| + 2|\Phi_2\rangle\langle\Phi_2|\Big)|\Phi_2\rangle \\
&= 2\langle\Phi_2|\Phi_1\rangle\langle\Phi_1|\Phi_2\rangle - i\langle\Phi_2|\Phi_1\rangle\langle\Phi_2|\Phi_2\rangle + i\langle\Phi_2|\Phi_2\rangle\langle\Phi_1|\Phi_2\rangle \\
&\quad + 2\langle\Phi_2|\Phi_2\rangle\langle\Phi_2|\Phi_2\rangle \\
&= 2
\end{aligned}$$

We can use these results to write teh matrix representation of A:

$$\Rightarrow \hat{A} = \begin{pmatrix} \langle\Phi_1|\hat{A}|\Phi_1\rangle\langle\Phi_1|\hat{A}|\Phi_2\rangle \\ \langle\Phi_2|\hat{A}|\Phi_1\rangle\langle\Phi_2|\hat{A}|\Phi_2\rangle \end{pmatrix} = \begin{pmatrix} 2 & -i \\ i & 2 \end{pmatrix}$$

We find that the eigenvalues are:

$$0 = \det\left|\hat{A} - \lambda I\right| = \det\left|\begin{pmatrix} 2 & -i \\ i & 2 \end{pmatrix} - \lambda\begin{pmatrix} 1 & 0 \\ 0 & 1 \end{pmatrix}\right| = \det\left|\begin{pmatrix} 2-\lambda & -i \\ i & 2-\lambda \end{pmatrix}\right|$$

$$\Rightarrow \lambda^2 - 4\lambda + 3 = 0$$

This leads to the eigenvalues:

$$\lambda_{1,2} = \{3, 1\}$$

We find the respective eigenvectors:

$$\begin{pmatrix} 2 & -i \\ i & 2 \end{pmatrix}\begin{pmatrix} a \\ b \end{pmatrix} = 3\begin{pmatrix} a \\ b \end{pmatrix}$$

This leads to:

$$\begin{aligned}
2a - ib &= 3a \\
ia + 2b &= 3b \\
\Rightarrow b &= ia
\end{aligned}$$

Normalizing, we have the following:

$$1 = \left(a^* \ -ia^*\right)\begin{pmatrix} a \\ ia \end{pmatrix} = |a|^2 + |a|^2 = 2\,|a|^2\,, \Rightarrow a = \frac{1}{\sqrt{2}}$$

So the first normalized eigenvector of $\hat{A}$ is:

$$|u_1\rangle = \frac{1}{\sqrt{2}}\begin{pmatrix} 1 \\ i \end{pmatrix}$$

Considering the second eigenvalue, we find:

$$\begin{pmatrix} 2 & -i \\ i & 2 \end{pmatrix}\begin{pmatrix} a \\ b \end{pmatrix} = \begin{pmatrix} a \\ b \end{pmatrix}$$
$$\Rightarrow 2a - ib = a \text{ or } b = -ia.$$

This leads to the second eigenvector:

$$|u_2\rangle = \frac{1}{\sqrt{2}}\begin{pmatrix} 1 \\ -i \end{pmatrix}$$

We verify that these eigenvectors are orthogonal:

$$\langle u_1\,|u_2\rangle = \left(\frac{1}{\sqrt{2}}\begin{pmatrix} 1 \\ i \end{pmatrix}\right)^{\dagger}\frac{1}{\sqrt{2}}\begin{pmatrix} 1 \\ -i \end{pmatrix}$$
$$= \frac{1}{2}\begin{pmatrix} 1 & -i \end{pmatrix}\begin{pmatrix} 1 \\ -i \end{pmatrix} = 0$$

Now we check the completeness relation:

$$|u_1\rangle\langle u_1| + |u_2\rangle\langle u_2| = \frac{1}{\sqrt{2}}\begin{pmatrix} 1 \\ i \end{pmatrix}\frac{1}{\sqrt{2}}(1 - i) + \frac{1}{\sqrt{2}}\begin{pmatrix} 1 \\ -i \end{pmatrix}\frac{1}{\sqrt{2}}\begin{pmatrix} 1 & i \end{pmatrix}$$
$$= \frac{1}{2}\begin{pmatrix} 1 \\ i \end{pmatrix}\begin{pmatrix} 1 & -i \end{pmatrix} + \frac{1}{2}\begin{pmatrix} 1 \\ -i \end{pmatrix}\begin{pmatrix} 1 & i \end{pmatrix}$$
$$= \frac{1}{2}\begin{pmatrix} 1(1) & 1(-i) \\ i(1) & i(-i) \end{pmatrix} + \frac{1}{2}\begin{pmatrix} 1\,(1) & 1\,(i) \\ -i\,(1) & -i\,(i) \end{pmatrix}$$
$$= \frac{1}{2}\left\{\begin{pmatrix} 1 & -i \\ i & 1 \end{pmatrix} + \begin{pmatrix} 1 & i \\ -i & 1 \end{pmatrix}\right\}$$
$$= \frac{1}{2}\begin{pmatrix} 2 & 0 \\ 0 & 2 \end{pmatrix} = \begin{pmatrix} 1 & 0 \\ 0 & 1 \end{pmatrix} = I$$

(c) The unitary transformation that diagnolizes $\hat{A}$ can be found from its eigenvectors. We now construct a unitary matrix from the basis vectors, with each column one of the basis vectors, i.e. $U = (|u_1\rangle\ |u_2\rangle)$:

$$U = \frac{1}{\sqrt{2}}\begin{pmatrix} 1 & 1 \\ i & -i \end{pmatrix}$$

We check that U is unitary:

$$U^\dagger = \frac{1}{\sqrt{2}}\begin{pmatrix} 1 & -i \\ 1 & i \end{pmatrix}$$

$$\Rightarrow U^\dagger U = \frac{1}{2}\begin{pmatrix} 1 & -i \\ 1 & i \end{pmatrix}\begin{pmatrix} 1 & 1 \\ i & -i \end{pmatrix} = \frac{1}{2}\begin{pmatrix} 1(1) + (-i)(i) & 1(1) + (-i)(-i) \\ 1(1) + i(i) & 1(1) + i(-i) \end{pmatrix}$$

$$= \frac{1}{2}\begin{pmatrix} 2 & 0 \\ 0 & 2 \end{pmatrix} = \begin{pmatrix} 1 & 0 \\ 0 & 1 \end{pmatrix} = 1$$

Therefore we have:

$$U^\dagger A U = \frac{1}{\sqrt{2}}\begin{pmatrix} 1 & -i \\ 1 & i \end{pmatrix}\begin{pmatrix} 2 & -i \\ i & 2 \end{pmatrix}\frac{1}{\sqrt{2}}\begin{pmatrix} 1 & 1 \\ i & -i \end{pmatrix}$$

$$= \frac{1}{2}\begin{pmatrix} 1 & -i \\ 1 & i \end{pmatrix}\begin{pmatrix} 2 & -i \\ i & 2 \end{pmatrix}\begin{pmatrix} 1 & 1 \\ i & -i \end{pmatrix}$$

$$= \frac{1}{2}\begin{pmatrix} 1 & -i \\ 1 & i \end{pmatrix}\begin{pmatrix} 2(1) - i(i) & 2(1) - i(-i) \\ i(1) + 2(i) & i(1) + 2(-i) \end{pmatrix} = \frac{1}{2}\begin{pmatrix} 1 & -i \\ 1 & i \end{pmatrix}\begin{pmatrix} 3 & 1 \\ 3i & -i \end{pmatrix}$$

$$= \frac{1}{2}\begin{pmatrix} 1(3) - i(3i) & 1(1) - i(-i) \\ 1(3) + i(3i) & 1(1) + i(-i) \end{pmatrix} = \frac{1}{2}\begin{pmatrix} 6 & 0 \\ 0 & 2 \end{pmatrix} = \begin{pmatrix} 3 & 0 \\ 0 & 1 \end{pmatrix} = \begin{pmatrix} \lambda_1 & 0 \\ 0 & \lambda_2 \end{pmatrix}$$

The Generalized Uncertainty Relation

Earlier we learned about the famous Hiesenberg uncertainty principle which relates the uncertainly in position to that of momentum via:

$$\Delta x \Delta p \geq \frac{\hbar}{2}$$

We now generalize this relation to any two arbitrary operators A and B. First, we recall that in a given state $|\psi\rangle$, the mean or expectation value of an operator O is found to be:

$$\langle O \rangle = \langle \psi | O | \psi \rangle$$

Now let's consider the standard deviation or uncertainty for two operators A and B:

$$(\Delta A)^2 = \left\langle (A - \langle A \rangle)^2 \right\rangle$$
$$(\Delta B)^2 = \left\langle (B - \langle B \rangle)^2 \right\rangle$$

Using $\langle O \rangle = \langle \psi | O | \psi \rangle$ we can rewrite these two equations as:

$$(\Delta A)^2 = \left\langle (A - \langle A \rangle)^2 \right\rangle = \left\langle \psi \left| (A - \langle A \rangle)^2 \right| \psi \right\rangle$$
$$(\Delta B)^2 = \left\langle (B - \langle B \rangle)^2 \right\rangle = \left\langle \psi \left| (B - \langle B \rangle)^2 \right| \psi \right\rangle$$

We now define the following kets:

$$|X\rangle = (A - \langle A \rangle) |\psi\rangle$$
$$|\Phi\rangle = (B - \langle B \rangle) |\psi\rangle$$

This allows us to write:

$$(\Delta A)^2 = \left\langle (A - \langle A \rangle)^2 \right\rangle = \left\langle \psi \left| (A - \langle A \rangle)^2 \right| \psi \right\rangle = \langle X | X \rangle$$
$$(\Delta B)^2 = \left\langle (B - \langle B \rangle)^2 \right\rangle = \langle \psi | (B - \langle B \rangle)^2 | \psi \rangle = \langle \Phi | \Phi \rangle$$

Now consider the product of these terms:

$$(\Delta A)^2 (\Delta B)^2 = \langle X | X \rangle \langle \Phi | \Phi \rangle$$

The Schwartz inequality tells us that:

$$\langle X | X \rangle \langle \Phi | \Phi \rangle \geq |\langle X | \Phi \rangle|^2 = \langle X | \Phi \rangle \langle \Phi | X \rangle$$

Remember that the inner product formed by a ket and a bra is just a complex number, so $|\langle X|\Phi\rangle|^2 = |z|^2 = zz^*$. For any complex number z, we have:

$$zz^* = Re(z)^2 + Im(z)^2 \geq Im(z)^2 = \left(\frac{z - \bar{z}}{2i} \right)^2$$

In this case we have:

$$\begin{aligned}
\langle X|\Phi\rangle &= \langle \psi|(A - \langle A \rangle)(B - \langle B \rangle)|\psi\rangle \\
&= \langle \psi|AB - A\langle B \rangle - \langle A \rangle B + \langle A \rangle\langle B \rangle|\psi\rangle \\
&= \langle \psi|AB|\psi\rangle - \langle \psi|A\langle B \rangle|\psi\rangle - \langle \psi|\langle A \rangle B|\psi\rangle + \langle \psi|\langle A \rangle\langle B \rangle|\psi\rangle
\end{aligned}$$

Now $\langle A \rangle$, the expectation value of an operator, is just a number. So we can pull it out of each term giving:

$$\begin{aligned}&\langle \psi | AB | \psi \rangle - \langle \psi | A | \psi \rangle \langle B \rangle - \langle A \rangle \langle \psi | B | \psi \rangle + \langle \psi | \langle A \rangle \langle B \rangle | \psi \rangle \\ &= \langle \psi | AB | \psi \rangle - \langle A \rangle \langle B \rangle - \langle A \rangle \langle B \rangle + \langle \psi | \langle A \rangle \langle B \rangle | \psi \rangle \\ &= \langle AB \rangle - 2 \langle A \rangle \langle B \rangle + \langle \psi | \langle A \rangle \langle B \rangle | \psi \rangle\end{aligned}$$

Now the expectation value of the mean, which is again just a number, is simply the mean back again, i.e.

$$\langle \psi | \langle A \rangle \langle B \rangle | \psi \rangle = \langle \langle A \rangle \langle B \rangle \rangle = \langle A \rangle \langle B \rangle$$

So, finally we have:

$$\begin{aligned}\langle X | \Phi \rangle &= \langle \psi | (A - \langle A \rangle)(B - \langle B \rangle) | \psi \rangle = \langle AB \rangle - 2 \langle A \rangle \langle B \rangle + \langle \psi | \langle A \rangle \langle B \rangle | \psi \rangle \\ &= \langle AB \rangle - 2 \langle A \rangle \langle B \rangle + \langle A \rangle \langle B \rangle = \langle AB \rangle - \langle A \rangle \langle B \rangle\end{aligned}$$

Following a similar procedure, we can show that:

$$\langle \Phi | X \rangle = \langle \psi | (B - \langle B \rangle)(A - \langle A \rangle) | \psi \rangle = \langle BA \rangle - \langle A \rangle \langle B \rangle$$

Putting everything together allows us to find an uncertainty relation for A and B. First we have:

$$(\Delta A)^2 (\Delta B)^2 = \langle X | X \rangle \langle \Phi | \Phi \rangle \geq |\langle X | \Phi \rangle|^2 = \langle X | \Phi \rangle \langle \Phi | X \rangle$$

Recalling that $zz^* = Re(z)^2 + Im(z)^2 \geq Im(z)^2 = \left(\frac{z - \bar{z}}{2i}\right)^2$, we set $z^* = \langle \Phi | X \rangle$. Then

$$\begin{aligned}(\Delta A)^2 (\Delta B)^2 \geq |\langle X | \Phi \rangle|^2 &= \left(\frac{\langle X | \Phi \rangle - \langle \Phi | X \rangle}{2i} \right) \\ &= \left(\frac{(\langle AB \rangle - \langle A \rangle \langle B \rangle) - (\langle BA \rangle - \langle A \rangle \langle B \rangle)}{2i} \right)^2 \\ &= \left(\frac{\langle AB \rangle - \langle A \rangle \langle B \rangle - \langle BA \rangle + \langle A \rangle \langle B \rangle}{2i} \right)^2 \\ &= \left(\frac{\langle AB \rangle - \langle BA \rangle}{2i} \right)^2 \\ &= \left(\frac{\langle AB - BA \rangle}{2i} \right)^2 \\ &= \left(\frac{\langle [A, B] \rangle}{2i} \right)^2\end{aligned}$$

Taking the square root of both sides gives us the generalized uncertainty relation, which applies to any two operators A and B.

Definition: The Uncertainty Relation

Given any two operators A, B:

$$\Delta A \Delta B \geq \frac{\langle [A, B] \rangle}{2i}$$

where $[A, B]$ is the commutator of the operators A and B.

e.g. **EXAMPLE 7.3**

Show that the commutation relation for the operators X and P leads to the Hiesenberg uncertainty principle.

✔ **SOLUTION**

Recalling that

$$[X, P] = i\hbar$$

we have:

$$\frac{[X, P]}{2i} = \frac{i\hbar}{2i} = \frac{\hbar}{2}$$

Therefore we obtain the famous Hiesenberg uncertainty principle:

$$\Delta X \Delta P \geq \frac{\hbar}{2}$$

Projection Operators

For an n-dimensional vector space V, let us denote a set of n orthonormal basis vectors by $|\Phi_1\rangle, |\Phi_2\rangle, \ldots, |\Phi_n\rangle$ where

$$\langle \Phi_i | \Phi_j \rangle = \delta_{ij}$$

A *subspace* W of V is a subset of V such that W is itself a vector space with respect to vector addition and scalar multiplication. We verify that W is a vector space in the usual way, but if W is a subset of V there are only two axioms that really need to be checked.

Definition: How to Check If a Subspace Is a Vector Space

Let V be a vector space and W be a subset of V, and let α be a scalar. W is also a vector space if:

1. The zero vector belongs to W
2. For every $|u\rangle$, $|v\rangle$ in W, $|u\rangle + |v\rangle$ and $\alpha|u\rangle$ also belong to W

Definition: Projection Operator

Suppose that a subspace is m-dimensional, where $m < n$ and is spanned by a set of orthonormal basis vectors $|u_1\rangle, |u_2\rangle, \ldots |u_m\rangle$. A *projection* operator P_m as given by:

$$P_m = \sum_{i=1}^{m} |u_i\rangle \langle u_i|$$

Projection operators have two important properties:

1. A projection operator P is Hermitian, i.e. $P = P^\dagger$
2. A projection operator P is equal to its own square, i.e. $P = P^2$

A projection operator can be formed from an individual basis vector that belongs to an orthonormal set, i.e. $P_i = |u_i\rangle \langle u_i|$ is a valid projection operator. This operator is obviously Hermitian, and is equal to its own square:

$$P_i^2 = (|u_i\rangle \langle u_i|)(|u_i\rangle \langle u_i|) = |u_i\rangle \langle u_i | u_i\rangle \langle u_i | = | u_i\rangle \langle u_i|$$

Since $\langle u_i | u_i\rangle = 1$, examining a projection operator constructed from an individual basis vector allows us to see what the fundamental action of a projection operator is. Consider the application of P_i on an arbitrary ket $|A\rangle$:

$$P_i |A\rangle = |u_i\rangle \langle u_i | A\rangle = (\langle u_i | A\rangle) |u_i\rangle$$

The operator P_i has "projected" the component of the ket $|A\rangle$ along the direction $|u_i\rangle$. This fits in with the completeness relation, which can now be expressed in the following way. We can represent the identity operator by a summation of projection operators, formed by the outer products of the basis vectors:

$$I = \sum_{i=1}^{n} |u_i\rangle \langle u_i| = \sum_{i=1}^{n} P_i$$

This makes sense, because it points a way to represent a vector in the given basis, i.e. we sum up the basis vectors multiplied by a weighting factor which is the component of that vector in that basis:

$$|A\rangle = I\,|A\rangle = \left(\sum_{i=1}^{n} |u_i\rangle\langle u_i|\right)|A\rangle = \sum_{i=1}^{n} |u_i\rangle \langle u_i | A\rangle = \sum_{i=1}^{n} (\langle u_i | A\rangle) |u_i\rangle$$

Is the sum of two projection operators a projector? Suppose that $|\Phi\rangle\langle\Phi|$ and $|\psi\rangle\langle\psi|$ are two projection operators. We then form their sum:

$$P = |\Phi\rangle \langle\Phi| + |\psi\rangle \langle\psi|$$

Then:

$$P^2 = (|\Phi\rangle\langle\Phi| + |\psi\rangle\langle\psi|)(|\Phi\rangle\langle\Phi| + |\psi\rangle\langle\psi|)$$
$$= |\Phi\rangle\langle\Phi|\Phi\rangle\langle\Phi| + |\Phi\rangle\langle\Phi|\psi\rangle\langle\psi| + |\psi\rangle\langle\psi|\Phi\rangle\langle\Phi| + |\psi\rangle\langle\psi|\psi\rangle\langle\psi|$$
$$= |\Phi\rangle\langle\Phi| + |\Phi\rangle\langle\Phi|\psi\rangle\langle\psi| + |\psi\rangle\langle\psi|\Phi\rangle\langle\Phi| + |\psi\rangle\langle\psi|$$

If $\langle\Phi|\psi\rangle = \langle\psi|\Phi\rangle = 0$, then we would have $P^2 = |\Phi\rangle\langle\Phi| + |\psi\rangle\langle\psi| = P$. Therefore we conclude that the sum of two projection operators P_1 and P_2 is a projection operator *if and only if* P_1 and P_2 are orthogonal. In fact we can extend this to any number of projection operators. The sum $P_1 + P_2 + \cdots + P_n$ is a projection operator if and only if the P_i, are all mutually orthogonal.

What about an arbitrary outer product, $|\Phi\rangle\langle\Phi|$? This operator is Hermitian. We form its square:

$$\left(|\Phi\rangle\langle\Phi|^2\right) = |\Phi\rangle\langle\Phi|\Phi\rangle\langle\Phi|$$

We see immediately that this outer product is equal to its own square if $\langle\Phi|\Phi\rangle = 1$, in other words $|\Phi\rangle$ is normalized

Definition: Forming a Projection Operator from an Arbitrary Ket $|\psi\rangle$

If a state $|\psi\rangle$ is normalized, then the operator formed from its bra and ket

$$|\psi\rangle\langle\psi|$$

is a projection operator.

e.g. EXAMPLE 7.4

Let P_1 and P_2 be two projection operators. Determine when their product, $P_1 P_2$, is also a projection operator.

✔ SOLUTION

We can find the conditions under which $P_1 P_2$ is a projection operator very easily. We check to see when it meets the two conditions that are required of a projection operator: whether the product is Hermitian and whether it is equal to its own square. We check the first condition. First note that since P_1 and P_2 are projection operators they are Hermitian, $P_1 = P_1^\dagger$ and $P_2 = P_2^\dagger$. Therefore:

$$(P_1 P_2)^\dagger = P_2^\dagger P_1^\dagger = P_2 P_1$$

So we see that a necessary condition for the product of two projection operators to be a projection operator is that:

$$P_2 P_1 = P_1 P_2$$

So we have $P_1 P_2 - P_2 P_1 = 0 = [P_1, P_2]$. Therefore P_1 and P_2 must commute. What about the square?

$$(P_1 P_2)^2 = (P_1 P_2)(P_1 P_2)$$

Again, if P_1 and P_2 commute, we have:

$$(P_1 P_2)^2 = (P_1 P_2)(P_1 P_2) = P_1 (P_2 P_1) P_2 = P_1 (P_1 P_2) P_2 = P_1^2 P_2^2 = P_1 P_2$$

So once again we see that the necessary and sufficient condition for a product of two projection operators to be itself a projection operator is that the two operators commute.

EXAMPLE 7.5

e.g.

Let $\{|a\rangle, |b\rangle\}$ be an orthonormal two-dimensional basis and let an operator A be given by:

$$A = |a\rangle\langle a| - i|a\rangle\langle b| + i|b\rangle\langle a| - |b\rangle\langle b|$$

(a) Is A a projection operator?

(b) Find the matrix representation of A and $Tr(A)$.

(c) Find the eigenvalues and eigenvectors of A.

SOLUTION

(a) First we find $A^\dagger$:

$$\begin{aligned} A^\dagger &= (|a\rangle \langle a| - i|a\rangle \langle b| + i|b\rangle \langle a| - |b\rangle \langle b|)^\dagger \\ &= |a\rangle \langle a| + i|b\rangle \langle a| - i|a\rangle \langle b| - |b\rangle \langle b| \\ &= |a\rangle \langle a| - i|a\rangle \langle b| + i|b\rangle \langle a| - |b\rangle \langle b| = A \end{aligned}$$

$\Rightarrow$ A is Hermitian. Now:

$$\begin{aligned} A^2 &= (|a\rangle \langle a| - i|a\rangle \langle b| + i|b\rangle \langle a| - |b\rangle \langle b|)(|a\rangle \langle a| - i|a \langle b| + i|b\rangle \langle a| - |b\rangle \langle b|) \\ &= |a\rangle \langle a| (|a\rangle \langle a|) + |a\rangle \langle a| (-i\,|a\rangle \langle b|) - i\,|a\rangle \langle b| (i\,|b\rangle \langle a|) - i\,|a\rangle \langle b| (-|b\rangle \langle b|) \\ &\quad + i\,|b\rangle \langle a| (|a\rangle \langle a|) + i\,|b\rangle \langle a| (-i\,|a\rangle \langle b|) - |b\rangle \langle b| (i\,|b\rangle \langle a|) - |b\rangle \langle b| (-|b\rangle \langle b|) \\ &= |a\rangle \langle a| - i\,|a\rangle \langle b| + |a\rangle \langle a| + i\,|a\rangle \langle b| + |b\rangle \langle b| + i\,|b\rangle \langle a| - i\,|b\rangle \langle a| + |b\rangle \langle b| \\ &= 2\,|a\rangle \langle a| + 2\,|b\rangle \langle b| \end{aligned}$$

Although A is Hermitian, since $A \neq A^2$, A is not a projection operator.

(b)

$$A = |a\rangle\langle a| - i\,|a\rangle\langle b| + i\,|b\rangle\langle a| - |b\rangle\langle b|$$

and so

$$\begin{aligned}\langle a|A|a\rangle &= \langle a|\left(|a\rangle\langle a| - i\,|a\rangle\langle b| + i\,|b\rangle\langle a| - |b\rangle\langle b|\right)|a\rangle \\ &= \langle a|a\rangle\langle a|a\rangle - i\langle a|a\rangle\langle b|a\rangle + i\langle a|b\rangle\langle a|a\rangle - \langle a|b\rangle\langle b|a\rangle \\ &= 1\end{aligned}$$

Similarly we find that:

$$\langle a|A|b\rangle = -i$$
$$\langle b|A|a\rangle = i$$
$$\langle b|A|b\rangle = -1$$

So the matrix representation of A is:

$$A = \begin{pmatrix} \langle a|A|a\rangle & \langle a|A|b\rangle \\ \langle b|A|a\rangle & \langle b|A|b\rangle \end{pmatrix} = \begin{pmatrix} 1 & -i \\ i & -1 \end{pmatrix}$$

The trace is the sum of the diagonal elements, and so: $Tr(A) = 1 - 1 = 0$

(c) Solving $\det|A - \lambda I| = 0$ we find:

$$\begin{aligned}0 &= \det\left|\begin{pmatrix} 1 & -i \\ i & -1 \end{pmatrix} - \lambda\begin{pmatrix} 1 & 0 \\ 0 & 1 \end{pmatrix}\right| = \det\left|\begin{pmatrix} 1-\lambda & -i \\ i & -1-\lambda \end{pmatrix}\right| \\ &= (1-\lambda)(-1-\lambda) + i^2\end{aligned}$$

This leads to the characteristic equation:

$$-2 + \lambda^2 = 0$$

So the eigenvalues of A are:

$$\lambda_{1,2} = \pm\sqrt{2}$$

To find the first eigenvector corresponding to $\lambda = \sqrt{2}$ we solve:

$$\begin{pmatrix} l & -i \\ i & -l \end{pmatrix}\begin{pmatrix} \alpha \\ \beta \end{pmatrix} = \sqrt{2}\begin{pmatrix} \alpha \\ \beta \end{pmatrix}$$

This leads to:

$$\alpha - i\beta = \sqrt{2}\alpha, \Rightarrow \beta = i\left(\sqrt{2} - 1\right)\alpha$$

and so we can write:

$$\begin{pmatrix} \alpha \\ \beta \end{pmatrix} = \begin{pmatrix} \alpha \\ i\left(\sqrt{2}-1\right)\alpha \end{pmatrix}$$

Normalizing we find:

$$1 = \begin{pmatrix} \alpha^* & -i(\sqrt{2}-1)\alpha^* \end{pmatrix} \begin{pmatrix} \alpha \\ -i\left(\sqrt{2}-1\right)\alpha \end{pmatrix}$$

and so the first eigenvector of A is:

$$|1\rangle = \begin{pmatrix} \frac{1}{2} \\ \frac{-i\left(\sqrt{2}-1\right)}{2} \end{pmatrix}$$

We leave it as an exercise to show that the other eigenvector of A is:

$$|2\rangle = \begin{pmatrix} \frac{1}{\sqrt{12}} \\ \frac{-i\left(\sqrt{2}+1\right)}{\sqrt{12}} \end{pmatrix}$$

Functions of Operators

An ordinary function can be expanded in a Taylor series:

$$f(x) = \sum_{n=0}^{\infty} a_n x^n$$

In a similar manner, we can define a function of an operator, $F(A)$:

$$F(A) = \sum_{n=0}^{\infty} a_n A^n$$

Note: In general, the expansion coefficients a_n can be complex.

e.g. EXAMPLE 7.6

Given that $[A, B^n] = n[A, B]B^{n-1}$, show that $[A, F(B)] = [A, B]F'(B)$, where $F'(B)$ is the ordinary derivative of F with respect to B.

✔ SOLUTION

This is easy to show by expanding $F(B)$ in a power series:

$$[A, F(B)] = [A, \sum_{n=0}^{\infty} b_n B^n]$$

Using the fact that $[A, B + C] = [A, B] + [A, C]$,

$$\left[A, \sum_{n=0}^{\infty} b_n B^n\right] = \sum_{n=0}^{\infty} b_n[A, B^n] = \sum_{n=0}^{\infty} b_n n[A, B]B^{n-1} = [A, B]\sum_{n=0}^{\infty} b_n n B^{b-1}$$

Given a power series expansion $g(x) = \sum a_n x^n$, then $g'(x) = \sum a_n n x^{n-1}$, and so $\sum_{n=0}^{\infty} b_n n B^{n-1} = F'(B)$. Therefore we have:

$$[A, F(B)] = [A, B]F'(B)$$

Generalization to Continuous Spaces

The bra-ket and linear algebra formalism that we have been developing can be generalized to the continuous case. There corresponds a ket $|\psi\rangle$ to every function $\psi(x)$ that belongs to the space of wavefunctions. To quickly review, this space is defined by functions that:

(a) Are defined everywhere, are continuous, and infinitely differentiable

(b) Are square-integrable, i.e. $\int |\psi|^2 \, dx < \infty$

The inner product between a bra and a ket is found by:

$$\langle \Phi | \psi \rangle = \int \Phi^* \psi \, dx$$

We recall that two ways that we can represent states include the position representation and the momentum representation, where the momentum representation is the Fourier transform of the position space wavefunction. Using the bra-ket formalism, the correspondence between a ket and its associated wavefunction is as follows:

$$\psi(x) = \langle x | \psi \rangle$$

$$\overline{\psi}(p) = \langle p|\psi\rangle$$

The bases of position and momentum space are defined as follows. For position space, we have:

$$\xi_{x_o}(x) = \delta(x - x_o)$$

where δ is the Dirac delta function. For momentum space:

$$v_{p_o}(x) = \frac{1}{\sqrt{2\pi\hbar}} e^{i\frac{v_o x}{\hbar}}$$

It should be noted that these basis functions do not belong to the space of wavefunctions. For position space, orthonormality is expressed in the following way:

$$\langle x | x' \rangle = \delta(x - x')$$

while completeness becomes:

$$1 = \int |x\rangle\langle x|\, dx$$

Similar results hold for the momentum space base kets:

$$\langle p | p' \rangle = \delta(p - p'),\ 1 = \int |p\rangle\langle p|\, dp$$

The inner product between the position and momentum space kets is:

$$\langle x | p \rangle = \langle p | x \rangle^* = \frac{1}{\sqrt{2\pi\hbar}} e^{i\frac{p_o x}{\hbar}}$$

Using the completeness relation, we can show that the position space and momentum space wavefunctions are related by the Fourier transform. For example:

$$\psi(x) = \langle x|\psi\rangle = \langle x|1|\psi\rangle = \langle x| \left(\int |p\rangle\langle p|dp \right) |\psi\rangle$$

$$= \int \langle x|p\rangle\langle p|\psi\rangle dp$$

$$= \frac{1}{\sqrt{2\pi\hbar}} \int e^{i\frac{px}{\hbar}} \overline{\psi}(p)\, dp$$

These ideas can be generalized to an arbitrary continuous basis. We recall that a ket can be expanded in terms of a basis by writing out a summation that includes the components of that ket in the given basis, i.e.

$$|\psi\rangle = \sum c_i |u_i\rangle$$

where $c_i = \langle u_i | \psi \rangle$. In an arbitrary continuous basis $|\alpha\rangle$, this becomes:

$$|\psi\rangle = \int c(\alpha) |\alpha\rangle \, d\alpha$$

The "components" in the given basis, $c(\alpha)$ are now *functions* of α. Closure is written as:

$$1 = \int |\alpha\rangle \langle\alpha| \, d\alpha$$

So, using $c(\alpha) = \langle \alpha | \psi \rangle$, we see that we obtain the representation of $|\Psi$ in the basis $|\alpha\rangle$ using the closure relation in the usual way:

$$|\psi\rangle = 1 |\psi\rangle = \left(\int |\alpha\rangle \langle\alpha| \, d\alpha \right) |\psi\rangle = \int |\alpha\rangle \langle\alpha | \psi \rangle \, d\alpha = \int c(\alpha) |\alpha\rangle \, d\alpha$$

The inner product of a ket and bra, which in the discrete case is obtained from the summation:

$$\langle \Phi | \psi \rangle = \sum b_i^* c_i$$

is generalized to the integral:

$$\langle \Phi | \psi \rangle = \int b(\alpha)^* c(\alpha) \, dx$$

We can also extend the representation of an operator to a continuous space in the obvious way; we represent an operator by a matrix that has an infinite number of rows and columns. So, if a component of an operator $\hat{A}$ is given by $A(\alpha, \alpha')$, we have:

$$\hat{A} \Rightarrow \begin{pmatrix} \ddots & \vdots & \iddots \\ \cdots & (A\alpha, \alpha') & \cdots \\ \iddots & \cdots & \ddots \end{pmatrix}$$

The expectation value of an operator is found from integration:

$$\langle \psi | A | \psi \rangle = \int_{-\infty}^{\infty} \psi(x)^* A \psi(x) \, dx$$

EXAMPLE 7.7

Show that the expansion of a ket in a continuous orthonormal basis $|\alpha\rangle$ is unique.

SOLUTION

We can expand any ket in a continuous basis by application of the closure relation:

$$|\psi\rangle = 1 |\psi\rangle = \int |\alpha\rangle \langle\alpha| \psi\rangle \, d\alpha = \int c(\alpha) |\alpha\rangle \, d\alpha$$

Let us assume that this expansion is not unique, so that:

$$|\psi\rangle = \int c(\alpha)\,|\alpha\rangle\,d\alpha \quad \text{and} \quad |\psi\rangle = \int d(\alpha)\,|\alpha\rangle\,d\alpha$$

Subtraction yields:

$$|\psi\rangle - |\psi\rangle = 0 = \int c(\alpha)\,|\alpha\rangle\,d\alpha - \int d(\alpha)\,|\alpha\rangle\,d\alpha = \int [c(\alpha) - d(\alpha)]\,|\alpha\rangle\,d\alpha$$

Since the $|\alpha\rangle$ form a basis, orthonormality tells us that $\langle\alpha'\,|\alpha\rangle = \delta\left(\alpha - \alpha'\right)$. Now we take the inner product with $\langle\alpha'|$:

$$\begin{aligned} 0 &= \langle\alpha'| \int [c(\alpha) - d(\alpha)]|\alpha\rangle d\alpha \\ &= \int [c(\alpha) - d(\alpha)]\langle\alpha'\,|\alpha\rangle\,d\alpha \\ &= \int [c(\alpha) - d(\alpha)]\,\delta\left(\alpha - \alpha'\right)d\alpha \end{aligned}$$

The sampling property of the δ function tells that this is:

$$\int [c(\alpha) - d(\alpha)]\,\delta\left(\alpha - \alpha'\right)d\alpha = c\left(\alpha'\right) - d\left(\alpha'\right)$$

But this is equal to zero. So, we have found that $c(\alpha') = d(\alpha')$, therefore the expansion is unique.

This generalization to continuous spaces allows us to connect wave and matrix mechanics. States, which in general are kets, can be represented as ordinary functions of position or momentum. Operators become mathematical actions on those functions. For example, momentum is represented by:

$$\hat{p} \rightarrow -i\hbar\frac{d}{dx}$$

EXAMPLE 7.8 e.g.

Let $\psi(x) = \left(\frac{\pi}{a}\right)^{-1/4} e^{-ax^2/2}$. Show that $\Delta x \Delta p = \hbar/2$

SOLUTION ✓

First we find $\langle x\rangle$ and $\langle x^2\rangle$:

$$\langle x\rangle = \int_{-\infty}^{\infty} x\,|\psi|^2\,dx = \sqrt{\frac{a}{\pi}}\int_{-\infty}^{\infty} xe^{-ax^2}\,dx = 0$$

To see this, let $u = -ax^2$. Then $du = -2ax\,dx$. Then

$$\int_{-\infty}^{\infty} xe^{-ax^2}\,dx = -\frac{1}{2a}\int e^u\,du = \frac{1}{2a}\,e^u$$

But, with $u = -ax^2$, we see that $u \to -\infty$ at both limits, so the integral is zero. Now,

$$\langle x^2\rangle = \int_{-\infty}^{\infty} x^2\,|\psi|^2\,dx = \sqrt{\frac{a}{\pi}}\int_{-\infty}^{\infty} x^2 e^{-ax^2}\,dx = 2\sqrt{\frac{a}{\pi}}\int_0^{\infty} x^2 e^{-ax^2}\,dx$$

This integral can be found from a table to be $\frac{\Gamma(1/2)}{4a^{3/2}}$ where Γ is the *factorial function*. You can look up the fact that $\Gamma(1/2) = \sqrt{\pi}$, and so:

$$\langle x^2\rangle = 2\sqrt{\frac{a}{\pi}}\int_0^{\infty} x^2 e^{-ax^2}\,dx = 2\sqrt{\frac{a}{\pi}}\frac{\Gamma\left(1/2\right)}{4a^{3/2}} = 2\sqrt{\frac{a}{\pi}}\frac{\sqrt{\pi}}{4a^{3/2}} = \frac{1}{2a}$$

$$\Rightarrow \Delta x = \sqrt{\langle x^2\rangle - \langle x\rangle^2} = \sqrt{\frac{1}{2a}}$$

$$\langle p\rangle = \int_{-\infty}^{\infty}\psi^* p\psi\,dx = \int_{-\infty}^{\infty}\left(\frac{\pi}{a}\right)^{-1/4} e^{-ax^2/2}\left(-i\hbar\frac{d}{dx}\right)\left(\left(\frac{\pi}{a}\right)^{-1/4} e^{-ax^2/2}\right)dx$$

$$= \sqrt{\frac{a}{\pi}}\,(-i\hbar)\int_{-\infty}^{\infty} e^{-ax^2/2}\frac{d}{dx}\left(e^{-ax^2/2}\right)dx$$

$$= \sqrt{\frac{a}{\pi}}\,(-i\hbar)\int_{-\infty}^{\infty} e^{-ax^2/2}\left[\left(-2a\frac{x}{2}\right)e^{-ax^2/2}\right]dx$$

$$= i\hbar\frac{a^{3/2}}{\sqrt{\pi}}\int_{-\infty}^{\infty} e^{-ax^2/2}\left[xe^{-ax^2/2}\right]dx$$

$$= i\hbar\frac{a^{3/2}}{\sqrt{\pi}}\int_{-\infty}^{\infty} xe^{-ax^2}\,dx$$

We saw above that $\int_{-\infty}^{\infty} xe^{-ax^2}\,dx = 0$, so $\langle p\rangle = 0$.

Now, $p^2 = \left(-i\hbar\frac{d}{dx}\right)^2 = -\hbar^2\frac{d^2}{dx^2}$. So:

$$p^2\psi = -\hbar^2\frac{d^2}{dx^2}\psi = -\hbar^2\frac{d^2}{dx^2}\left(\frac{\pi}{a}\right)^{-1/4} e^{-ax^2/2}$$

$$= -\hbar^2\left(\frac{\pi}{a}\right)^{-1/4}\frac{d}{dx}\left(-axe^{-ax^2/2}\right)$$

$$= -\hbar^2 \left(\frac{\pi}{a}\right)^{-1/4} \left[-ae^{-ax^2/2} - ax\left(-axe^{-ax^2/2}\right)\right]$$
$$= -\hbar^2 \left(\frac{\pi}{a}\right)^{-1/4} \left[-ae^{-ax^2/2} + a^2x^2e^{-ax^2/2}\right]$$
$$= -\hbar^2 \left(\frac{\pi}{a}\right)^{-1/4} \left[a^2x^2 - a\right]e^{-ax^2/2}$$

So we obtain:

$$\langle p\rangle^2 = \int_{-\infty}^{\infty} \psi^* p^2 \psi\, dx = \int_{-\infty}^{\infty} \left(\frac{\pi}{a}\right)^{-1/4} e^{-ax^2/2}\left[-\hbar^2\left(\frac{\pi}{a}\right)^{-1/4}\left[a^2x^2 - a\right]e^{-ax^2/2}\right] dx$$
$$= -\sqrt{\frac{a}{\pi}}\hbar^2 \int_{-\infty}^{\infty} a^2x^2e^{-ax^2} - ae^{-ax^2}\, dx$$
$$= -\sqrt{\frac{a}{\pi}}\hbar^2 \int_{-\infty}^{\infty} a^2x^2e^{-ax^2}\, dx - \sqrt{\frac{a}{\pi}}\hbar^2 \int_{-\infty}^{\infty} ae^{-ax^2}\, dx$$
$$= -\sqrt{\frac{a}{\pi}}\hbar^2\left(a^2\frac{\Gamma(1/2)}{2a^{3/2}}\right) - \sqrt{\frac{a}{\pi}}\hbar^2\left(a\sqrt{\frac{\pi}{a}}\right)$$
$$= -\sqrt{\frac{a}{\pi}}\hbar^2a^2\frac{\sqrt{\pi}}{2a^{3/2}} - \hbar^2 a$$
$$= -\frac{\hbar^2 a}{2} - \hbar^2 a = \frac{\hbar^2 a}{2}$$
$$\Rightarrow \Delta p = \sqrt{\langle p^2\rangle - \langle p\rangle^2} = \sqrt{\frac{a}{2}}\hbar$$

Finally we can write the uncertainty relation:

$$\Delta x\,\Delta p = \sqrt{\frac{1}{2a}}\sqrt{\frac{a}{2}}\hbar = \frac{\hbar}{2}$$

EXAMPLE 7.9 *e.g.*

Let an operator $\hat{O} = -i\dfrac{d}{d\varphi}$ where φ is the usual azimuthal angle in spherical coordinates.

(a) Find the eigenfunctions $f(\varphi)$ and eigenvalues λ subject to the constraint that $f(0) = f(2\pi) = \dfrac{1}{\sqrt{2\pi}}$ and that λ must be positive.

(b) Let $\hat{\varphi}$ act as the position operator in cartesian coordinates, i.e. $\hat{\varphi} f = \varphi f$. Find $[\hat{O}, \hat{\varphi}]$.

(c) Is $\hat{O}$ Hermitian?

✔ SOLUTION

(a) In a continuous space the eigenvalue equation for an operator $\hat{O}$ can be written as:

$$\hat{O} f = \lambda f$$

In this case we have the familiar equation:

$$-i\frac{df}{d\varphi} = \lambda f$$

Rearranging terms and integrating:

$$\frac{df}{f} = i\lambda d\varphi \Rightarrow ln = i\lambda\varphi + C$$

where C is the constant of integration. Taking the exponential of both sides:

$$f = Ce^{i\lambda\varphi}$$

To find C, we apply the normalization condition $1 = \int f^* f \, d\varphi$:

$$1 = \int_0^{2\pi} f^* f \, d\varphi = \int_0^{2\pi} \left(Ce^{i\lambda\varphi}\right)^* Ce^{i\lambda\varphi} \, d\varphi$$

$$= \int_0^{2\pi} C^* e^{-i\lambda\varphi} Ce^{i\lambda\varphi} \, d\varphi = \int_0^{2\pi} |C|^2 \, d\varphi$$

$$= C^2 \int_0^{2\pi} d\varphi = C^2\varphi \,|_0^{2\pi} = C^2 2\pi$$

$$\Rightarrow C = \frac{1}{\sqrt{2\pi}}$$

A quick check shows that λ is in fact the eigenvalue for $\hat{O}$:

$$\hat{O} f = -i\frac{d}{d\varphi}\left(\frac{1}{\sqrt{2\pi}}e^{i\lambda\varphi}\right) = -i\,(i\lambda)\,\frac{1}{\sqrt{2\pi}}e^{i\lambda\varphi} = \lambda\frac{1}{\sqrt{2\pi}}e^{i\lambda\varphi} = \lambda f$$

(b) To find the eigenvalues, we first apply Euler's formula:

$$f = \frac{1}{\sqrt{2\pi}} e^{i\lambda\varphi} = \frac{1}{\sqrt{2\pi}} [\cos \lambda\varphi + i \sin \lambda\varphi]$$

The boundary condition at $\varphi = 0$ is automatically satisfied:

$$f\,(0) = \frac{1}{\sqrt{2\pi}} [\cos\,(0) + i \sin(0)] = \frac{1}{\sqrt{2\pi}} [1 + i0] = \frac{1}{\sqrt{2\pi}}$$

At $\varphi = 2\pi$ we have:

$$f(2\pi) = \frac{1}{\sqrt{2\pi}} [\cos \lambda 2\pi + i \sin \lambda 2\pi]$$

The function f is complex and has real and imaginary parts. For $f(2\pi) = \frac{1}{\sqrt{2\pi}}$ to be satisfied, the imaginary part of f which is given by the sin function must be zero. Therefore we must have:

$$\sin \lambda 2\pi = 0$$

But let's ignore that for the moment, and focus on the *cos* term. To have $f(2\pi) = \frac{1}{\sqrt{2\pi}} = \frac{1}{\sqrt{2\pi}} \cos \lambda 2\pi$, the condition $\cos \lambda 2\pi = +1$ must be met. Notice that:

$$\cos(0) = +1$$

$$\cos(\pi) = -1$$

$$\cos(2\pi) = +1$$

$$\cos(3\pi) = -1$$

$$\cos(4\pi) = +1$$

So we see that the argument to the *cos* function must be an even integral multiple of π. Since we have $\cos \lambda 2\pi$, this is automatically satisfied for any λ that is an integer. For example:

$$\lambda = 0 \Rightarrow [\cos(0)(2\pi)] = \cos(0) = +1$$

$$\lambda = 1 \Rightarrow [\cos(1)(2\pi)] = \cos(2\pi) = +1$$

$$\lambda = 2 \Rightarrow [\cos(2)(2\pi)] = \cos(4\pi) = +1$$

$$\lambda = 3 \Rightarrow [\cos(3)(2\pi)] = \cos(6\pi) = +1$$

With the further requirement that λ be strictly positive, we omit $\lambda = 0$ and find that the eigenvalues of $\hat{O}$ are:

$$\lambda = 1, 2, 3, \ldots \cdot$$

We see that the condition $\sin \lambda 2\pi = 0$ is automatically satisfied:

$$\lambda = 1 \Rightarrow \sin[(1)2\pi] = \sin[2\pi] = 0$$
$$\lambda = 2 \Rightarrow \sin[(2)2\pi] = \sin[4\pi] = 0$$
$$\lambda = 3 \Rightarrow \sin[(3)2\pi] = \sin[6\pi] = 0$$

and so on.

To find $[\hat{O}, \hat{\varphi}]$, we apply the commutator to a test function g. Remember the operator $\hat{\varphi}$ results in multiplication, so $\hat{\varphi} g = \varphi g$

$$\begin{aligned}
[\hat{O}, \hat{\varphi}]g &= (\hat{O}\hat{\varphi} - \hat{\varphi}\hat{O})g \\
&= \hat{O}\left(\hat{\phi} g\right) - \hat{\varphi}\left(\hat{O} g\right) \\
&= -i\frac{d}{d\varphi}(\varphi g) - \varphi\left(-i\frac{dg}{d\varphi}\right) \\
&= -i\frac{d}{d\varphi}(\varphi g) + i\varphi\frac{dg}{d\varphi} \\
&= -ig - i\varphi\frac{dg}{d\varphi} + i\varphi\frac{dg}{d\varphi} \\
&= -ig \\
[\hat{O}, \hat{\varphi}]g &= -ig, \Rightarrow [\hat{O}, \hat{\varphi}] = -i
\end{aligned}$$

(c) To check if $\hat{O}$ is Hermitian, we need to see if $\langle g|\hat{O} f\rangle = \langle g\hat{O}|f\rangle$:

$$\langle g|\hat{O} f\rangle = \int_0^{2\pi} g^* \left(-i\frac{df}{d\varphi}\right) d\varphi$$

Integrating by parts, we find that:

$$\langle g|\hat{O} f\rangle = \int_0^{2\pi} g^* \left(-i\frac{df}{d\varphi}\right) d\varphi = fg|_0^{2\pi} - \int_0^{2\pi} (-i)\,\frac{dg^*}{d\varphi} f\, d\varphi$$

Since $f(2\pi) = f(0)$ and $g(2\pi) = g(0)$, the boundary term vanishes and we have:

$$\begin{aligned}\langle g|\hat{O} f\rangle &= -\int_0^{2\pi} (-i)\frac{dg^*}{d\varphi} f\,d\varphi \\ &= \int_0^{2\pi} i\frac{dg^*}{d\varphi} f\,d\varphi \\ &= \int_0^{2\pi} \left(-i\frac{dg}{d\varphi}\right)^* f\,d\varphi \\ &= \langle g|\hat{O}|f\rangle, \Rightarrow \hat{O} \text{ is Hermitian}\end{aligned}$$

Quiz

1. Consider the Pauli matrix $\sigma_x = \begin{pmatrix} 0 & 1 \\ 1 & 0 \end{pmatrix}$. Use the eigenvectors of this matrix to construct a unitary transformation such that:

$$U^\dagger \sigma_x U = \begin{pmatrix} \lambda_1 & 0 \\ 0 & \lambda_2 \end{pmatrix}$$

where $\lambda_{1,2}$ are the eigenvalues of σ_x.

2. For the following matrix:

$$X = \begin{pmatrix} 2 & i & 0 \\ -i & -2 & 0 \\ 0 & 0 & 0 \end{pmatrix}$$

(a) Show that the eigenvalues are $(2, i, -i)$.

(b) Find the eigenvectors of the matrix and normalize them.

(c) Construct a transformation matrix S, find its inverse and show that it diagonalizes X.

3. Using the completeness relation for $|x\rangle$ and $\langle x|p\rangle$ find a representation $\overline{\psi}(p)$ in terms $\psi'(x)$.

4. If A and B are operators, show that $e^A e^B = e^{A+B} e^{[A,B]/2}$ Baker-Campbell-Hausdorff formula

5. If A is a Hermitian operator, under what conditions with $F(A)$ be Hermitian?

6. Show that
   (a) $\left[\hat{p}, \hat{x}^n\right] = -i\hbar n\hat{x}^{n-1}$,
   (b) $\left[\hat{p}, f\left(\hat{x}\right)\right] = -i\hbar\frac{df}{dx}$

7. Let an operator

$$\hat{A} = \frac{\partial}{\partial x} \text{ and } \hat{B} = -i\frac{\partial}{\partial x} + \frac{\partial}{\partial y}.$$

   (a) Show $[\hat{A}, \hat{B}] = 0$ by applying the commutator to a test function $f(x, y)$.
   (b) Show that $f(x, y) = e^{-x+y}$ is an eigenfunction of $\hat{A}$ with eigenvalue $\lambda = -1$. Is it an eigenfunction of $\hat{B}$? If so, what is the eigen value?

8. Suppose that $\{|+\rangle, |-\rangle\}$, Form an orthonormal basis for a two-dimensional vector space.
   (a) Find the matrix representations of the operators $|+\rangle\langle+|$ and $|-\rangle\langle-|$.
   (b) Are these operators projection operators?
   (c) Let $|\psi\rangle = 2i|+\rangle - 4|-\rangle$. Is $|\psi\rangle$ normalized? If not, normalize the state. Find the representation of $|\psi\rangle$ as a column vector.
   (d) Find the action of $|+\rangle\langle+|$ and $|-\rangle\langle-|$ on the normalized state $|\psi\rangle$, using both the outer product notation and by the action of the matrices representing these operators.

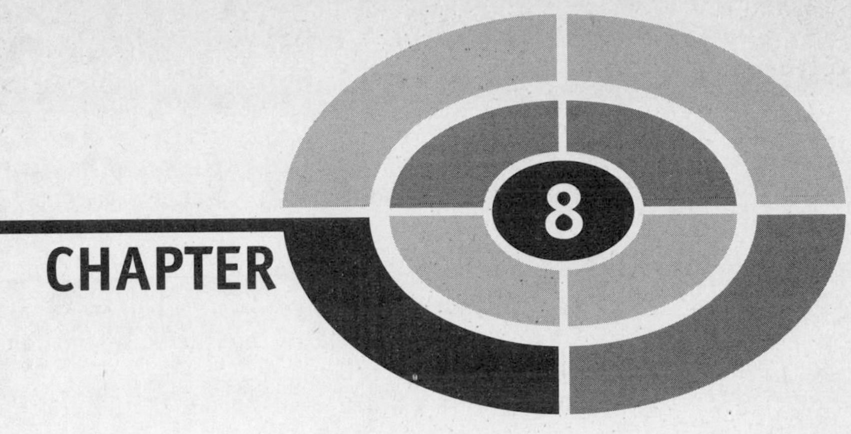

The Foundations of Quantum Mechanics

The postulates of quantum mechanics are a mathematical prescription for using the theory to predict the results of experiments. These postulates can be presented in terms of the state vector, or using the density operator. We discuss both topics and also cover material related to the representation of composite quantum states and the Heisenberg picture. We also introduce a useful tool for two-level systems called the Bloch vector.

The Postulates of Quantum Mechanics

In this section the postulates of quantum mechanics are described in terms of the state vector. This formalism works for an isolated physical system.

Postulate 1: States of physical systems are represented by vectors

The state of a physical system is described by a state vector $|\psi\rangle$ that belongs to a complex Hilbert space. The superposition principle holds, meaning that if $|\phi_1\rangle, |\phi_2\rangle, \ldots, |\phi_n\rangle$ are kets belonging to the Hilbert space, the linear combination

$$|\chi\rangle = \alpha_1 |\phi_1\rangle + \alpha_2 |\phi_2\rangle + \cdots + \alpha_n |\phi_n\rangle$$

is also a valid state that belongs to the Hilbert space. States are normalized to conform to the Born probability interpretation, meaning

$$\langle \psi | \psi \rangle = 1$$

If a state is formed from a superposition of other states, normalization implies that the squares of the expansion coefficients must add up to 1:

$$\langle \chi | \chi \rangle = |\alpha_1|^2 + |\alpha_2|^2 + \cdots + |\alpha_n|^2 = 1$$

TWO-LEVEL SYSTEM EXAMPLE: STATE VECTORS

A quantum-bit or *qubit* is a quantum state that is the basic unit of information in a quantum computer. The state space is two-dimensional with orthonormal basis vectors $\{|0\rangle, |1\rangle\}$:

$$|0\rangle = \begin{pmatrix} 1 \\ 0 \end{pmatrix}, \quad |1\rangle = \begin{pmatrix} 0 \\ 1 \end{pmatrix}$$

An arbitrary state $|\psi\rangle$ that belongs to this vector space can be written as a superposition of the basis states

$$|\psi\rangle = \alpha |0\rangle + \beta |1\rangle$$

where α and β are generally complex numbers. Normalization of the state implies that

$$|\alpha|^2 + |\beta|^2 = 1$$

The *Born rule* tells us that the coefficients in the expansion are related to the probability of obtaining a given measurement as follows:

The probability that measurement finds the system in state $|0\rangle$

$$|\langle 0 | \psi \rangle|^2 = |\alpha|^2$$

The probability that measurement finds the system in state $|1\rangle$

$$|\langle 1 | \psi \rangle|^2 = |\beta|^2$$

Postulate 2: Physical observables are represented by operators

Physically measureable quantities like energy and momentum are known as *observables*. Mathematacally, an observable is a Hermitian operator that acts on state vectors in the Hilbert space. The eigenvectors of Hermitian operators form an orthonormal basis of the state space for the system.

TWO-LEVEL SYSTEM EXAMPLE: THE PAULI OPERATORS

The Pauli operators X, Y, and Z correspond to the measurement of spin (intrinsic angular momentum of a particle) along the x, y, and z axes, respectively. They act on the basis states $\{|0\rangle, |1\rangle\}$ as follows

$$X|0\rangle = |1\rangle, \quad X|1\rangle = |0\rangle$$

$$Y|0\rangle = -i|1\rangle, \quad Y|1\rangle = i|0\rangle$$

$$Z|0\rangle = |0\rangle, \quad Z|1\rangle = -|1\rangle$$

We see that the basis states are eigenvectors of Z. Let's remind ourselves how to construct the matrix representation of an operator.

Reminder: The Matrix Representation of an Operator

In a discrete basis $|u_i\rangle$ an operator is represented by a set of numbers

$$A_{ij} = \langle u_i|A|u_j\rangle$$

These numbers are arranged into a square matrix such that A_{ij} is the element at the ith row and jth column.

EXAMPLE 8.1

e.g.

The matrix representation of the Pauli operators can be derived using the $\{|0\rangle, |1\rangle\}$ basis (remember, we could equally well choose a different basis, giving a different matrix representation). In this basis we form the matrix representation of an operator A by calculating

$$\begin{pmatrix} \langle 0| A |0\rangle & \langle 0| A |1\rangle \\ \langle 1| A |0\rangle & \langle 1| A |1\rangle \end{pmatrix}$$

The basis is orthonormal, and so

$$\langle 0|0\rangle = \langle 1|1\rangle = 1$$
$$\langle 0|1\rangle = \langle 1|0\rangle = 0$$

Using the action of the X operator described earlier we find

$$X = \begin{pmatrix} \langle 0|X|0\rangle & \langle 0|X|1\rangle \\ \langle 1|X|0\rangle & \langle 1|X|1\rangle \end{pmatrix} = \begin{pmatrix} \langle 0|1\rangle & \langle 0|0\rangle \\ \langle 1|1\rangle & \langle 1|0\rangle \end{pmatrix} = \begin{pmatrix} 0 & 1 \\ 1 & 0 \end{pmatrix}$$

A similar procedure applied to the other operators shows that

$$Y = \begin{pmatrix} 0 & -i \\ i & 0 \end{pmatrix}, \quad Z = \begin{pmatrix} 1 & 0 \\ 0 & -1 \end{pmatrix}$$

WRITING AN OPERATOR IN TERMS OF OUTER PRODUCTS

Consider two kets $|\psi\rangle$ and $|\phi\rangle$. The quantity formed by their outer product

$$|\psi\rangle\langle\phi|$$

is an operator. Any operator can be written in this fashion. In terms of some basis $|u_i\rangle$ an operator A can be written as

$$A = \sum_{i,j=1}^{n} A_{ij}|u_i\rangle\langle u_j|$$

Looking at the matrix representation of the Pauli X operator

$$X = \begin{pmatrix} 0 & 1 \\ 1 & 0 \end{pmatrix} = \begin{pmatrix} 0 & X_{01} \\ X_{10} & 0 \end{pmatrix}$$

We see that with respect to the $\{|0\rangle, |1\rangle\}$ basis, the operator can be written as

$$X = X_{01}|0\rangle\langle 1| + X_{10}|1\rangle\langle 0| = |0\rangle\langle 1| + |1\rangle\langle 0|$$

e.g. EXAMPLE 8.2

A useful operator in quantum information theory is known as the *Hadamard gate*. This operator can be used to construct superposition states from $\{|0\rangle, |1\rangle\}$. In this basis the Hadamard gate has the representation

$$H = \frac{1}{\sqrt{2}}\begin{pmatrix} 1 & 1 \\ 1 & -1 \end{pmatrix} = \begin{pmatrix} H_{00} & H_{01} \\ H_{10} & H_{11} \end{pmatrix}$$

In outer product notation, the operator is written as

$$H = \sum_{i,j} H_{ij}|i\rangle\langle j| = H_{00}|0\rangle\langle 0| + H_{01}|0\rangle\langle 1| + H_{10}|1\rangle\langle 0| + H_{11}|1\rangle\langle 1|$$

$$\Rightarrow H = \frac{1}{\sqrt{2}}(|0\rangle\langle 0| + |0\rangle\langle 1| + |1\rangle\langle 0| - |1\rangle\langle 1|)$$

We can use this representation to quickly find the action of H on a state. For example:

$$\begin{aligned} H|0\rangle &= \frac{1}{\sqrt{2}}\Big(|0\rangle\langle 0| + |0\rangle\langle 1| + |1\rangle\langle 0| - |1\rangle\langle 1|\Big)|0\rangle \\ &= \frac{1}{\sqrt{2}}\Big(|0\rangle\langle 0|0\rangle + |0\rangle\langle 1|0\rangle + |1\rangle\langle 0|0\rangle - |1\rangle\langle 1|0\rangle\Big) \\ &= \frac{1}{\sqrt{2}}\Big(|0\rangle\langle 0|0\rangle + |1\rangle\langle 0|0\rangle\Big) = \frac{1}{\sqrt{2}}\Big(|0\rangle + |1\rangle\Big) \end{aligned}$$

We also have

$$\begin{aligned} H|1\rangle &= \frac{1}{\sqrt{2}}\Big(|0\rangle\langle 0| + |0\rangle\langle 1| + |1\rangle\langle 0| - |1\rangle\langle 1|\Big)|1\rangle \\ &= \frac{1}{\sqrt{2}}\Big(|0\rangle\langle 0|1\rangle + |0\rangle\langle 1|1\rangle + |1\rangle\langle 0|1\rangle - |1\rangle\langle 1|1\rangle\Big) \\ &= \frac{1}{\sqrt{2}}\Big(|0\rangle\langle 1|1\rangle - |1\rangle\langle 1|1\rangle\Big) = \frac{1}{\sqrt{2}}(|0\rangle - |1\rangle) \end{aligned}$$

Using these results we find that for an arbitrary state

$$\begin{aligned} H|\psi\rangle = \alpha H|0\rangle + \beta H|1\rangle &= \frac{\alpha}{\sqrt{2}}(|0\rangle + |1\rangle) + \frac{\beta}{\sqrt{2}}(|0\rangle - |1\rangle) \\ &= \frac{\alpha+\beta}{\sqrt{2}}|0\rangle + \frac{\alpha-\beta}{\sqrt{2}}|1\rangle \end{aligned}$$

An operator can be written in terms of projection operators formed by its own eigenvectors. This is called the *spectral decomposition*.

Spectral Decomposition

Suppose that an operator A has eigenvalues λ_i and eigenvectors $|a_i\rangle$:

$$A|a_i\rangle = \lambda_i|a_i\rangle$$

The *spectral decomposition* of the operator A is a representation of the operator in terms of its eigenvalues and the projector

$$P_i = |a_i\rangle\langle a_i|$$

are given by

$$A = \sum_{i=1}^{n}\lambda_i|a_i\rangle\langle a_i|$$

TWO-LEVEL SYSTEM EXAMPLE: SPECTRAL DECOMPOSITION OF Z

A quick exercise shows that the eigenvectors of Z are $\{|0\rangle, |1\rangle\}$ with eigenvalues ± 1

$$Z|0\rangle = |0\rangle, \quad Z|1\rangle = -|1\rangle$$

The spectral decomposition of Z is given by

$$Z = \sum_{i=0}^{1} \lambda_i \left| i \right\rangle \left\langle i \right| = \left| 0 \right\rangle \left\langle 0 \right| - \left| 1 \right\rangle \left\langle 1 \right|$$

Postulate 3 : The possible results of a measurement are the eigenvalues of an operator

The possible results of a measurement of a physical quantity are the eigenvalues of its corresponding operator.

e.g. EXAMPLE 8.3

Elementary particles like electrons carry an intrinsic angular momentum called spin. Measurement of spin $1/2$ along the x-axis is represented by the Pauli operator

$$S_x = \frac{\hbar}{2} \begin{pmatrix} 0 & 1 \\ 1 & 0 \end{pmatrix}$$

If spin is measured along the x-axis, what are the possible results of measurement?

✔ SOLUTION

The possible results of measurement are found from the eigenvalues of the matrix. The characteristic polynomial is found from

$$0 = \det \left| S_x - \lambda I \right| = \det \left| \frac{\hbar}{2} \begin{pmatrix} 0 & 1 \\ 1 & 0 \end{pmatrix} - \begin{pmatrix} \lambda & 0 \\ 0 & \lambda \end{pmatrix} \right| = \det \begin{vmatrix} \lambda & \frac{\hbar}{2} \\ \frac{\hbar}{2} & \lambda \end{vmatrix}$$

$$\Rightarrow \lambda^2 - \frac{\hbar^2}{4} = 0$$

This equation is satisfied by

$$\lambda = \pm \frac{\hbar}{2}$$

Therefore the possible results of measurment of spin along the x-axis are

$$\lambda_1 = \frac{\hbar}{2}$$

$$\lambda_2 = -\frac{\hbar}{2}$$

According to the measurement postulate, these are the *only* values of measurement that will ever be found.

Projective Measurements

Projective measurements, the type encountered in standard quantum mechanics, deal with mutually exclusive measurement results. For example, we can ask if a particle is located at position x_1 or position x_2. Another example might be asking if a trapped electron has ground state energy E_1 or excited state energy E_2. Mutually exclusive measurement outcomes like these are represented by orthogonal projection operators:

$$P_1, P_2, \ldots, P_n$$

The statement that two projection operators are orthogonal means that if $i \neq j$ then

$$P_i P_j = 0$$

We can see this using the spectral decomposition of some operator A:

$$A = \sum_{i=1}^{n} \lambda_i \left|a_i\right\rangle \left\langle a_i\right| = \sum_{i=1}^{n} \lambda_i P_i$$

The eigenvectors of a Hermitian operator are orthogonal, and in fact the practice is to normalize them. Assuming the eigenvectors are orthonormal, then we have

$$P_i P_j = |a_i\rangle\langle a_i|a_j\rangle\langle a_j| = |a_i\rangle\langle a_j|\delta_{ij}$$

Therefore if $i \neq j$, we obtain $P_i P_j = 0$. However, if $i = j$ then

$$P_i P_i = P_i^2 = \left|a_i\right\rangle \left\langle a_i \left|a_i\right.\right\rangle \left\langle a_i\right| = \left|a_i\right\rangle \left\langle a_i\right| = P_i$$

TWO-LEVEL SYSTEM EXAMPLE: PROJECTION OPERATORS

In the standard basis, there are two projection operators that correspond to measurements of $+1$ and -1:

$$P_0 = |0\rangle\langle 0|, \quad P_1 = |1\rangle\langle 1|$$

The operator P_0 projects any state onto $|0\rangle$ while P_1 projects any state onto $|1\rangle$. For some arbitrary state $|\psi\rangle = \alpha|0\rangle + \beta|1\rangle$, we have

$$P_0|\psi\rangle = \alpha\left(|0\rangle\langle 0|\right)|0\rangle + \beta\left(|0\rangle\langle 0|\right)|1\rangle = \alpha|0\rangle\langle 0|0\rangle = \alpha|0\rangle$$

$$P_1|\psi\rangle = \alpha\left(|1\rangle\langle 1|\right)|0\rangle + \beta\left(|1\rangle\langle 1|\right)|1\rangle = \beta|1\rangle\langle 1|1\rangle = \beta|1\rangle$$

since $\langle 0|1\rangle = \langle 1|0\rangle = 0$. Now we consider the sum of a set of projection operators. This gives the identity.

The Completeness Relation

The sum of the projection operators gives the identity:

$$\sum_{i=1}^{n} P_i = \sum_{i=1}^{n} |a_i\rangle \langle a_i| = I$$

For the two-level system we have been studying, this means that

$$P_0 + P_1 = |0\rangle \langle 0| + |1\rangle \langle 1| = I$$

We can show this by calculating the matrix representations of the outer products

$$|0\rangle \langle 0| = \begin{pmatrix} 1 \\ 0 \end{pmatrix} \begin{pmatrix} 1 & 0 \end{pmatrix} = \begin{pmatrix} 1 & 0 \\ 0 & 0 \end{pmatrix}$$

$$|1\rangle \langle 1| = \begin{pmatrix} 0 \\ 1 \end{pmatrix} \begin{pmatrix} 0 & 1 \end{pmatrix} = \begin{pmatrix} 0 & 0 \\ 0 & 1 \end{pmatrix}$$

$$\Rightarrow |0\rangle \langle 0| + |1\rangle \langle 1| = \begin{pmatrix} 1 & 0 \\ 0 & 0 \end{pmatrix} + \begin{pmatrix} 0 & 0 \\ 0 & 1 \end{pmatrix} = \begin{pmatrix} 1 & 0 \\ 0 & 1 \end{pmatrix}$$

Postulate 4 : The probability of obtaining a given measurement result

Suppose that an operator A has eigenvalues λ_i and eigenvectors $|a_i\rangle$

$$A\,|a_i\rangle = \lambda_i\,|a_i\rangle$$

We can expand a state $|\psi\rangle$ in terms of the eigenvectors of A

$$|\psi\rangle = \alpha_1\,|a_1\rangle + \alpha_2\,|a_2\rangle + \cdots + \alpha_n\,|a_n\rangle$$

The probability of obtaining measurement result λ_i, which is associated with eigenvector $|a_i\rangle$, is given by the *Born Rule*

$$\text{Probability of finding } \lambda_i = |\langle a_i | \psi \rangle|^2 = |\alpha_i|^2$$

Remember, the inner product is a complex number:

$$|\langle a_i | \psi \rangle|^2 = \langle a_i | \psi \rangle \left(\langle a_i | \psi \rangle\right)^* = \langle a_i | \psi \rangle \langle \psi | a_i \rangle$$

Since these are *numbers* we can change their order, and write the probability in terms of a projection operator:

$$\langle a_i \mid \psi\rangle \langle \psi | a_i \rangle = \langle \psi | a_i \rangle \langle a_i | \psi \rangle = \langle \psi | P_i | \psi \rangle$$

EXAMPLE 8.4

A quantum system is in the state

$$|\psi\rangle = \frac{\sqrt{2}}{3}|\Phi_1\rangle + \frac{\sqrt{3}}{3}|\Phi_2\rangle + \frac{2}{3}|\Phi_3\rangle$$

where the $|\Phi_1\rangle$ constitute an orthonormal basis. These states are eigenvectors of the Hamiltonian operator such that

$$H|\Phi_1\rangle = E|\Phi_1\rangle\,,\ H|\Phi_2\rangle = 2E|\Phi_2\rangle\,,\ H|\Phi_3\rangle = 3E|\Phi_3\rangle$$

(a) Is $|\psi\rangle$ normalized?

(b) If the energy is measured, what are the probabilities of obtaining E, $2E$, and $3E$?

SOLUTION

(a) The bra that corresponds to the state is given by

$$\langle\psi| = \frac{\sqrt{2}}{3}\langle\Phi_1| + \frac{\sqrt{3}}{3}\langle\Phi_2| + \frac{2}{3}\langle\Phi_3|$$

Therefore the norm of the state is found to be

$$\begin{aligned}\langle\psi|\psi\rangle &= \left(\frac{\sqrt{2}}{3}\langle\Phi_1| + \frac{\sqrt{3}}{3}\langle\Phi_2| + \frac{2}{3}\langle\Phi_3|\right)\left(\frac{\sqrt{2}}{3}|\Phi_1\rangle + \frac{\sqrt{3}}{3}|\Phi_2\rangle + \frac{2}{3}|\Phi_3\rangle\right)\\ &= \frac{2}{9}\langle\Phi_1|\Phi_1\rangle + \frac{3}{9}\langle\Phi_2|\Phi_2\rangle + \frac{4}{9}\langle\Phi_3|\Phi_3\rangle\\ &= \frac{2}{9} + \frac{3}{9} + \frac{4}{9} = 1\end{aligned}$$

Notice that we have used the fact that the basis is orthonormal, and so

$$\langle\Phi_1|\Phi_2\rangle = \langle\Phi_1|\Phi_3\rangle = \langle\Phi_2|\Phi_3\rangle = 0$$

Since $\langle\psi|\psi\rangle = 1$ we conclude the state is normalized.

(b) To find the probability of obtaining each measurement, we use the Born rule. We have

$$\begin{aligned}\langle\Phi_1|\psi\rangle &= \frac{\sqrt{2}}{3}\langle\Phi_1|\Phi_1\rangle + \frac{\sqrt{3}}{3}\langle\Phi_1|\Phi_2\rangle + \frac{2}{3}\langle\Phi_1|\Phi_3\rangle\\ &= \frac{\sqrt{2}}{3}\end{aligned}$$

The Born rule tells us that the probability of obtaining measurement result E is

$$|\langle\Phi_1|\psi\rangle|^2 = \left(\frac{\sqrt{2}}{3}\right)^2 = \frac{2}{9}$$

For the next eigenvector, we find

$$\begin{aligned}\langle\Phi_2|\psi\rangle &= \frac{\sqrt{2}}{3}\langle\Phi_2|\Phi_1\rangle + \frac{\sqrt{3}}{3}\langle\Phi_2|\Phi_2\rangle + \frac{2}{3}\langle\Phi_2|\Phi_3\rangle \\ &= \frac{\sqrt{3}}{3}\end{aligned}$$

The Born rule tells us that the probability of obtaining measurement result $2E$ is

$$|\langle\Phi_2|\psi\rangle|^2 = \left(\frac{\sqrt{3}}{3}\right)^2 = \frac{3}{9}$$

Finally, we have

$$\begin{aligned}\langle\Phi_3|\psi\rangle &= \frac{\sqrt{2}}{3}\langle\Phi_3|\Phi_1\rangle + \frac{\sqrt{3}}{3}\langle\Phi_3|\Phi_2\rangle + \frac{2}{3}\langle\Phi_3|\Phi_3\rangle \\ &= \frac{2}{3}\end{aligned}$$

The Born rule tells us that the probability of finding $3E$ upon measurement is

$$|\langle\Phi_3|\psi\rangle|^2 = \left(\frac{2}{3}\right)^2 = \frac{4}{9}$$

Although we have already verified that the state is normalized, it is always a good idea to sum the probabilities and verify that they sum to one, as they do in this case:

$$|\langle\Phi_1|\psi\rangle|^2 + |\langle\Phi_2|\psi\rangle|^2 + |\langle\Phi_3|\psi\rangle|^2 = \frac{2}{9} + \frac{3}{9} + \frac{4}{9} = \frac{9}{9} = 1$$

---

e.g. EXAMPLE 8.5

An orthonormal basis of a Hamiltonian operator in four dimensions is defined as follows

$$H|1\rangle = E|1\rangle,\ H|2\rangle = 2E|2\rangle,\ H|3\rangle = 3E|3\rangle,\ H|4\rangle = 4E|4\rangle$$

A system is in the state

$$|\psi\rangle = 3|1\rangle + |2\rangle - |3\rangle + 7|4\rangle$$

(a) If a measurement of the energy is made, what results can be found and with what probabilities?
(b) Find the average energy of the system.

SOLUTION

First we check to see if the state is normalized. We have

$$\langle\psi \mid \psi\rangle = |3|^2 + |1|^2 + |1|^2 + |7|^2 = 9 + 1 + 1 + 49 = 60$$

Therefore it is necessary to normalize the state. The normalized state is found by dividing by $\sqrt{\langle\psi|\psi\rangle}$. Calling the normalized state $|\chi\rangle$ we obtain

$$|\chi\rangle = \frac{1}{\sqrt{\langle\psi\,|\psi\,\rangle}}\,|\psi\rangle = \frac{3}{\sqrt{60}}\,|1\rangle + \frac{1}{\sqrt{60}}\,|2\rangle - \frac{1}{\sqrt{60}}\,|3\rangle + \frac{7}{\sqrt{60}}\,|4\rangle$$

Since the state is expanded in the eigenbasis of the Hamiltonian, the only possible values of measurment are the eigenvalues of the Hamiltonian, which have been given to us as $(E, 2E, 3E, 4E)$. The probabilties of obtaining each measurement are found by application of the Born rule, squaring the coefficients for the normalized state.

(a) The probability of obtaining E is

$$\left|\frac{3}{\sqrt{60}}\right|^2 = \frac{9}{60}$$

The probability of obtaining $2E$ is

$$\left|\frac{1}{\sqrt{60}}\right|^2 = \frac{1}{60}$$

The probability of obtaining $3E$ is

$$\left|\frac{-1}{\sqrt{60}}\right|^2 = \frac{1}{60}$$

The probability of finding $4E$ is

$$\left|\frac{7}{\sqrt{60}}\right|^2 = \frac{49}{60}$$

A quick check shows these probabilities add up to one

$$\frac{9}{60} + \frac{1}{60} + \frac{1}{60} + \frac{49}{60} = \frac{60}{60} = 1$$

(b) We can find the average energy from

$$\langle H \rangle = \sum p(E_i) E_i$$

where $p(E_i)$ is the probability of obtaining measurement result E_i. We find that the average energy is

$$\langle H \rangle = E\left(\frac{9}{60}\right) + 2E\left(\frac{1}{60}\right) + 3E\left(\frac{1}{60}\right) + 4E\left(\frac{49}{60}\right) = \frac{210}{60}E = \frac{7}{2}E$$

ASIDE ON DEGENERACY

Suppose that an operator A has degenerate eigenvalues corresponding to eigenvectors

$$\left\{ \left|a_m^1\right\rangle, \left|a_m^2\right\rangle, \ldots, \left|a_m^{g_m}\right\rangle \right\}$$

Where each of these eigenvectors corresponds to the same eigenvector

$$A\left|a_m^1\right\rangle = \lambda_m \left|a_m^1\right\rangle$$
$$A\left|a_m^2\right\rangle = \lambda_m \left|a_m^2\right\rangle$$
$$\vdots$$
$$A\left|a_m^{g_m}\right\rangle = \lambda_m \left|a_m^{g_m}\right\rangle$$

This set of eigenvectors constitutes a subspace of the vector space ε. In the case of degeneracy, the probability of obtaining measurement result λ_m is found by summing over the inner products of all of the eigenvectors that belong to this subspace:

$$\textit{Probability of finding } \lambda_m = \sum_{i=1}^{g_m} \left|\langle a_m^i | \psi \rangle\right|^2$$

Postulate 5 : The state of a system after measurement

The reader who studies quantum theory more extensively will find that not all measurements can be described as projective measurements as we have done here. Nonetheless, in a first exposure to quantum theory it is good to restrict our attention to measurements of this type. We consider the non-degenerate case first. *Immediately* after a measurement, the state of the system is given by the eigenvector corresponding to the eigenvalue that has been measured. If the state of the system is given by

$$|\psi\rangle = \alpha_1 |a_1\rangle + \alpha_2 |a_2\rangle + \cdots + \alpha_n |a_n\rangle$$

suppose a measurement is made and result λ_j is obtained. The state of the system immediately after measurement is

$$|\psi\rangle \overset{measurement}{\rightarrow} |a_j\rangle$$

In terms of projection operators, the state of the system after measurement is

$$|\psi\rangle \overset{measurement}{\rightarrow} \frac{1}{\sqrt{\langle\psi| P_j |\psi\rangle}} P_j |\psi\rangle$$

TWO-LEVEL SYSTEM EXAMPLE: STATE OF THE SYSTEM AFTER MEASUREMENT

Recalling the projection operators for the two-state system

$$P_0 = |0\rangle \langle 0|, \quad P_1 = |1\rangle \langle 1|$$

And the action on a state $|\psi\rangle = \alpha|0\rangle + \beta|1\rangle$ that we found to be

$$P_0 |\psi\rangle = \alpha |0\rangle$$
$$P_1 |\psi\rangle = \beta |1\rangle$$

the state after measurement is found by dividing these quantities by $\sqrt{\langle\psi|P_i|\psi\rangle}$. We have

$$\langle\psi| P_0 |\psi\rangle = \alpha \langle\psi| 0\rangle = \alpha \left(\alpha^* \langle 0| + \beta^* \langle 1|\right) |0\rangle = |\alpha|^2 \langle 0 |0\rangle + \alpha\beta^* \langle 1 |0\rangle = |\alpha|^2$$

and

$$\langle\psi| P_1 |\psi\rangle = \beta \langle\psi| 1\rangle = \beta \left(\alpha^* \langle 0| + \beta^* \langle 1|\right) |1\rangle = \beta\alpha^* \langle 0 |1\rangle + |\beta|^2 \langle 1 |1\rangle = |\beta|^2$$

Therefore, if measurement result $+1$ is obtained, the state of the system after measurement is

$$\left|\psi_{after}\right\rangle = \frac{\alpha}{|\alpha|} |0\rangle$$

and if -1 is obtained it is

$$\left|\psi_{after}\right\rangle = \frac{\beta}{|\beta|} |1\rangle$$

EXAMPLE 8.6

Suppose a Hamiltonian operator has a basis $|u_i\rangle$ such that

$$H|u_1\rangle = E|u_1\rangle$$
$$H|u_2\rangle = E|u_2\rangle$$
$$H|u_3\rangle = 2E|u_3\rangle$$
$$H|u_4\rangle = 4E|u_4\rangle$$

the system is in the state

$$|\psi\rangle = \frac{1}{\sqrt{6}}|u_1\rangle + \frac{1}{\sqrt{2}}|u_2\rangle + \frac{1}{2}|u_3\rangle + \frac{1}{\sqrt{12}}|u_4\rangle$$

(a) Write the Hamiltonian operator in outer product notation.

(b) Write the projection operator P that projects a state onto the subspace spanned by $\{|u_1\rangle, |u_2\rangle\}$.

(c) The energy is measured. What values can be found, and with what probabilities?

(d) Suppose the energy is measured and found to be E. What is the state of the system after measurement?

SOLUTION

(a) Using the spectral decomposition of H, we find

$$H = E|u_1\rangle\langle u_1| + E|u_2\rangle\langle u_2| + 2E|u_3\rangle\langle u_3| + 4E|u_4\rangle\langle u_4|$$

(b) The projection operator for the subspace spanned by $\{|u_1\rangle, |u_2\rangle\}$ is found by summing over the individual projection operators for each state

$$P_E = |u_1\rangle\langle u_1| + |u_2\rangle\langle u_2|$$

(c) The reader should verify that the state is normalized. The possible results of measurement are E, $2E$, and $4E$, corresponding to eigenvectors $\{|u_1\rangle, |u_2\rangle\}$, $|u_3\rangle$, and $|u_4\rangle$, respectively. To calculate the probability of obtaining the measurement result E, which is degenerate, we use

$$p_E = \sum_{i=1}^{2} |\langle u_i|\psi\rangle|^2 = |\langle u_1|\psi\rangle|^2 + |\langle u_2|\psi\rangle|^2 = \left(\frac{1}{\sqrt{6}}\right)^2 + \left(\frac{1}{\sqrt{2}}\right)^2 = \frac{1}{6} + \frac{1}{2} = \frac{2}{3}$$

The other probabilities are for non-degenerate eigenvalues and can be calculated immediately using the Born rule:

$$p_{2E} = |\langle u_3 | \psi \rangle|^2 = \left(\frac{1}{2}\right)^2 = \frac{1}{4}$$

$$p_{4E} = |\langle u_4 | \psi \rangle|^2 = \left(\frac{1}{\sqrt{12}}\right)^2 = \frac{1}{12}$$

(d) If the energy is measured and is found to be E, the state after measurement is:

$$|\psi_{after}\rangle = \frac{1}{\sqrt{\langle\psi| P_E |\psi\rangle}} P_E |\psi\rangle$$

$$P_E |\psi\rangle = \left(|u_1\rangle \langle u_1| + |u_2\rangle \langle u_2| \right)\left(\frac{1}{\sqrt{6}} |u_1\rangle + \frac{1}{\sqrt{2}} |u_2\rangle + \frac{1}{2} |u_3\rangle + \frac{1}{\sqrt{12}} |u_4\rangle\right)$$

$$= \frac{1}{\sqrt{6}} |u_1\rangle + \frac{1}{\sqrt{2}} |u_2\rangle$$

It is easy to show that

$$\langle\psi| P_E |\psi\rangle = \left(\frac{1}{\sqrt{6}}\right)^2 + \left(\frac{1}{\sqrt{2}}\right)^2 = \frac{1}{6} + \frac{1}{2} = \frac{2}{3}$$

and so we have

$$\frac{1}{\sqrt{\langle\psi| P_E |\psi\rangle}} = \sqrt{\frac{3}{2}}$$

Therefore, the state after measurement is

$$|\psi_{after}\rangle = \frac{1}{\sqrt{\langle\psi| P_E |\psi\rangle}} P_E |\psi\rangle = \sqrt{\frac{3}{2}}\left(\frac{1}{\sqrt{6}} |u_1\rangle + \frac{1}{\sqrt{2}} |u_2\rangle\right)$$

$$= \frac{1}{2} |u_1\rangle + \frac{\sqrt{3}}{2} |u_2\rangle$$

Note: The state of the system cannot be further distinguished; with the information given there is nothing we can do to distinguish $|u_1\rangle$ from $|u_2\rangle$.

Postulate 6: The time evolution of a quantum system is governed by the Schrödinger equation

The Schrödinger equation determines how a quantum system changes with time according to

$$i\hbar \frac{\partial}{\partial t} |\psi\rangle = H |\psi (t_o)\rangle$$

where H is the Hamiltonian and t_o is some initial time. For an isolated system the Hamiltonian is independent of time. The equation can be integrated in that case, giving

$$|\psi (t)\rangle = \exp\left(-\frac{i}{\hbar} H (t - t_o)\right) |\psi (t_o)\rangle$$

The exponential term defines a time evolution operator; which we denote U because this operator is unitary:

$$U (t, t_o) = \exp\left(-\frac{i}{\hbar} H (t - t_o)\right)$$

The *infinitesimal* operator is found by expanding I in a Taylor series and keeping the first two terms:

$$U (t + \varepsilon, t) = I - \frac{i}{\hbar} H\varepsilon$$

Completely Specifying a State with a CSCO

In a previous example, we found that when two energy eigenstates had the same eigenvalue E, and a measurement result turned out to be E, the state of the system after measurement was:

$$\frac{1}{2} |u_1\rangle + \frac{\sqrt{3}}{2} |u_2\rangle$$

There was no way to distinguish between the states $|u_1\rangle$ and $|u_2\rangle$. There are many situations like this that can arise. However, many eigenfunctions of one operator with degenerate eigenvalues turn out to be eigenfunctions of another operator with eigenvalues that can be used to distinguish the states. For example, free particle energy states have degenerate energy eigenvalues, but can be distinguished by momentum $\pm p$. So, to completely distinguish the state, the energy E is not enough; we must also specify a momentum of $+p$ or $-p$. This notion is formalized by the idea of a CSCO. A CSCO is a *complete set of commuting observables* and is the *minimum* set of operators required to completely specify a state. Two or more operators that commute can be simultaneously measured. Suppose that:

$$A\,|\psi\rangle = \alpha\,|\psi\rangle\,, \quad B\,|\psi\rangle = \beta\,|\psi\rangle$$

Now notice what happens when we apply the product AB to the state

$$AB\,|\psi\rangle = A\,(\beta\,|\psi\rangle) = \beta\,(A\,|\psi\rangle) = \alpha\beta\,|\psi\rangle$$

and the application of BA gives

$$BA\,|\psi\rangle = B\,(\alpha\,|\psi\rangle) = \alpha\,(B\,|\psi\rangle) = \alpha\beta\,|\psi\rangle$$

and so

$$(AB - BA)\,|\psi\rangle = 0$$

Since the state $|\psi\rangle \neq 0$ then we must have

$$AB - BA = [A, B] = 0$$

Now suppose that a state $|\Phi\rangle$ is also an eigenstate of A with the same eigenvalue α, so that this eigenvalue is degenerate. Then a measurement of A cannot distinguish the states. However, if $|\Phi\rangle$ is also an eigenvector of B such that

$$B\,|\Phi\rangle = \gamma\,|\Phi\rangle \text{ but } B\,|\psi\rangle = \beta\,|\psi\rangle$$

then we can distinguish the states by measuring B. In this example, A and B form a CSCO. What if the eigenvalues of B are also degenerate? Then we must find an operator C that commutes with both A and B, but has eigenvalues such that the states can be distinguished. If it turns out that no such set of operators can be found, then the states are physically indistinguishable.

The Heisenberg versus Schrödinger Pictures

In the Schrödinger picture, the time evolution of the system is contained in the state. However, since the physical predictions of quantum mechanics are determined by inner products, the time evolution can be moved from states to operators:

$$\langle\psi\,(t)|\,A\,|\psi\,(t)\rangle = \langle\psi\,(0)|\,U^{\dagger}AU\,|\psi\,(0)\rangle$$

We can then describe time evolution in terms of evolving the operator forward in time according to

$$A\,(t) = U^{\dagger}A\,(0)\,U$$

The governing equation for time evolution is given by

$$\frac{d}{dt}\langle A\rangle = \frac{1}{i\hbar}[A, H] + \left\langle\frac{\partial A}{\partial t}\right\rangle$$

where H is the system's Hamiltonian.

Describing Composite Systems in Quantum Mechanics

Sometimes it is necessary to describe a system that is composed of multiple particles, each particle being a system in its own right described by a vector space. A mathematical technique called the tensor product allows us to bring together vector spaces that describe individual systems together into a larger vector space. The tensor product between vector spaces V and W is denoted by writing

$$V \otimes W$$

DIMENSION OF A VECTOR SPACE FORMED BY A TENSOR PRODUCT

If V is a vector space of dimension p and W is a vector space of dimension q, then the vector space

$$V \otimes W$$

has dimension pq.

Let $|v\rangle \in V$ and $|w\rangle \in W$ be two vectors. Then the vector $|v\rangle \otimes |w\rangle \in V \otimes W$. Alternative shorthand notations for the vector $|v\rangle \otimes |w\rangle$ include $|v\rangle|w\rangle$ and $|vw\rangle$. If A is an operator that acts on the vector space V and B is an operator that acts on the vector space W, then the operator formed by the tensor product given by

$$A \otimes B$$

acts on vectors $|v\rangle \otimes |w\rangle$ in the following way:

$$(A \otimes B) |v\rangle |w\rangle = (A |v\rangle) (B |w\rangle)$$

We now consider the norm of a state that belongs to $V \otimes W$. Suppose that

$$|\psi\rangle = |v\rangle |w\rangle$$

Then the norm is

$$\langle \psi | \psi \rangle = \langle v | v \rangle \langle w | w \rangle$$

The Matrix Representation of a Tensor Product

Let A be an $m \times n$ matrix and B be a $p \times q$ matrix. The tensor product $A \otimes B$ is a matrix with mp rows and nq columns. The elements of this matrix are constructed by the submatrices:

$$A_{ij} B$$

that is, multiply B by each component of A and then arrange these submatrices into a larger matrix. The entire matrix representing the tensor product is:

$$A = \begin{pmatrix} A_{11}B & A_{12}B & \dots & A_{1n}B \\ A_{21}B & A_{22}B & \dots & \vdots \\ \vdots & \vdots & \ddots & \vdots \\ A_{m1}B & \dots & \dots & A_{mn}B \end{pmatrix}$$

A column vector in two dimensions has two rows and one column. Since the tensor product between an $(m \times n)$ matrix and a $(p \times q)$ matrix has mp rows and nq columns, the tensor product between two (2×1) column vectors has $(2 \times 2) = 4$ rows and $(1 \times 1) = 1$ columns. The way to compute this is:

$$\begin{pmatrix} a \\ b \end{pmatrix} \otimes \begin{pmatrix} c \\ d \end{pmatrix} = \begin{pmatrix} ac \\ ad \\ bc \\ bd \end{pmatrix}$$

Now consider the tensor product of two 2×2 matrices. This tensor product will be a matrix with $(2 \times 2) = 4$ rows and $(2 \times 2) = 4$ columns. If we let

$$A = \begin{pmatrix} a & b \\ c & d \end{pmatrix}, \quad B = \begin{pmatrix} w & x \\ y & z \end{pmatrix}$$

then the tensor product of these two matrices is

$$A \otimes B = \begin{pmatrix} aB & bB \\ cB & dB \end{pmatrix} = \begin{pmatrix} aw & ax & bw & bx \\ ay & az & by & bz \\ cw & cx & dw & dx \\ cy & cz & dy & dz \end{pmatrix}$$

EXAMPLE 8.7 e.g.

Consider the two Pauli matrices Y and Z. Compute the tensor product

$$Y \otimes Z$$

SOLUTION
Recalling that

$$Y = \begin{pmatrix} 0 & -i \\ i & 0 \end{pmatrix}, \quad Z = \begin{pmatrix} 1 & 0 \\ 0 & -1 \end{pmatrix}$$

we have

$$Y \otimes Z = \begin{pmatrix} (0)Z & (-i)Z \\ (i)Z & (0)Z \end{pmatrix} = \begin{pmatrix} 0 & 0 & -i & 0 \\ 0 & 0 & 0 & i \\ i & 0 & 0 & 0 \\ 0 & -i & 0 & 0 \end{pmatrix}$$

The Tensor Product of State Vectors

We can use the tensor product to compute the tensor product of a state with itself n times. This is represented by the notation

$$|\psi\rangle^{\otimes n} = |\psi\rangle \otimes |\psi\rangle \otimes \cdots \otimes |\psi\rangle \ (n\text{times})$$

EXAMPLE 8.8
Let

$$|\psi\rangle = \sqrt{\frac{2}{3}}\,|+\rangle + \frac{1}{\sqrt{3}}\,|-\rangle$$

where $|+\rangle = \begin{pmatrix} 1 \\ 0 \end{pmatrix}$ and $|-\rangle = \begin{pmatrix} 0 \\ 1 \end{pmatrix}$. Find $|\psi\rangle^{\otimes 2}$ and $|\psi\rangle^{\otimes 3}$.

SOLUTION
It is helpful to write the state as a column vector:

$$|\psi\rangle = \sqrt{\frac{2}{3}}\,|+\rangle + \frac{1}{\sqrt{3}}\,|-\rangle = \sqrt{\frac{2}{3}}\begin{pmatrix} 1 \\ 0 \end{pmatrix} + \frac{1}{\sqrt{3}}\begin{pmatrix} 0 \\ 1 \end{pmatrix} = \frac{1}{\sqrt{3}}\begin{pmatrix} \sqrt{2} \\ 1 \end{pmatrix}$$

So we obtain:

$$|\psi\rangle^{\otimes 2} = |\psi\rangle \otimes |\psi\rangle = \frac{1}{\sqrt{3}}\begin{pmatrix} \sqrt{2} \\ 1 \end{pmatrix} \otimes \frac{1}{\sqrt{3}}\begin{pmatrix} \sqrt{2} \\ 1 \end{pmatrix}$$

$$= \begin{pmatrix} \left(\sqrt{\frac{2}{3}}\right)\left(\sqrt{\frac{2}{3}}\right) \\ \left(\sqrt{\frac{2}{3}}\right)\left(\frac{1}{\sqrt{3}}\right) \\ \left(\frac{1}{\sqrt{3}}\right)\left(\sqrt{\frac{2}{3}}\right) \\ \left(\frac{1}{\sqrt{3}}\right)\left(\frac{1}{\sqrt{3}}\right) \end{pmatrix} = \begin{pmatrix} \frac{2}{3} \\ \frac{\sqrt{2}}{3} \\ \frac{\sqrt{2}}{3} \\ \frac{1}{3} \end{pmatrix}$$

The tensor product $|\psi\rangle^{\otimes 3} = |\psi\rangle \otimes |\psi\rangle^{\otimes 2}$

$$|\psi\rangle^{\otimes 3} = \frac{1}{\sqrt{3}}\begin{pmatrix} \sqrt{2} \\ 1 \end{pmatrix} \otimes \begin{pmatrix} \frac{2}{3} \\ \frac{\sqrt{2}}{3} \\ \frac{\sqrt{2}}{3} \\ \frac{1}{3} \end{pmatrix}$$

$$= \begin{pmatrix} \left(\sqrt{\frac{2}{3}}\right)\left(\frac{2}{3}\right) \\ \left(\sqrt{\frac{2}{3}}\right)\left(\frac{\sqrt{2}}{3}\right) \\ \left(\sqrt{\frac{2}{3}}\right)\left(\frac{\sqrt{2}}{3}\right) \\ \left(\sqrt{\frac{2}{3}}\right)\left(\frac{1}{3}\right) \\ \left(\frac{1}{\sqrt{3}}\right)\left(\frac{2}{3}\right) \\ \left(\frac{1}{\sqrt{3}}\right)\left(\frac{\sqrt{2}}{3}\right) \\ \left(\frac{1}{\sqrt{3}}\right)\left(\frac{\sqrt{2}}{3}\right) \\ \left(\frac{1}{\sqrt{3}}\right)\left(\frac{1}{3}\right) \end{pmatrix} = \begin{pmatrix} \frac{2\sqrt{2}}{3\sqrt{3}} \\ \frac{2}{3\sqrt{3}} \\ \frac{2}{3\sqrt{3}} \\ \frac{\sqrt{2}}{3\sqrt{3}} \\ \frac{2}{3\sqrt{3}} \\ \frac{\sqrt{2}}{3\sqrt{3}} \\ \frac{\sqrt{2}}{3\sqrt{3}} \\ \frac{1}{3\sqrt{3}} \end{pmatrix}$$

EXAMPLE 8.9

e.g.

Consider a composite system of qubits. The first qubit is denoted by A and the second by B. The *spin-singlet state* is formed from a linear superposition:

$$|\psi\rangle = \frac{|0\rangle_A |1\rangle_B - |1\rangle_A |0\rangle_B}{\sqrt{2}}$$

The operator X_A applies the X operator only to states belonging to subsystem A. Suppose that we act the operator X_A on the state. What is the resulting state of the system?

SOLUTION

✔

The operator X is

$$X = |0\rangle\langle 1| + |1\rangle\langle 0|$$

and it acts as follows:

$$X|0\rangle = |1\rangle, \quad X|1\rangle = |0\rangle$$

Therefore, when X_A acts on the composite state we find

$$X_A \left|\psi\right\rangle = X_A \left(\frac{\left|0\right\rangle_A \left|1\right\rangle_B - \left|1\right\rangle_A \left|0\right\rangle_B}{\sqrt{2}} \right) = \frac{X_A \left|0\right\rangle_A \left|1\right\rangle_B - X_A \left|1\right\rangle_A \left|0\right\rangle_B}{\sqrt{2}}$$

$$= \frac{\left|1\right\rangle_A \left|1\right\rangle_B - \left|0\right\rangle_A \left|0\right\rangle_B}{\sqrt{2}}$$

EXAMPLE 8.10

Suppose that a two-qubit system is in the state:

$$\left|\psi\right\rangle = \frac{\left|0\right\rangle \left|0\right\rangle + \left|1\right\rangle \left|1\right\rangle}{\sqrt{2}}$$

Find $(Z_A \otimes X_B) \left|\psi\right\rangle$.

SOLUTION

Recalling that $(A \otimes B) \left|v\right\rangle \left|w\right\rangle = (A \left|v\right\rangle)(B \left|w\right\rangle)$, we have

$$(Z_A \otimes X_B) \left|0\right\rangle \left|0\right\rangle = (Z_A \left|0\right\rangle)(X_B \left|0\right\rangle) = \left|0\right\rangle \left|1\right\rangle$$

$$(Z_A \otimes X_B) \left|1\right\rangle \left|1\right\rangle = (Z_A \left|1\right\rangle)(X_B \left|1\right\rangle) = -\left|1\right\rangle \left|0\right\rangle$$

and so

$$(Z_A \otimes X_B) \left|\psi\right\rangle = \frac{(Z_A \otimes X_B) \left|0\right\rangle \left|0\right\rangle + (Z_A \otimes X_B) \left|1\right\rangle \left|1\right\rangle}{\sqrt{2}}$$

$$= \frac{\left|0\right\rangle \left|1\right\rangle - \left|1\right\rangle \left|0\right\rangle}{\sqrt{2}}$$

The Density Operator

So far our studies of quantum systems have involved the examination of a *single* state. This state can be expanded in terms of basis kets:

$$\left|\psi\right\rangle = c_1 \left|u_1\right\rangle + c_2 \left|u_2\right\rangle + \cdots + c_n \left|u_n\right\rangle$$

Such an expansion is referred to as a coherent superposition of the $\left|u_i\right\rangle$ states. The task has been to find out what the possible measurement results are on the system, and to calculate the probability of obtaining each possible result.

Now consider an ensemble of quantum states that is prepared as a *classical statistical mixture*. The members of the ensemble are the states $|\psi_1\rangle, |\psi_2\rangle, \ldots, |\psi_n\rangle$, and the probability of finding each state is given by $p_1, p_2, \ldots, p_n$. We can describe this type of mixture of states with a *density operator*, which is usually represented by the symbol ρ.

The density operator representing a collection of states $|\psi_1\rangle, |\psi_2\rangle, \ldots, |\psi_n\rangle$ is

$$\rho = \sum_{i=1}^{n} p_i |\psi_i\rangle \langle\psi_i|$$

If the state of a system is known exactly, we say it is a *pure state*. This is a state where a single $p_i = 1$ and all others are zero. This means that the density operator for a pure state $|\psi\rangle$ can be written as

$$\rho = |\psi\rangle \langle\psi|$$

This is a projection operator, so in the case of a pure state, the density operator satisfies $\rho^2 = \rho$. A *mixed state* is a collection of different pure states, each occurring with a given probability.

PROPERTIES OF THE DENSITY OPERATOR

- The density operator is Hermitian, $\rho = \rho^\dagger$
- The trace of any density matrix is equal to one, $Tr(\rho) = 1$
- For a pure state, since $\rho^2 = \rho$, $Tr(\rho^2) = 1$
- For a mixed state $Tr(\rho^2) < 1$
- The eigenvalues of a density operator satisfy $0 \le \lambda_i \le 1$
- The expectation value of an operator A can be calculated using $\langle A\rangle = Tr(\rho A)$

We can use a density operator to make all of the physical predictions of quantum mechanics. Let a measurement result m be represented by a projection operator $P_m = |m\rangle\langle m|$. The probability of obtaining measurement result m is

$$Tr(\rho P_m) = Tr(\rho|m\rangle\langle m|) = \langle m|\rho|m\rangle$$

Density operators can be represented by matrices. The off-diagonal elements of a density matrix represent the ability of the system to exhibit quantum interference. Consider, once again, a state formed by a superposition:

$$|\psi\rangle = c_1 |u_1\rangle + c_2 |u_2\rangle + \cdots + c_n |u_n\rangle$$

This type of state is a pure state. In this case the density operator is found from the outer product

$$\rho = |\psi\rangle \langle\psi|$$

It is not too hard to show that this is

$$\rho = \sum_{i=1}^{n} |c_i|^2 |u_i\rangle \langle u_i| + \sum_{i \neq j}^{n} c_i c_j^* |u_i\rangle \langle u_j|$$

We see that the density operator for this pure state can be split into two parts. The terms in the first part tell us

$$\langle u_i| \rho |u_i\rangle = |c_i|^2$$

which is the probability of finding the system in state $|u_i\rangle$.

To determine the meaning of the second part of the density operator, given by the sum

$$\sum_{i \neq j}^{n} c_i c_j^* |u_i\rangle\langle u_j|$$

first recall that the expansion coefficients are complex numbers. We can write any complex number in polar form, i.e. $z = re^{i\phi}$. In this case we have

$$c_j = |c_j| e^{i\phi_j}$$

So, with respect to the second summation, we have

$$\langle u_i| \rho |u_j\rangle = c_i c_j^* = |c_i| |c_j| e^{i(\phi_i - \phi_j)}$$

The phase difference in the exponential expresses the *coherence* or capability of terms in the state to interfere with each other. This is a characteristic of a pure quantum state and these terms are represented by off-diagonal elements in the density matrix. Mixed states are classical statistical mixtures and therefore have no terms of this form.

The off-diagonal elements of a mixed state will be zero, while those of a pure state—a state with coherences—will be non-zero. Note, however, that the presence of the off-diagonal terms is basis-dependent. Always check purity by computing the trace $Tr(\rho^2)$.

A *completely mixed* state is one for which the probability of each state is equal to all others. For example,

$$\rho = \frac{1}{2} |0\rangle \langle 0| + \frac{1}{2} |1\rangle \langle 1|$$

is a completely mixed state with a 50% probability of finding the system in state $|0\rangle$ or $|1\rangle$.

The Density Operator for a Completely Mixed State

In an n-dimensional Hilbert space the density operator can be expressed in terms of the identity matrix, and:

$$\rho = \frac{1}{n} I, \quad Tr(\rho^2) = \frac{1}{n}$$

A completely mixed state represents one extreme possibility, a statistical mixture with no interference terms. A pure state represents the other extreme, a quantum state in a coherent superposition. A measure of purity can be expressed by computing the trace of ρ^2 and setting it in between these bounds:

$$\frac{1}{n} \leq Tr(\rho^2) \leq 1$$

A state that falls in this range without being pure $Tr(\rho^2) = 1$ or completely mixed for which

$$Tr(\rho^2) = \frac{1}{n}$$

is called *partially coherent*. Testing purity by checking the trace of the square of the density matrix is a basis independent concept, unlike the values of the off-diagonal elements of the matrix, which depend on the basis you choose to express the density operator.

EXAMPLE 8.11

In this example we compare pure *vs* mixed states for a three-dimensional quantum system. We consider two orthonormal bases:

$$\{|1\rangle, |2\rangle, |3\rangle\} \quad \text{and} \quad \{|A\rangle, |B\rangle, |C\rangle\}$$

The second basis can be expressed in terms of the first using the relations

$$|A\rangle = \frac{1}{2}\left(|1\rangle + \sqrt{2}\,|2\rangle + |3\rangle\right)$$

$$|B\rangle = \frac{1}{\sqrt{2}}\left(|1\rangle - |3\rangle\right)$$

$$|C\rangle = \frac{1}{2}\left(|1\rangle - \sqrt{2}\,|2\rangle + |3\rangle\right)$$

Now consider the pure state

$$\rho_A = |A\rangle \langle A|$$

The state $|A\rangle$ is a coherent superposition of the states $|1\rangle, |2\rangle, |3\rangle$. Notice that in this state the probabilities of finding the states $|1\rangle, |2\rangle, |3\rangle$ are

$$prob\,(|1\rangle) = \frac{1}{4}, \quad prob\,(|2\rangle) = \frac{1}{2}, \quad prob\,(|3\rangle) = \frac{1}{4}$$

The state $|A\rangle$ is pure because if we make a measurement in the $\{|A\rangle, |B\rangle, |C\rangle\}$ basis the probabilities of obtaining each measurement result are

$$prob\,(|A\rangle) = 1, \quad prob\,(|B\rangle) = 0, \quad prob\,(|C\rangle) = 0$$

Now, using the expansion of $|A\rangle$ in the $|1\rangle, |2\rangle, |3\rangle$ basis, we can write the density operator $\rho_A = |A\rangle\langle A|$ as

$$\begin{aligned}
\rho_A &= |A\rangle\langle A| \\
&= \frac{1}{2}\left(|1\rangle + \sqrt{2}\,|2\rangle + |3\rangle\right)\frac{1}{2}\left(\langle 1| + \sqrt{2}\,\langle 2| + \langle 3|\right) \\
&= \frac{1}{4}\left[|1\rangle\langle 1| + 2\,|2\rangle\langle 2| + |3\rangle\langle 3| + |1\rangle\langle 3| + |3\rangle\langle 1|\right] \\
&\quad + \left[\sqrt{2}\,(|1\rangle\langle 2| + |2\rangle\langle 1| + |2\rangle\langle 3| + |3\rangle\langle 2|)\right]
\end{aligned}$$

Now we consider another state, a state that is a statistical mixture of the $|1\rangle, |2\rangle, |3\rangle$ states. Let

$$\rho_m = \frac{1}{4}|1\rangle\langle 1| + \frac{1}{2}|2\rangle\langle 2| + \frac{1}{4}|3\rangle\langle 3|$$

Recalling that a density operator is a summation of the form

$$\rho = \sum p_i\,|\psi_i\rangle\langle\psi_i|$$

where p_i is the probability of finding state $|\psi_i\rangle$, we see that in the state ρ_m the probabilities of finding each of the states is

$$prob\,(|1\rangle) = \frac{1}{4}$$

$$prob\,(|2\rangle) = \frac{1}{2}$$

$$prob\,(|3\rangle) = \frac{1}{4}$$

These are the same probabilities we found for the state $|A\rangle$, and so this seems to describe the same state. However, what are the probabilities of finding the system in the states $\{|A\rangle, |B\rangle, |C\rangle\}$ when the system is in the state described by ρ_m? We can answer this question by considering the projection operators

$|A\rangle\langle A|, |B\rangle\langle B|, |C\rangle\langle C|$ and using $Tr(\hat{O}|\Phi\rangle\langle\Phi|) = \langle\Phi|\hat{O}|\Phi\rangle$. The probability of finding $|A\rangle$ when the system is described by ρ_m is

$$prob(|A\rangle) = tr(\rho_m |A\rangle\langle A|) = \langle A|\rho_m|A\rangle$$

Now

$$\begin{aligned}\rho_m|A\rangle &= \left(\frac{1}{4}|1\rangle\langle 1| + \frac{1}{2}|2\rangle\langle 2| + \frac{1}{4}|3\rangle\langle 3|\right)\left(\frac{1}{2}\left(|1\rangle + \sqrt{2}|2\rangle + |3\rangle\right)\right)\\ &= \frac{1}{8}|1\rangle + \frac{\sqrt{2}}{4}|2\rangle + \frac{1}{8}|3\rangle\end{aligned}$$

and so

$$\begin{aligned}\langle A|\rho_m|A\rangle &= \left(\frac{1}{2}\langle 1| + \frac{\sqrt{2}}{2}\langle 2| + \frac{1}{2}\langle 3|\right)\left(\frac{1}{8}|1\rangle + \frac{\sqrt{2}}{4}|2\rangle + \frac{1}{8}|3\rangle\right)\\ &= \frac{1}{16}\langle 1|1\rangle + \frac{1}{4}\langle 2|2\rangle + \frac{1}{16}\langle 3|3\rangle\\ &= \frac{1}{16} + \frac{1}{4} + \frac{1}{16} = \frac{3}{8}\end{aligned}$$

This is quite different from the pure state where we found $prob(|A\rangle) = 1$. A similar procedure using the expansion of the state $|B\rangle$ in the $|1\rangle, |2\rangle, |3\rangle$ basis shows that the probability of finding $|B\rangle$ in the state ρ_m is

$$prob(|B\rangle) = Tr(\rho_m|B\rangle\langle B|) = \langle B|\rho_m|B\rangle = \frac{1}{4}$$

This is certainly different than what we found for ρ_A, where the probability was zero. In conclusion, the predictions of ρ_m are identical to those of ρ_A for measurements in the $|1\rangle, |2\rangle, |3\rangle$ basis, but they are quite different for the $\{|A\rangle, |B\rangle, |C\rangle\}$ basis.

Now let's consider the matrix representations of these density operators in the $|1\rangle, |2\rangle, |3\rangle$ basis. We have

$$\rho_m = \begin{pmatrix} \langle 1|\rho_m|1\rangle & \langle 1|\rho_m|2\rangle & \langle 1|\rho_m|3\rangle \\ \langle 2|\rho_m|1\rangle & \langle 2|\rho_m|2\rangle & \langle 2|\rho_m|3\rangle \\ \langle 3|\rho_m|1\rangle & \langle 3|\rho_m|2\rangle & \langle 3|\rho_m|3\rangle \end{pmatrix} = \begin{pmatrix} 1/4 & 0 & 0 \\ 0 & 1/2 & 0 \\ 0 & 0 & 1/4 \end{pmatrix}$$

Notice that

$$Tr(\rho_m) = \frac{1}{4} + \frac{1}{2} + \frac{1}{4} = 1$$

as it should for a density matrix.

An exercise shows that the density operator representing ρ_A in this basis is

$$\rho_A = \begin{pmatrix} 1/4 & \sqrt{2}/4 & 1/4 \\ \sqrt{2}/4 & 1/2 & \sqrt{2}/4 \\ 1/4 & \sqrt{2}/4 & 1/4 \end{pmatrix}$$

Notice that the mixed state has zero off-diagonal elements and the pure state has non-zero off-diagonal elements. While this is a characteristic property, it depends on the basis chosen to represent the density operator (try writing the matrix for ρ_A in the $\{|A\rangle, |B\rangle, |C\rangle\}$ basis.) We also have

$$Tr(\rho_A) = \frac{1}{4} + \frac{1}{2} + \frac{1}{4} = 1$$

Now consider the square of this matrix. A calculation shows

$$\rho_A^2 = \begin{pmatrix} \frac{1}{4} & \frac{1}{2\sqrt{2}} & \frac{1}{4} \\ \frac{1}{2\sqrt{2}} & \frac{1}{2} & \frac{1}{2\sqrt{2}} \\ \frac{1}{4} & \frac{1}{2\sqrt{2}} & \frac{1}{4} \end{pmatrix} \Rightarrow Tr(\rho_A^2) = \frac{1}{4} + \frac{1}{2} + \frac{1}{4} = 1$$

telling us that this is a pure state. Now consider

$$\rho_m^2 = \begin{pmatrix} \frac{1}{16} & 0 & 0 \\ 0 & \frac{1}{4} & 0 \\ 0 & 0 & \frac{1}{16} \end{pmatrix} \quad \Rightarrow \quad Tr\left(\rho_m^2\right) = \frac{1}{16} + \frac{1}{4} + \frac{1}{16} = \frac{3}{8} < 1$$

Therefore, this density operator represents a mixed state (as we knew already). Recalling that the dimension $n = 3$ of this Hilbert space, the completely mixed state would be represented by the density operator:

$$\rho = \frac{1}{3}|1\rangle\langle 1| + \frac{1}{3}|2\rangle\langle 2| + \frac{1}{3}|3\rangle\langle 3|$$

The matrix representation in this basis is

$$\rho = \begin{pmatrix} \frac{1}{3} & 0 & 0 \\ 0 & \frac{1}{3} & 0 \\ 0 & 0 & \frac{1}{3} \end{pmatrix} = \frac{1}{3}\begin{pmatrix} 1 & 0 & 0 \\ 0 & 1 & 0 \\ 0 & 0 & 1 \end{pmatrix} = \frac{1}{3}I$$

We see that

$$\rho^2 = \frac{1}{9}I$$

and so

$$Tr(\rho^2) = \frac{1}{9}Tr(I) = \frac{3}{9} = \frac{1}{3}$$

This is the completely mixed state for three dimensions. And so we have the bounds

$$\frac{1}{3} < Tr(\rho_m^2) = \frac{3}{8} < 1 = Tr(\rho_A^2)$$

where ρ_A is a pure state. The state ρ_m is a partially coherent state.

EXAMPLE 8.12

e.g.

We work in a two-dimensional vector space with basis $\{|0\rangle, |1\rangle\}$. Consider an ensemble of systems prepared in the following way. We have two quantum states

$$|A\rangle = \frac{2}{\sqrt{5}}|0\rangle + \frac{1}{\sqrt{5}}|1\rangle, \quad |B\rangle = \frac{\sqrt{3}}{2}|0\rangle + \frac{1}{2}|1\rangle$$

In the preparation of the ensemble, 3/4 of the states are prepared in state $|A\rangle$ and 1/4 of the states are prepared in state $|B\rangle$.

(a) Write down the density operator for each individual pure state $|A\rangle$, and $|B\rangle$.

(b) Write down the density operator for the ensemble.

(c) Express the density operator as a matrix with respect to the $\{|0\rangle, |1\rangle\}$ basis.

(d) Compute the trace.

(e) If a particle is drawn from the ensemble and a measurement is performed, what is the probability it is found in state $|0\rangle$? What is the probability it is found in state $|1\rangle$?

(f) Suppose there is another measurement with basis states $|+\rangle = 1/\sqrt{2}(|0\rangle + |1\rangle)$ and $|-\rangle = 1/\sqrt{2}(|0\rangle - |1\rangle)$. A particle is drawn from the ensemble and a measurement is made in this basis. What is the probability of finding $|+\rangle$ and what is the probability of finding $|-\rangle$?

SOLUTION

(a) The density operators for each pure state are

$$\rho_A = |A\rangle\langle A| = \frac{4}{5}|0\rangle\langle 0| + \frac{2}{5}|0\rangle\langle 1| + \frac{2}{5}|1\rangle\langle 0| + \frac{1}{5}|1\rangle\langle 1|$$

$$\rho_B = |B\rangle\langle B| = \frac{3}{4}|0\rangle\langle 0| + \frac{\sqrt{3}}{4}|0\rangle\langle 1| + \frac{\sqrt{3}}{4}|1\rangle\langle 0| + \frac{1}{4}|1\rangle\langle 1|$$

(b) The density operator for the ensemble can be found from the formula

$$\rho = \sum_{i=1}^{n} p_i \left|\Phi_i\right\rangle \left\langle\Phi_i\right|$$

We are told that 3/4 of the systems are prepared in state $|A\rangle$ and 1/4 of the systems are prepared in state $|B\rangle$. Therefore in this case we have

$$\rho = \frac{3}{4}\left|A\right\rangle\left\langle A\right| + \frac{1}{4}\left|B\right\rangle\left\langle B\right| = \frac{3}{4}\rho_A + \frac{1}{4}\rho_B$$

Now

$$\frac{3}{4}\rho_A = \frac{3}{4}\left(\frac{4}{5}\left|0\right\rangle\left\langle 0\right| + \frac{2}{5}\left|0\right\rangle\left\langle 1\right| + \frac{2}{5}\left|1\right\rangle\left\langle 0\right| + \frac{1}{5}\left|1\right\rangle\left\langle 1\right|\right)$$

$$= \frac{3}{5}\left|0\right\rangle\left\langle 0\right| + \frac{3}{10}\left|0\right\rangle\left\langle 1\right| + \frac{3}{10}\left|1\right\rangle\left\langle 0\right| + \frac{3}{20}\left|1\right\rangle\left\langle 1\right|$$

$$\frac{1}{4}\rho_B = \frac{1}{4}\left(\frac{3}{4}\left|0\right\rangle\left\langle 0\right| + \frac{\sqrt{3}}{4}\left|0\right\rangle\left\langle 1\right| + \frac{\sqrt{3}}{4}\left|1\right\rangle\left\langle 0\right| + \frac{1}{4}\left|1\right\rangle\left\langle 1\right|\right)$$

$$= \frac{3}{16}\left|0\right\rangle\left\langle 0\right| + \frac{\sqrt{3}}{16}\left|0\right\rangle\left\langle 1\right| + \frac{\sqrt{3}}{16}\left|1\right\rangle\left\langle 0\right| + \frac{1}{16}\left|1\right\rangle\left\langle 1\right|$$

Adding these terms together gives

$$\rho = \frac{3}{4}\left|A\right\rangle\left\langle A\right| + \frac{1}{4}\left|B\right\rangle\left\langle B\right| = \frac{3}{4}\rho_A + \frac{1}{4}\rho_B$$

$$= \frac{63}{80}\left|0\right\rangle\left\langle 0\right| + \frac{24+5\sqrt{3}}{80}\left|0\right\rangle\left\langle 1\right| + \frac{24+5\sqrt{3}}{80}\left|1\right\rangle\left\langle 0\right| + \frac{17}{80}\left|1\right\rangle\left\langle 1\right|$$

(c) The matrix that represents this density operator in the $\{|0\rangle, |1\rangle\}$ basis is

$$\rho = \begin{pmatrix} \left\langle 0\right|\rho\left|0\right\rangle & \left\langle 0\right|\rho\left|1\right\rangle \\ \left\langle 1\right|\rho\left|0\right\rangle & \left\langle 1\right|\rho\left|1\right\rangle \end{pmatrix} = \begin{pmatrix} \frac{63}{80} & \frac{24+5\sqrt{3}}{80} \\ \frac{24+5\sqrt{3}}{80} & \frac{17}{80} \end{pmatrix}$$

(d) We find the trace by summing the diagonal elements

$$Tr(\rho) = \frac{63}{80} + \frac{17}{80} = \frac{80}{80} = 1$$

as it should be for a density matrix.

(e) The probabilities that a particle drawn is found in the state $|0\rangle$ or the state $|1\rangle$ are found from the elements along the diagonal of the matrix.

The probability the particle is found to be in state $|0\rangle$ is $63/80 = 0.79$

The probability the particle is found to be in state $|1\rangle$ is $17/80 = 0.21$

(f) We can find the probability that a particle is found in the state $|+\rangle = 1/\sqrt{2}\,(|0\rangle + |1\rangle)$ by calculating $\langle +|\,\rho\,|+\rangle$. The column vector representation of $|+\rangle$ is

$$|+\rangle = \frac{|0\rangle + |1\rangle}{\sqrt{2}} = \frac{1}{\sqrt{2}}\left[\begin{pmatrix}1\\0\end{pmatrix} + \begin{pmatrix}0\\1\end{pmatrix}\right] = \frac{1}{\sqrt{2}}\begin{pmatrix}1\\1\end{pmatrix}$$

We can calculate the probability using matrix multiplication:

$$\langle +|\,\rho\,|+\rangle = \begin{pmatrix}\frac{1}{\sqrt{2}} & \frac{1}{\sqrt{2}}\end{pmatrix}\begin{pmatrix}\frac{63}{80} & \frac{24+5\sqrt{3}}{80}\\ \frac{24+5\sqrt{3}}{80} & \frac{17}{80}\end{pmatrix}\begin{pmatrix}\frac{1}{\sqrt{2}}\\ \frac{1}{\sqrt{2}}\end{pmatrix}$$

$$= \begin{pmatrix}\frac{1}{\sqrt{2}} & \frac{1}{\sqrt{2}}\end{pmatrix}\begin{pmatrix}\frac{63}{80\sqrt{2}} & \frac{24+5\sqrt{3}}{80\sqrt{2}}\\ \frac{24+5\sqrt{3}}{80\sqrt{2}} & \frac{17}{80\sqrt{2}}\end{pmatrix} = 0.91$$

The probability the particle is found in the state $|-\rangle = \frac{1}{\sqrt{2}}\,(|0\rangle - |1\rangle)$ is

$$\langle -|\,\rho\,|-\rangle = \begin{pmatrix}\frac{1}{\sqrt{2}} & -\frac{1}{\sqrt{2}}\end{pmatrix}\begin{pmatrix}\frac{63}{80} & \frac{24+5\sqrt{3}}{80}\\ \frac{24+5\sqrt{3}}{80} & \frac{17}{80}\end{pmatrix}\begin{pmatrix}\frac{1}{\sqrt{2}}\\ -\frac{1}{\sqrt{2}}\end{pmatrix} = 0.09$$

EXAMPLE 8.13

e.g.

A density matrix is given by

$$\rho = \frac{1}{4}\begin{pmatrix}3 & 1+i\\ 1-i & 1\end{pmatrix}$$

In quantum information theory, a measurement of information content is given by the entropy. The entropy is given by:

$$S = Tr(\rho \log \rho) = -\sum \lambda_i \log \lambda_i$$

where the λ_i are the eigenvalues of the density matrix and the log is base 2. Higher entropy means more disorder (and less information) in a system. Find the entropy for the system described by this density matrix.

SOLUTION

The eigenvalues of this matrix are $\{0.933, 0.067\}$. The entropy is:

$$S = -(0.933)\log(0.933) - 0.067\log(0.067) = 0.35$$

The maximum entropy for the basic unit of quantum information theory, a qubit, is $S = 1$.

EXAMPLE 8.14

Suppose the density matrix for some system is:

$$\rho = \begin{pmatrix} \frac{3}{4} & \frac{1+i}{4} \\ \frac{1-i}{4} & \frac{1}{4} \end{pmatrix}$$

(a) Show that this is a mixed state.

(b) Find $\langle X \rangle$ and $\langle Z \rangle$ for this state.

SOLUTION

(a) We calculate the square of the matrix

$$\rho^2 = \begin{pmatrix} \frac{3}{4} & \frac{1+i}{4} \\ \frac{1-i}{4} & \frac{1}{4} \end{pmatrix}\begin{pmatrix} \frac{3}{4} & \frac{1+i}{4} \\ \frac{1-i}{4} & \frac{1}{4} \end{pmatrix} = \begin{pmatrix} \frac{11}{16} & \frac{1+i}{4} \\ \frac{1-i}{4} & \frac{3}{16} \end{pmatrix}$$

The trace is

$$Tr(\rho^2) = \frac{11}{16} + \frac{3}{16} = \frac{14}{16} = \frac{7}{8} < 1$$

Therefore this is a mixed state.

(b) We calculate the matrix product

$$\rho X = \begin{pmatrix} \frac{3}{4} & \frac{1+i}{4} \\ \frac{1-i}{4} & \frac{1}{4} \end{pmatrix}\begin{pmatrix} 0 & 1 \\ 1 & 0 \end{pmatrix} = \begin{pmatrix} \frac{1+i}{4} & \frac{3}{4} \\ \frac{1}{4} & \frac{1-i}{4} \end{pmatrix}$$

The expectation value can be found from

$$Tr(\rho X) = Tr\begin{pmatrix} \frac{1+i}{4} & \frac{3}{4} \\ \frac{1}{4} & \frac{1-i}{4} \end{pmatrix} = \frac{1+i}{4} + \frac{1-i}{4} = \frac{1}{4} + \frac{1}{4} = \frac{1}{2}$$

For Z we have

$$\rho Z = \begin{pmatrix} \frac{3}{4} & \frac{1+i}{4} \\ \frac{1-i}{4} & \frac{1}{4} \end{pmatrix}\begin{pmatrix} 1 & 0 \\ 0 & -1 \end{pmatrix} = \begin{pmatrix} \frac{3}{4} & -\left(\frac{1+i}{4}\right) \\ \frac{1-i}{4} & -\frac{1}{4} \end{pmatrix}$$

and the expectation value is given by

$$\langle Z \rangle = Tr(\rho Z) = \frac{3}{4} - \frac{1}{4} = \frac{1}{2}$$

A Brief Introduction to the Bloch Vector

For a two-level system, we can write the density operator in the following form

$$\rho = \frac{I + \vec{r}.\vec{\sigma}}{2}$$

The vector $\vec{r}$ is called the *Bloch vector*. We can use the Bloch vector to determine if a state is a pure state or a mixed state.

The magnitude of the Bloch vector is less than or equal to one (with equality only for a pure state):

$$\|\vec{r}\| \leq 1$$

The components of the Bloch vector are calculated from

$$r_x = Tr(\rho X), \quad r_y = Tr(\rho Y), \quad r_z = Tr(\rho Z)$$

where X, Y, Z are the Pauli operators.

EXAMPLE 8.15

A density operator for some system is given by

$$\rho = \begin{pmatrix} \frac{2}{3} & \frac{1}{6} - \frac{1}{3}i \\ \frac{1}{6} + \frac{1}{3}i & \frac{1}{3} \end{pmatrix}$$

(a) Find the components of the Bloch vector.
(b) Is this state mixed or is it a pure state?
(c) A measurement of spin is made in the z-direction. What is the probability that the measurement result is spin-down? What is the probility that the measurement is spin-up?

SOLUTION

(a) The components of the Bloch vector are

$$r_x = Tr\,(\rho X) = Tr\left[\begin{pmatrix} \frac{2}{3} & \frac{1}{6} - \frac{1}{3}i \\ \frac{1}{6} + \frac{1}{3}i & \frac{1}{3} \end{pmatrix}\begin{pmatrix} 0 & 1 \\ 1 & 0 \end{pmatrix}\right]$$

$$= Tr\left[\begin{pmatrix} \frac{1}{6} - \frac{1}{3}i & \frac{2}{3} \\ \frac{1}{3} & \frac{1}{6} + \frac{1}{3}i \end{pmatrix}\right] = \frac{1}{6} - \frac{1}{3}i + \frac{1}{6} + \frac{1}{3}i = \frac{1}{3}$$

$$r_y = Tr\,(\rho Y) = Tr\left[\begin{pmatrix} \frac{2}{3} & \frac{1}{6} - \frac{1}{3}i \\ \frac{1}{6} + \frac{1}{3}i & \frac{1}{3} \end{pmatrix}\begin{pmatrix} 0 & -i \\ i & 0 \end{pmatrix}\right]$$

$$= Tr\left[\begin{pmatrix} \frac{1}{3} + \frac{1}{6}i & -i\frac{2}{3} \\ i\frac{1}{3} & \frac{1}{3} - \frac{1}{6}i \end{pmatrix}\right] = \frac{1}{3} + \frac{1}{6}i + \frac{1}{3} - \frac{1}{6}i = \frac{2}{3}$$

$$r_z = Tr\,(\rho Z) = Tr\left[\begin{pmatrix} \frac{2}{3} & \frac{1}{6} - \frac{1}{3}i \\ \frac{1}{6} + \frac{1}{3}i & \frac{1}{3} \end{pmatrix}\begin{pmatrix} 1 & 0 \\ 0 & -1 \end{pmatrix}\right]$$

$$= Tr\left[\begin{pmatrix} \frac{2}{3} & -\frac{1}{6} + \frac{1}{3}i \\ \frac{1}{6} + \frac{1}{3}i & -\frac{1}{3} \end{pmatrix}\right] = \frac{2}{3} - \frac{1}{3} = \frac{1}{3}$$

(b) The magnitude of the Bloch vector is

$$||\vec{r}|| = \sqrt{r_x^2 + r_y^2 + r_z^2} = \sqrt{\frac{1}{9} + \frac{4}{9} + \frac{1}{9}} = \sqrt{\frac{2}{3}} < 1$$

Since the magnitude is less than one, this density operator represents a mixed state.

(c) The probability of obtaining a given result is given by $Tr(\rho P)$ where P is a projection operator corresponding to the given measurement. The projection operator for spin-down is

$$P_1 = |1\rangle\langle 1| = \begin{pmatrix} 0 & 0 \\ 0 & 1 \end{pmatrix}$$

$$P_1\rho = \begin{pmatrix} 0 & 0 \\ 0 & 1 \end{pmatrix}\begin{pmatrix} \frac{2}{3} & \frac{1}{6} - \frac{1}{3}i \\ \frac{1}{6} + \frac{1}{3}i & \frac{1}{3} \end{pmatrix} = \begin{pmatrix} 0 & 0 \\ \frac{1}{6} + \frac{1}{3}i & \frac{1}{3} \end{pmatrix}$$

The probability is found from the trace, which is

$$Tr(P_1\rho) = 0 + \frac{1}{3} = \frac{1}{3}$$

For spin-up, the projection operator is

$$P_0 = |0\rangle\langle 0| = \begin{pmatrix} 1 & 0 \\ 0 & 0 \end{pmatrix}$$

The product of the matrices is

$$P_0\rho = \begin{pmatrix} 1 & 0 \\ 0 & 0 \end{pmatrix}\begin{pmatrix} \frac{2}{3} & \frac{1}{6} - \frac{1}{3}i \\ \frac{1}{6} + \frac{1}{3}i & \frac{1}{3} \end{pmatrix} = \begin{pmatrix} \frac{2}{3} & \frac{1}{6} - \frac{1}{3}i \\ 0 & 0 \end{pmatrix}$$

and so the probability of finding spin-up for this state is

$$Tr(P_0\rho) = \frac{2}{3}$$

Quiz

1. Using matrix representation, show that the Hadamard matrix

$$H = \frac{1}{\sqrt{2}}\begin{pmatrix} 1 & 1 \\ 1 & -1 \end{pmatrix}$$

produces superposition states out of

$$|0\rangle = \begin{pmatrix} 1 \\ 0 \end{pmatrix}, \quad |1\rangle = \begin{pmatrix} 0 \\ 1 \end{pmatrix}$$

2. For the system described by the density matrix

$$\rho = \begin{pmatrix} \frac{63}{80} & \frac{24+5\sqrt{3}}{80} \\ \frac{24+5\sqrt{3}}{80} & \frac{17}{80} \end{pmatrix}$$

show that $\langle Z \rangle = 23/40$.

3. Using the basis states

$$|+\rangle = \frac{|0\rangle + |1\rangle}{\sqrt{2}}, \quad |-\rangle = \frac{|0\rangle - |1\rangle}{\sqrt{2}}$$

(a) Verify that the new basis is orthonormal. Write down the density operator for each state, and then derive the density matrices.

(b) Suppose that an ensemble of systems is prepared with 50% of the systems in state $|0\rangle$ and 50% of the systems in state $|1\rangle$. Write down the density operator that describes this state.

(c) For each density matrix in this problem, is the state pure or mixed? In each state, what is the probability of finding $|+\rangle$?

4. The density matrix for a given state is:

$$\rho = \begin{pmatrix} \frac{3}{4} & -\frac{i}{4} \\ \frac{i}{4} & \frac{1}{4} \end{pmatrix}$$

(a) Is this a pure state?

(b) Show that the components of the Bloch vector are $\{0, 1/2, 1/2\}$. What can you learn about the purity of the state from the Bloch vector?

(c) If the Z operator is measured, show that the probability of finding $|1\rangle$ is $1/4$.

5. Consider an orthonormal basis $\{|u_1\rangle, |u_2\rangle, |u_3\rangle\}$. The Hamiltonian acts on this basis as

$$H|u_1\rangle = \hbar\omega|u_1\rangle$$

$$H|u_2\rangle = 3\hbar\omega|u_2\rangle$$

$$H|u_3\rangle = 5\hbar\omega|u_3\rangle$$

A system is in the state $|\psi\rangle = A|u_1\rangle + 1/2|u_2\rangle - i/\sqrt{3}|u_3\rangle$.

(a) Write down the bra that corresponds to the state $|\psi\rangle$.

(b) Find A so that the state is normalized.

(c) Write down the outer product and matrix representations of the Hamiltonian.

(d) Write $|\psi\rangle$ as a column vector.

(e) Using the Born rule, find the probabilities of obtaining each of the possible measurement results when the energy is measured.

(f) Find the average energy for this system.

(g) How does this state evolve in time?

The Harmonic Oscillator

The harmonic oscillator for a particle of mass m is described by the potential $V = 1/2kx^2$ where $k = m\omega^2$. Solutions to the Schrödinger equation for this potential are given in terms of Hermite polynomials, and can be obtained by either working in the position representation or by using an algebraic method based on the raising and lowering operators.

The Solution of the Harmonic Oscillator in the Position Representation

The Hamiltonian for the harmonic oscillator in one dimension is

$$H = \frac{p^2}{2m} + \frac{1}{2}kx^2 = \frac{p^2}{2m} + \frac{1}{2}m\omega^2 x^2$$

where ω is the angular frequency and m is the mass of the oscillator. The time-independent Schrödinger equation takes the form

$$-\frac{\hbar^2}{2m}\frac{d^2\psi}{dx^2} + \frac{1}{2}m\omega^2 x^2 \psi = E\psi$$

Typically, dimensionless parameters are introduced for the position coordinate and the energy

$$u = \sqrt{\frac{m\omega}{\hbar}}x, \quad \varepsilon = \frac{2E}{\hbar\omega}$$

The Schrödinger equation can then be rewritten as

$$\frac{d^2\psi}{du^2} + \left(\varepsilon - u^2\right)\psi = 0$$

where $\psi = \psi(u)$. Solutions to this equation can be obtained using the series method; we simply summarize them here. They are a product of an exponential and a *Hermite polynomial*

$$\psi_n(u) = H_n(u)\, e^{-u^2/2}$$

In terms of the position coordinate, the solution takes the form

$$\psi_n(x) = A_n H_n\left(\sqrt{\frac{m\omega}{\hbar}}x\right) e^{-m\omega x^2/2\hbar}$$

Here A_n is a normalization constant. The following plots (Figs. 9-1, 9-2, and 9-3) show the form of the first three wave functions.

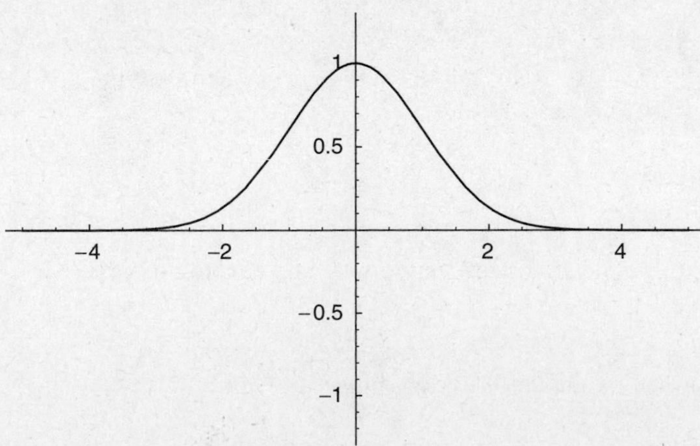

Fig. 9-1

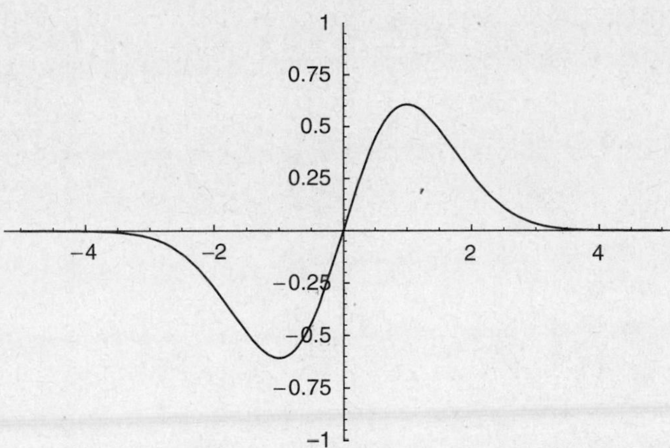

Fig. 9-2

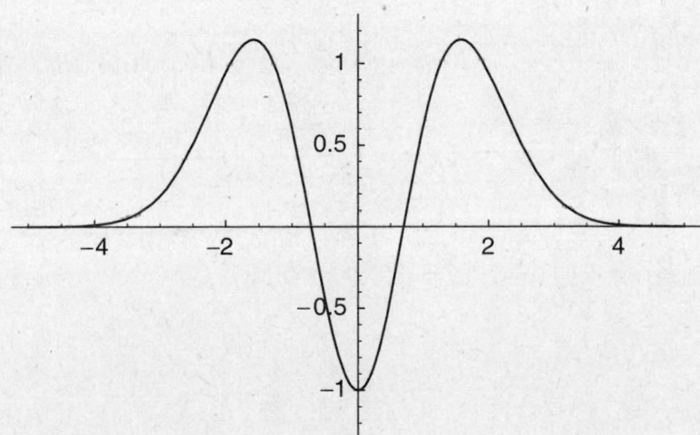

Fig. 9-3

EXAMPLE 9.1

e.g.

Show that the ground state wavefunction of the harmonic oscillator

$$\psi_0(x) = \left(\frac{m\omega}{\pi\hbar}\right)^{\frac{1}{4}} \exp\left(-\frac{m\omega}{2\hbar}x^2\right)$$

is normalized. If a harmonic oscillator is in this state, find the probability that the particle can be found in the range $0 \leq x \leq 1$.

SOLUTION

To check normalization, we begin by squaring the wavefunction

$$\psi_0^2 = \sqrt{\frac{m\omega}{\pi\hbar}} \exp\left(-\frac{m\omega}{\hbar}x^2\right)$$

Recalling that

$$\int_{-\infty}^{\infty} e^{-z^2}\, dz = \sqrt{\pi}$$

we now set $z^2 = \frac{m\omega}{\hbar}x^2$. Then we have

$$z = \sqrt{\frac{m\omega}{\hbar}}x, \quad \Rightarrow dz = \sqrt{\frac{m\omega}{\hbar}}\, dx$$

We invert this relation to give

$$dx = \sqrt{\frac{\hbar}{m\omega}}\, dz$$

Using these substitutions, the normalization integral becomes

$$\begin{aligned}\int_{-\infty}^{\infty} \psi_0^2(x)\, dx &= \int_{-\infty}^{\infty} \sqrt{\frac{m\omega}{\pi\hbar}} \exp\left(-\frac{m\omega}{\hbar}x^2\right) dx \\ &= \sqrt{\frac{\hbar}{m\omega}} \int_{-\infty}^{\infty} \sqrt{\frac{m\omega}{\pi\hbar}} \exp\left(-z^2\right) dz \\ &= \frac{1}{\sqrt{\pi}} \int_{-\infty}^{\infty} e^{-z^2}\, dz = \frac{\sqrt{\pi}}{\sqrt{\pi}} = 1\end{aligned}$$

Therefore the state is normalized. The probability that the particle is found in the range $0 \leq x \leq 1$ is given by

$$\int_0^1 \psi_0^2(x)\, dx = \sqrt{\frac{m\omega}{\pi\hbar}} \int_0^1 \exp\left(-\frac{m\omega}{\hbar}x^2\right) dx$$

This integral is nearly in the form of the error function

$$erf(z) = \frac{2}{\sqrt{\pi}} \int_0^z e^{-u^2}\, du$$

Following the procedure used in checking normalization, we set

$$u^2 = \frac{m\omega}{\hbar}x^2$$

and obtain

$$\sqrt{\frac{m\omega}{\pi\hbar}}\int_0^1 \exp\left(-\frac{m\omega}{\hbar}x^2\right)dx = \sqrt{\frac{m\omega}{\pi\hbar}}\sqrt{\frac{\hbar}{m\omega}}\int_0^1 \exp\left(-u^2\right)du$$

$$= \frac{1}{\sqrt{\pi}}\int_0^1 \exp\left(-u^2\right)du = \frac{1}{2}erf(1) \approx \frac{1}{2}(0.84) = 0.42$$

Definition: Normalization of the Hermite Polynomials

The normalization of the wavefunctions comes from that of the Hermite polynomials. The orthonormality of the Hermite polynomials is written as

$$\int_{-\infty}^{\infty} H_m(u)\, H_n(u)\, e^{-u^2}\, du = \sqrt{\pi}2^n n!\delta_{mn}$$

Using this relationship, we can normalize the wavefunctions by integrating with the normalization constant

$$\int_{-\infty}^{\infty} A_n^2 \psi_n^2\, du = A_n^2 \int_{-\infty}^{\infty} H_n^2(u)\, e^{-u^2}\, du$$

$$= A_n^2\left(\sqrt{\frac{\hbar}{m\omega}}\right)\int_{-\infty}^{\infty} H_n^2\left(\sqrt{\frac{m\omega}{\hbar}}x\right)e^{-m\omega x^2/\hbar}\, dx$$

$$= A_n^2\left(\sqrt{\frac{\hbar}{m\omega}}\right)\sqrt{\pi}2^n n!$$

To have a normalized wavefunction, this must be equal to unity and so we have

$$A_n^2 = \frac{1}{\sqrt{\pi}2^n n!}\sqrt{\frac{m\omega}{\hbar}}, \quad \Rightarrow A_n = \frac{1}{\sqrt{2^n n!}}\left(\frac{m\omega}{\pi\hbar}\right)^{\frac{1}{4}}$$

We can then write the normalized wavefunction as

$$\psi_n(x) = \frac{1}{\sqrt{2^n n!}}\left(\frac{m\omega}{\pi\hbar}\right)^{\frac{1}{4}} H_n\left(\sqrt{\frac{m\omega}{\hbar}}x\right)e^{-m\omega x^2/2\hbar}$$

The energy is found from the series solution technique applied to the Schrödinger equation. The termination condition for this solution dictates that the energy of state n is given by

$$E_n = \left(n + \frac{1}{2}\right)\hbar\omega, \quad n = 0, 1, 2, \ldots$$

Helpful recursion relationships exist that can be used to derive higher-order Hermite polynomials. These include

$$H_{n+1}(u) = 2u\, H_n(u) - 2n\, H_{n-1}(u)$$

$$\frac{dH_n}{du} = 2n\, H_{n-1}(u)$$

The first few Hermite polynomials are given by

$$H_0(u) = 1$$

$$H_1(u) = 2u$$

$$H_2(u) = 4u^2 - 2$$

$$H_3(u) = 8u^3 - 12u$$

The recursion relations can be useful for determining expectation values.

e.g. **EXAMPLE 9.2**

Find $H_4(u)$.

✔ **SOLUTION**

Using the recursion formula

$$H_{n+1}(u) = 2uH_n(u) - 2nH_{n-1}(u)$$

and setting $n + 1 = 4$, we obtain

$$\begin{aligned} H_4(u) &= 2uH_3(u) - 2(3)H_2(u) \\ &= 2u\left(8u^3 - 12u\right) - 6\left(4u^2 - 2\right) \\ &= 16u^4 - 48u^2 + 12 \end{aligned}$$

---

e.g. **EXAMPLE 9.3**

Show that

$$\psi(u) = \left(8u^3 - 12u\right)e^{-u^2/2}$$

is an eigenfunction of the dimensionless equation

$$\frac{d^2\psi}{du^2} + \left(\varepsilon - u^2\right)\psi = 0$$

and find the corresponding eigenvalue. Use the relationships used to derive the dimensionless parameters to find the energy that this represents for a particle in the harmonic oscillator potential, and find the energy level.

SOLUTION

$$\frac{d\psi}{du} = \frac{d}{du}\left[\left(8u^3 - 12u\right)e^{-u^2/2}\right] = \left(24u^2 - 12\right)e^{-u^2/2} - u\left(8u^3 - 12u\right)e^{-u^2/2}$$

$$= \left(-12 + 36u^2 - 8u^4\right)e^{-u^2/2}$$

$$\Rightarrow$$

$$\frac{d^2\psi}{du^2} = \left(72u - 32u^3\right)e^{-u^2/2} - u\left(-12 + 36u^2 - 8u^4\right)e^{-u^2/2}$$

$$= \left(84u - 68u^3 + 8u^5\right)e^{-u^2/2}$$

We rewrite the dimensionless equation by moving the energy term to the other side:

$$\frac{d^2\psi}{du^2} - u^2\psi = -\varepsilon\psi$$

Using the result obtained for the second derivative, the left-hand side is

$$\frac{d^2\psi}{du^2} - u^2\psi = \left(84u - 68u^3 + 8u^5\right)e^{-u^2/2} - u^2\left(8u^3 - 12u\right)e^{-u^2/2}$$

$$= \left(-56u^3 + 84u\right)e^{-u^2/2} = -7\left(8u^3 - 12u\right)e^{-u^2/2} = -7\psi$$

and so we have

$$-7\psi = -\varepsilon\psi, \quad \Rightarrow \varepsilon = 7$$

Now we recall that

$$\varepsilon = \frac{2E}{\hbar\omega}$$

Solving for E, we obtain

$$E = \frac{\varepsilon\hbar\omega}{2} = \frac{7\hbar\omega}{2}$$

To determine the energy level, we recall that the energy of the harmonic oscillator is

$$E_n = \left(n + \frac{1}{2}\right)\hbar\omega$$

So we rewrite the expression we have derived in this form

$$E = \frac{7\hbar\omega}{2} = \left(\frac{6}{2} + \frac{1}{2}\right)\hbar\omega = \left(3 + \frac{1}{2}\right)\hbar\omega$$

Therefore we see that this is the $n = 3$ excited state of the harmonic oscillator.

EXAMPLE 9.4

Suppose that a particle is in the state

$$\psi\,(x, 0) = \frac{1}{\sqrt{2}}\psi_0\,(x) + \frac{1}{\sqrt{2}}\psi_1\,(x)$$

Write down the state at time t and show that $\langle x \rangle$ oscillates in time.

SOLUTION

If the initial wavefunction is in some superposition of basis states

$$\psi\,(x, 0) = \sum c_n \Phi_n\,(x)$$

setting $E = \hbar\omega$, the time evolution of the state can be written as

$$\psi\,(x, t) = \sum c_n \Phi_n\,(x)\, e^{-i\omega_n t}$$

We apply this procedure to the wavefunction as stated in the problem

$$\begin{aligned}\psi\,(x, t) &= \frac{1}{\sqrt{2}}\psi_0\,(x)\, e^{-iE_1 t/\hbar} + \frac{1}{\sqrt{2}}\psi_1\,(x)\, e^{-iE_2 t/\hbar} \\ &= \frac{1}{\sqrt{2}}\psi_0\,(x)\, e^{-i\omega t/2} + \frac{1}{\sqrt{2}}\psi_1\,(x)\, e^{-i3\omega t/2}\end{aligned}$$

Now we turn to the problem of finding the expectation value. The exact form of the basis states is

$$\psi_0\,(x) = \left(\frac{m\omega}{\pi\hbar}\right)^{\frac{1}{4}} \exp\left(-\frac{m\omega}{2\hbar}x^2\right), \quad \psi_1\,(x) = \sqrt{2}\pi^{-1/4}\left(\frac{m\omega}{\hbar}\right)^{\frac{3}{4}} x \exp\left(-\frac{m\omega}{2\hbar}x^2\right)$$

For simplicity we denote

$$A_0 = \left(\frac{m\omega}{\pi\hbar}\right)^{\frac{1}{4}}, \quad A_1 = \sqrt{2}\pi^{-1/4}\left(\frac{m\omega}{\hbar}\right)^{\frac{3}{4}}$$

We will also use the frequently seen integrals

$$\int_{-\infty}^{\infty} z^{2n} e^{-z^2}\, dz = \sqrt{\pi}\frac{1.3.5\ldots(2n-1)}{2^n}, \qquad n = 1, 2, \ldots$$

$$\int_{-\infty}^{\infty} z^{n} e^{-z^2}\, dz = 0 \quad (n \quad \text{odd})$$

So the expectation value of position for the given state is

$$\langle x\rangle = \int_{-\infty}^{\infty}\left\{\left[\frac{1}{\sqrt{2}}\psi_0(x)\exp\left(\frac{i\omega}{2}t\right) + \frac{1}{\sqrt{2}}\psi_1(x)\exp\left(\frac{i3\omega}{2}t\right)\right](x)\right.$$
$$\left.\left[\frac{1}{\sqrt{2}}\psi_0(x)\exp\left(-\frac{i\omega}{2}t\right) + \frac{1}{\sqrt{2}}\psi_1(x)\exp\left(-\frac{i3\omega}{2}t\right)\right]\right\} dx$$

First we simplify the integrand

$$\left[\frac{1}{\sqrt{2}}\psi_0(x)\exp\left(\frac{i\omega}{2}t\right) + \frac{1}{\sqrt{2}}\psi_1(x)\exp\left(\frac{i3\omega}{2}t\right)\right]$$
$$\left[\frac{1}{\sqrt{2}}\psi_0(x)\exp\left(-\frac{i\omega}{2}t\right) + \frac{1}{\sqrt{2}}\psi_1(x)\exp\left(-\frac{i3\omega}{2}t\right)\right]$$
$$= \frac{1}{2}\left[\psi_0^2(x) + \psi_0(x)\psi_1(x)e^{-i\omega t} + \psi_1(x)\psi_0(x)e^{i\omega t} + \psi_1^2(x)\right]$$

and so we obtain

$$\langle x\rangle = \int_{-\infty}^{\infty}\frac{x}{2}\left[\psi_0^2(x) + \psi_0(x)\psi_1(x)e^{-i\omega t} + \psi_1(x)\psi_0(x)e^{i\omega t} + \psi_1^2(x)\right] dx$$
$$= \frac{1}{2}\int_{-\infty}^{\infty} x\psi_0^2(x)\, dx + \frac{1}{2}\int_{-\infty}^{\infty} x\psi_1^2(x)\, dx + \frac{e^{-i\omega t}}{2}\int_{-\infty}^{\infty} x\psi_0(x)\psi_1(x)\, dx$$
$$+ \frac{e^{i\omega t}}{2}\int_{-\infty}^{\infty} x\psi_1(x)\psi_0(x)\, dx$$

The first two terms vanish because $\int_{-\infty}^{\infty} z^n e^{-z^2}\, dz = 0$. To see this recall that

$$\psi_0(x) = A_0\exp\left(-\frac{m\omega}{2\hbar}x^2\right)$$

and perform a substitution. The second integral vanishes for the same reason. Therefore we are left with

$$\langle x\rangle = \frac{e^{-i\omega t}}{2}\int_{-\infty}^{\infty} x\psi_0(x)\psi_1(x)\, dx + \frac{e^{i\omega t}}{2}\int_{-\infty}^{\infty} x\psi_1(x)\psi_0(x)\, dx$$

$$= \frac{e^{i\omega t} + e^{-i\omega t}}{2} \int_{-\infty}^{\infty} x\psi_0(x)\,\psi_1(x)\,dx$$

$$= \cos(\omega t) \int_{-\infty}^{\infty} x\left(A_0 e^{-\frac{m\omega}{\hbar}x^2}\right)\left(A_1 x e^{-\frac{m\omega}{\hbar}x^2}\right) dx$$

$$= \cos(\omega t)\, A_0 A_1 \int_{-\infty}^{\infty} x^2 e^{-\frac{2m\omega}{\hbar}x^2}\, dx$$

This can be rewritten as

$$\cos(\omega t)\, A_0 A_1 \int_{-\infty}^{\infty} x^2 e^{-\frac{2m\omega}{\hbar}x^2}\, dx =$$

$$\cos(\omega t)\left(\frac{m\omega}{\pi\hbar}\right)^{\frac{1}{4}} \sqrt{2}\pi^{\frac{-1}{4}} \left(\frac{m\omega}{\hbar}\right)^{\frac{3}{4}} \left(\frac{\hbar}{m\omega}\right)^{\frac{3}{2}} \int_{-\infty}^{\infty} z^2 e^{-z^2}\, dz$$

But

$$\int_{-\infty}^{\infty} z^2 e^{-z^2}\, dz = \frac{\sqrt{\pi}}{2}$$

and so we find that the expectation value is

$$\langle x\rangle = \cos(\omega t)\left(\frac{m\omega}{\pi\hbar}\right)^{\frac{1}{4}} \sqrt{2}\pi^{-\frac{1}{4}} \left(\frac{m\omega}{\hbar}\right)^{\frac{3}{4}} \left(\frac{\hbar}{m\omega}\right)^{\frac{3}{2}} \frac{\sqrt{\pi}}{2} = \cos(\omega t)\sqrt{\frac{\pi\hbar}{2m\omega}}$$

We see that the expectation value oscillates in time with frequency ω.

The Operator Method for the Harmonic Oscillator

We now proceed to solve the harmonic oscillator problem using an entirely different method based on operators and algebra alone. Consider the following operators defined in terms of the position and momentum operators

$$a = \sqrt{\frac{m\omega}{2\hbar}}\left(x + \frac{ip}{m\omega}\right)$$

$$a^\dagger = \sqrt{\frac{m\omega}{2\hbar}}\left(x - \frac{ip}{m\omega}\right)$$

We can rewrite the Hamiltonian using these operators and then solve the eigenvector/eigenvalue problem in an algebraic way. An important part of working with these operators is to determine their commutator.

EXAMPLE 9.5

Derive the commutator $[a, a^\dagger]$.

SOLUTION

To find this commutator we rely on $[x, p] = i\hbar$

$$\begin{aligned}
[a, a^\dagger] &= \left[\sqrt{\frac{m\omega}{2\hbar}}\left(x + \frac{ip}{m\omega}\right), \sqrt{\frac{m\omega}{2\hbar}}\left(x - \frac{ip}{m\omega}\right)\right] \\
&= \frac{m\omega}{2\hbar}\left[\left(x + \frac{ip}{m\omega}\right), \left(x - \frac{ip}{m\omega}\right)\right] \\
&= \frac{m\omega}{2\hbar}\left\{[x, x] - \frac{i}{m\omega}[x, p] + \frac{i}{m\omega}[p, x] + \frac{1}{m^2\omega^2}[p, p]\right\}
\end{aligned}$$

Since $[x, x] = [p, p] = 0$, this simplifies to

$$\begin{aligned}
[a, a^\dagger] &= \frac{m\omega}{2\hbar}\left\{-\frac{i}{m\omega}[x, p] + \frac{i}{m\omega}[p, x]\right\} \\
&= \frac{i}{2\hbar}\{-[x, p] + [p, x]\} \\
&= \frac{i}{2\hbar}\{-[x, p] - [x, p]\} \\
&= \frac{-i}{\hbar}[x, p] = \frac{-i}{\hbar}(i\hbar) = 1 \\
\Rightarrow [a, a^\dagger] &= aa^\dagger - a^\dagger a = 1
\end{aligned}$$

EXAMPLE 9.6

Show that the harmonic oscillator Hamiltonian can be written in the form

$$H = \hbar\omega\left(a^\dagger a + \frac{1}{2}\right)$$

SOLUTION

We begin by writing the position and momentum operators in terms of $a, a^\dagger$. Notice that

$$a + a^\dagger = \sqrt{\frac{m\omega}{2\hbar}}\left(x + \frac{ip}{m\omega}\right) + \sqrt{\frac{m\omega}{2\hbar}}\left(x - \frac{ip}{m\omega}\right) = \sqrt{\frac{2m\omega}{\hbar}}x$$

and so the position operator can be written as

$$x = \sqrt{\frac{\hbar}{2m\omega}}\left(a + a^{\dagger}\right)$$

The harmonic oscillator Hamiltonian contains the square of x. Squaring this term we find

$$x^2 = \frac{\hbar}{2m\omega}\left(a + a^{\dagger}\right)^2 = \frac{\hbar}{2m\omega}\left(a^2 + aa^{\dagger} + a^{\dagger}a + \left(a^{\dagger}\right)^2\right)$$

Now we write the momentum operator in terms of $a, a^{\dagger}$. Consider

$$\begin{aligned} a - a^{\dagger} &= \sqrt{\frac{m\omega}{2\hbar}}\left(x + \frac{ip}{m\omega}\right) - \sqrt{\frac{m\omega}{2\hbar}}\left(x - \frac{ip}{m\omega}\right) \\ &= \sqrt{\frac{m\omega}{2\hbar}}\left(\frac{ip}{m\omega}\right) = i\frac{1}{\sqrt{2\hbar m\omega}}p \end{aligned}$$

And so we can write momentum as

$$p = -i\sqrt{\frac{\hbar m\omega}{2}}\left(a - a^{\dagger}\right), \Rightarrow$$

$$\frac{p^2}{2m} = -\frac{\hbar m\omega}{4m}\left(a^2 - aa^{\dagger} - a^{\dagger}a + \left(a^{\dagger}\right)^2\right)$$

Now we can insert these terms into the Hamiltonian

$$\begin{aligned} H &= \frac{p^2}{2m} + \frac{1}{2}m\omega^2 x^2 \\ &= -\frac{\hbar\omega}{4}\left[a^2 - aa^{\dagger} - a^{\dagger}a + \left(a^{\dagger}\right)^2\right] + \frac{m\omega^2}{2}\left(\frac{\hbar}{2m\omega}\right)\left[a^2 + aa^{\dagger} + a^{\dagger}a + \left(a^{\dagger}\right)^2\right] \end{aligned}$$

Notice that

$$\frac{m\omega^2}{2}\left(\frac{\hbar}{2m\omega}\right) = \frac{\hbar\omega}{4}$$

Therefore the a^2, $\left(a^{\dagger}\right)^2$ terms cancel. This leaves

$$H = \frac{\hbar\omega}{4}\left[2aa^{\dagger} + 2a^{\dagger}a\right] = \frac{\hbar\omega}{2}\left[aa^{\dagger} + a^{\dagger}a\right]$$

Now we use the commutation relation $\left[a, a^{\dagger}\right] = aa^{\dagger} - a^{\dagger}a = 1$ to write $aa^{\dagger} = 1 + a^{\dagger}a$, and we have

$$H = \frac{\hbar\omega}{4}\left[1 + 2a^{\dagger}a\right] = \hbar\omega\left(a^{\dagger}a + \frac{1}{2}\right)$$

Some other important commutation relations are

$$[H, a] = -\hbar\omega a, \quad \left[H, a^{\dagger}\right] = \hbar\omega a^{\dagger}$$

Number States of the Harmonic Oscillator

Now that we have expressed the Hamiltonian in terms of the operators $a, a^{\dagger}$, we can derive the energy eigenstates. We begin by stating the eigenvalue equation

$$H\left|E_n\right\rangle = E_n\left|E_n\right\rangle$$

To simplify notation, we set $\left|E_n\right\rangle = |n\rangle$. We have already seen that

$$E_n = \hbar\omega\left(n + \frac{1}{2}\right), \quad n = 0, 1, 2, \ldots$$

Using the form of the Hamiltonian written in terms of $a, a^{\dagger}$, we find that

$$H|n\rangle = \hbar\omega\left(a^{\dagger}a + \frac{1}{2}\right)|n\rangle = \hbar\omega\left(a^{\dagger}a|n\rangle\right) + \frac{\hbar\omega}{2}|n\rangle$$

However, we know that

$$H|n\rangle = E_n|n\rangle = \hbar\omega\left(n + \frac{1}{2}\right)|n\rangle$$

Equating this to the above, we have

$$\hbar\omega\left(a^{\dagger}a|n\rangle\right) + \frac{\hbar\omega}{2}|n\rangle = \hbar\omega n|n\rangle + \frac{\hbar\omega}{2}|n\rangle$$

Now divide through by $\hbar\omega$ and subtract the common term $1/2\,|n\rangle$ from both sides, giving

$$a^{\dagger}a|n\rangle = n|n\rangle$$

This shows that the energy eigenstate is an eigenstate of $a^{\dagger}a$ with eigenvalue n. The operator $a^{\dagger}a$ is called the *number operator.*

SUMMARY

The number operator N is defined in the following way

$$N = a^{\dagger} a$$

$$N \left| n \right\rangle = n \left| n \right\rangle$$

The state $\left| n \right\rangle$ is sometimes referred to as the number state. The lowest possible state for the harmonic oscillator is the state $\left| n \right\rangle = \left| 0 \right\rangle$ and is called the ground state. Energy levels of the harmonic oscillator are equally spaced, and we move up and down the ladder of energy states using the operators $a^{\dagger}, a$.

EXAMPLE 9.7

Using $[H, a] = -\hbar\omega a$, $[H, a^{\dagger}] = \hbar\omega a^{\dagger}$, show that $a \left| n \right\rangle$ is an eigenvector of H with eigenvalue $E_n - \hbar\omega$ and that $a^{\dagger} \left| n \right\rangle$ is an eigenvector of H with eigenvalue $E_n + \hbar\omega$.

SOLUTION

First we write the commutator explicitly:

$$[H, a] = Ha - aH, \Rightarrow Ha = aH - \hbar\omega a$$

Now we apply this to the state $a \left| n \right\rangle$

$$H \left(a \left| n \right\rangle \right) = (Ha) \left| n \right\rangle = (aH - \hbar\omega a) \left| n \right\rangle = aH \left| n \right\rangle - \hbar\omega a \left| n \right\rangle$$

On the first term, we use

$$H \left| n \right\rangle = \hbar\omega \left(n + \frac{1}{2} \right) \left| n \right\rangle$$

giving

$$aH \left| n \right\rangle - \hbar\omega a \left| n \right\rangle = a\hbar\omega \left(n + \frac{1}{2} \right) \left| n \right\rangle - \hbar\omega a \left| n \right\rangle = (E_n - \hbar\omega)\, a \left| n \right\rangle$$

From this we see that a steps the energy down by $\hbar\omega$. Because of this, this operator is called the *lowering operator*. We now follow the same procedure for $a^{\dagger} \left| n \right\rangle$. Beginning with the commutator

$$\left[H, a^{\dagger} \right] = Ha^{\dagger} - a^{\dagger} H = \hbar\omega a^{\dagger}$$

we obtain

$$H\left(a^{\dagger}|n\rangle\right) = \left(Ha^{\dagger}\right)|n\rangle = \left(a^{\dagger}H + \hbar\omega a^{\dagger}\right)|n\rangle = a^{\dagger}H|n\rangle + \hbar\omega a^{\dagger}|n\rangle$$

$$a^{\dagger}H|n\rangle = a^{\dagger}(E_n|n\rangle) = E_n a^{\dagger}|n\rangle \Rightarrow$$

$$a^{\dagger}H|n\rangle + \hbar\omega a^{\dagger}|n\rangle = E_n a^{\dagger}|n\rangle + \hbar\omega a^{\dagger}|n\rangle = (E_n + \hbar\omega)\left(a^{\dagger}|n\rangle\right)$$

From this we see that $a^{\dagger}$ steps up the energy by one unit of $\hbar\omega$. This gives it its name, the *raising operator.* Together these operators are sometimes known as *ladder operators.*

EXAMPLE 9.8

e.g.

At time $t = 0$, a wavefunction is in the state

$$|\psi\rangle = \frac{1}{\sqrt{2}}|0\rangle + \frac{1}{\sqrt{3}}|1\rangle + \frac{1}{\sqrt{6}}|2\rangle$$

(a) If the energy is measured, what values can be found and with what probabilities?
(b) Find the average value of the energy, $\langle H\rangle$.
(c) Find the explicit forms of the $\Phi_i(x)$, the basis functions for this expansion, and write the form of the wavefunction at time t.

SOLUTION

(a) Using the energy of the nth eigenstate $E_n = (n + 1/2)\hbar\omega$, we make a table of possible energies for the basis states found in this wavefunction (Table 9-1).

Table 9-1 Ground State of the Harmonic Oscillator

Basis State	Energy	Probability
$\|0\rangle$	$\hbar\omega\left(0+\frac{1}{2}\right) = \frac{\hbar\omega}{2}$	$\left(\frac{1}{\sqrt{2}}\right)^2 = \frac{1}{2}$
$\|1\rangle$	$\hbar\omega\left(1+\frac{1}{2}\right) = \frac{3\hbar\omega}{2}$	$\left(\frac{1}{\sqrt{3}}\right)^2 = \frac{1}{3}$
$\|2\rangle$	$\hbar\omega\left(2+\frac{1}{2}\right) = \frac{5\hbar\omega}{2}$	$\left(\frac{1}{\sqrt{6}}\right)^2 = \frac{1}{6}$

Notice that these probabilities sum to one:

$$\frac{1}{2} + \frac{1}{3} + \frac{1}{6} = \frac{3}{6} + \frac{2}{6} + \frac{1}{6} = 1$$

(b) The average energy is found to be

$$\langle H\rangle = \sum p_i E_i = \frac{1}{2}\left(\frac{\hbar\omega}{2}\right) + \frac{1}{3}\left(\frac{3\hbar\omega}{2}\right) + \frac{1}{6}\left(\frac{5\hbar\omega}{2}\right)$$
$$= \frac{\hbar\omega}{4} + \frac{\hbar\omega}{2} + \frac{5\hbar\omega}{12} = \frac{7\hbar\omega}{6}$$

(c) The nth state wavefuntion of the harmonic oscillator is

$$\Phi_n(x) = \left(\frac{m\omega}{\pi\hbar}\right)^{\frac{1}{4}} \frac{1}{\sqrt{2^n n!}} H_n\left(\sqrt{\frac{m\omega}{\hbar}}x\right) \exp\left(-\frac{m\omega}{2\hbar}x^2\right)$$

Therefore we have

$$\Phi_0(x) = \langle x|0\rangle = \left(\frac{m\omega}{\pi\hbar}\right)^{\frac{1}{4}} \exp\left(-\frac{m\omega}{2\hbar}x^2\right)$$

$$\Phi_1(x) = \langle x|1\rangle = \sqrt{2}\pi^{-1/4}\left(\frac{m\omega}{\hbar}\right)^{\frac{3}{4}} x \exp\left(-\frac{m\omega}{2\hbar}x^2\right)$$

$$\Phi_2(x) = \langle x|2\rangle = \left(\frac{m\omega}{\pi\hbar}\right)^{\frac{1}{4}} \frac{1}{\sqrt{8}}\left(4\frac{m\omega}{\hbar}x^2 - 2\right) \exp\left(-\frac{m\omega}{2\hbar}x^2\right)$$

The wavefunction at time $t = 0$ is

$$\psi(x,0) = \frac{1}{\sqrt{2}}\Phi_0(x) + \frac{1}{\sqrt{3}}\Phi_1(x) + \frac{1}{\sqrt{6}}\Phi_2(x)$$

and so at time t, the wavefunction is given by

$$\psi(x,t) = \frac{1}{\sqrt{2}}\Phi_0(x)e^{-iE_0t/\hbar} + \frac{1}{\sqrt{3}}\Phi_1(x)e^{-iE_1t/\hbar} + \frac{1}{\sqrt{6}}\Phi_2(x)e^{-iE_2t/\hbar}$$
$$= \frac{1}{\sqrt{2}}\Phi_0(x)e^{-i\omega t/2} + \frac{1}{\sqrt{3}}\Phi_1(x)e^{-i3\omega t/2} + \frac{1}{\sqrt{6}}\Phi_2(x)e^{-i5\omega t/2}$$

More on the Action of the Raising and Lowering Operators

We now work out the action of the raising and lowering operators on the eigenstates of the Hamiltonian. We being by applying the commutator $[H, a^\dagger]$ to an arbitary number state

$$[H, a^\dagger]|n\rangle = \hbar\omega a^\dagger|n\rangle$$

Now, on the left-hand side we expand the commutator

$$[H, a^\dagger]\,|n\rangle = \left(Ha^\dagger - a^\dagger H\right)|n\rangle = Ha^\dagger\,|n\rangle - a^\dagger H\,|n\rangle$$
$$= Ha^\dagger\,|n\rangle - a^\dagger\left(n + \frac{1}{2}\right)\hbar\omega\,|n\rangle$$

Now we equate this to $[H, a^\dagger]\,|n\rangle = \hbar\omega a^\dagger\,|n\rangle$

$$Ha^\dagger\,|n\rangle - na^\dagger\hbar\omega\,|n\rangle - \frac{\hbar\omega}{2}a^\dagger\,|n\rangle = \hbar\omega a^\dagger\,|n\rangle$$

Now we move terms over to the right side and combine, giving

$$Ha^\dagger\,|n\rangle = \hbar\omega\left(n + \frac{3}{2}\right)a^\dagger\,|n\rangle$$

From this we conclude that $a^\dagger\,|n\rangle$ is an eigenvector of H with eigenvalue $\hbar\omega\,(n + 3/2)$. Now if

$$H\,|n\rangle = \left(n + \frac{1}{2}\right)\hbar\omega\,|n\rangle\,, \quad \Rightarrow H\,|n+1\rangle = \left(n + \frac{3}{2}\right)\hbar\omega\,|n+1\rangle$$

Therefore we conclude that

$$a^\dagger\,|n\rangle = (n+1)\,|n+1\rangle$$

The operator $a^\dagger$ raises the state $|n\rangle$ to $|n+1\rangle$ (this is why it is called the raising operator). A similar exercise shows that

$$a\,|n\rangle = \sqrt{n}\,|n-1\rangle$$

Since the lowest state of the harmonic oscillator is the ground state, we cannot lower below $|0\rangle$. To avoid going lower than this state, the lowering operator annihilates the state

$$a\,|0\rangle = 0$$

The nth eigenstate can be obtained from the ground state by application of $a^\dagger n$ times to the ground state

$$|n\rangle = \frac{1}{\sqrt{n!}}\left(a^\dagger\right)^n|0\rangle$$

Quiz

1. A harmonic oscillator is in the state

$$\psi(x,0) = \frac{1}{\sqrt{8}}\Phi_0(x) + \frac{1}{\sqrt{2}}\Phi_1(x) + A\Phi_2(x)$$

   (a) Find A so that the state is normalized.

   (b) A measurement is made of the energy. What energies can be found? What is the probability of obtaining each value of the energy?

   (c) Find the state of the system at a later time t.

2. Show that the harmonic oscillator Hamiltonian can be written as

$$H = \hbar\omega\left(aa^\dagger - \frac{1}{2}\right)$$

3. Use $H = \hbar\omega(aa^\dagger - 1/2)$ to show that $aa^\dagger|n\rangle = (n+1)|n\rangle$.

4. Use the fact that $\langle\psi|H|\psi\rangle \geq 0$ to explain why $a|0\rangle = 0$.

5. Show that

$$[N, a] = -a, \quad [N, a^\dagger] = a^\dagger$$

6. Use the recursion relations for the Hermite polynomials to find $\langle x\rangle$ and $\langle p\rangle$ for the ground state of the harmonic oscillator in the coordinate representation.

Angular Momentum

In classical physics angular momentum is defined as

$$\vec{L} = \vec{r} \times \vec{p}$$

where $\vec{r}$ is the displacement vector from the origin and $\vec{p}$ is the linear momentum. The components of angular momentum, in Cartesian coordinates, can be found by computing the cross product explicitly:

$$\vec{L} = \vec{r} \times \vec{p} = \begin{vmatrix} \hat{x} & \hat{y} & \hat{z} \\ x & y & z \\ p_x & p_y & p_z \end{vmatrix} = \hat{x}\left(y\, p_z - z\, p_y\right) + \hat{y}\left(z\, p_x - x\, p_z\right) + \hat{z}\left(x\, p_y - y\, p_x\right)$$

Therefore we have

$$L_x = y\, p_z - z\, p_y$$

$$L_y = z\, p_x - x\, p_z$$

$$L_z = x\, p_y - y\, p_x$$

By using the coordinate representation of the momentum operator, we can write out the components of the angular momentum operator in quantum mechanics. We make the usual substitution

$$p_x \rightarrow -i\hbar \frac{\partial}{\partial x}$$

et cetera for each component of linear momentum. For example, L_z is given by

$$L_z = x\,p_y - y\,p_x = x\left(-i\hbar\frac{\partial}{\partial y}\right) - y\left(-i\hbar\frac{\partial}{\partial x}\right) = -i\hbar\left(x\frac{\partial}{\partial y} - y\frac{\partial}{\partial x}\right)$$

Given that the angular momentum operator is defined in terms of position and linear momentum operators, and these operators obey the commutation relation $\left[x_i, p_j\right] = i\hbar\delta_{ij}$, we expect the components of angular momentum to obey certain commutation relations. We now summarize them.

The Commutation Relations of Angular Momentum

The components of angular momentum satisfy the following commutation relations:

$$\left[L_x, L_y\right] = i\hbar L_z \quad \left[L_y, L_z\right] = i\hbar L_x \quad \left[L_z, L_x\right] = i\hbar L_y$$

EXAMPLE 10.1

Show that

$$\left[L_x,\ L_y\right] = i\hbar L_z$$

SOLUTION

We begin by writing each angular momentum operator in terms of the position and linear momentum. Doing this for L_y first and using the fact that $[A,\ B + C] = [A, B] + [A, C]$ we have

$$\left[L_x, L_y\right] = \left[L_x,\ z\,p_x - x\,p_z\right] = \left[L_x,\ z\,p_x\right] - \left[L_x,\ x\,p_z\right]$$

Now we recall that

$$[A, BC] = [A, B]\,C + B\,[A, C]$$

and use this to expand each of the terms

$$\left[L_x,\ zp_x\right] = \left[L_x,\ z\right]p_x + z\left[L_x,\ p_x\right]$$

$$\left[L_x,\ xp_z\right] = \left[L_x,\ x\right]p_z + x\left[L_x,\ p_z\right]$$

Now we use the definition of angular momentum to work out each term in turn.

$$\begin{aligned}[L_x, z] &= \left[y\, p_z - z\, p_y, z\right] = \left[y\, p_z, z\right] - \left[z\, p_y, z\right] \\ &= -\left[z, y\right] p_z - y\left[z, p_z\right] + [z, z]\, p_y + z\left[z, p_y\right] \\ &= -y\left[z, p_z\right] = -i\hbar\, y\end{aligned}$$

The next term vanishes, because there are no matching coordinates in the commutators:

$$\left[L_x, p_x\right] = \left[y\, p_z - z\, p_y,\ p_x\right] = 0$$

We get a similar result for the third term:

$$[L_x,\ x] = \left[y\, p_z - z\, p_y, x\right] = 0$$

For the last term we find

$$\begin{aligned}\left[L_x, p_z\right] &= \left[y\, p_z - z\, p_y, p_z\right] = \left[y\, p_z, p_z\right] - \left[z\, p_y, p_z\right] \\ &= -\left[z, p_z\right] p_y = -i\hbar\, p_y\end{aligned}$$

Putting these results together we obtain the final result:

$$\begin{aligned}\left[L_x, L_y\right] &= [L_x,\ z\,]\, p_x + z\left[L_x, p_x\right] - [L_x,\ x]\, p_z - x\left[L_x,\ p_z\right] \\ &= -i\hbar y p_x + i\hbar x p_y = i\hbar\left(x p_y - y p_x\right) = i\hbar L_z\end{aligned}$$

The fact that the components of angular momentum do not commute tells us that we can only specify one component at a time. Also, note that the components of angular momentum cannot have common eigenvectors.

While we can't simultaneously specify all of the individual components of angular momentum, there is more information available about the state. This comes from the square of the angular momentum operator:

$$\vec{L} \cdot \vec{L} = L^2 = L_x^2 + L_y^2 + L_z^2$$

We can choose one component of angular momentum; it is customary to choose L_z. Notice that

$$\left[L_z, L^2\right] = \left[L_z, L_x^2\right] + \left[L_z, L_y^2\right] + \left[L_z, L_z^2\right]$$

Since every operator commutes with itself, the last term vanishes. This leaves

$$\begin{aligned}\left[L_z, L_x^2\right] + \left[L_z, L_y^2\right] &= \left[L_z, L_x\right] L_x + L_x\left[L_z, L_x\right] + \left[L_z, L_y\right] L_y + L_y\left[L_z, L_y\right] \\ &= i\hbar L_y L_x + i\hbar L_x L_y - i\hbar L_x L_y - i\hbar L_y L_x = 0\end{aligned}$$

A similar exercise shows that every component of angular momentum commutes with L^2. Therefore, to specify a state of angular momentum, we use L^2 and one of $L_x,\ L_y,\ L_z$.

The Uncertainty Relations for Angular Momentum

Recalling the generalized uncertainty relation for two operators A and B,

$$(\Delta A)^2 (\Delta B)^2 \geq \left(\left| \frac{\langle [A, B] \rangle}{2i} \right| \right)^2$$

we can write down uncertainty relations for the components of angular momentum using the commutators. For example, we find

$$\Delta L_x \, \Delta L_y \geq \left| \frac{\langle [L_x, L_y] \rangle}{2i} \right| = \left| \frac{\langle i\hbar L_z \rangle}{2i} \right| = \frac{\hbar}{2} \langle L_z \rangle$$

Generalized Angular Momentum and the Ladder Operators

In quantum mechanics we find that there are two types of angular momentum. These include *orbital angular momentum* **L** and *spin angular momentum* **S**. In addition, we can form total angular momentum, the combination **L** + **S**. It turns out that all types of angular momentum have certain properties in common; therefore we can study a "generalized" angular momentum, which we label **J**. We have already seen the commutation relations satisfied by the components of orbital angular momentum. It turns out these are characteristic of all types of angular momentum, so we restate them here for **J**:

$$\left[J_x, J_y\right] = i\hbar J_z, \quad \left[J_y, J_z\right] = i\hbar J_x, \quad \left[J_z, J_x\right] = i\hbar J_y$$

A very useful tool that can be used in working out the states of angular momentum are the *ladder operators*.

LADDER OPERATORS FOR ANGULAR MOMENTUM

The ladder operators are constructed from J_x and J_y and are defined as follows:

$$J_+ = J_x + i\,J_y$$

$$J_- = J_x - i\,J_y$$

Notice that

$$\frac{1}{2}(J_+ + J_-) = \frac{1}{2}\left(J_x + i\,J_y + J_x - i\,J_y\right) = J_x$$

$$\frac{1}{2i}(J_+ - J_-) = \frac{1}{2i}\left(J_x + i\,J_y - J_x + i\,J_y\right) = J_y$$

We can also work out some important commutation relations involving these operators:

$$\begin{aligned}[J_z, J_+] &= [J_z, J_x + i\,J_y] = [J_z, J_x] + i\,[J_z, J_y] = i\hbar\,J_y + i\,(-i\hbar\,J_x) \\ &= \hbar\,(J_x + i\,J_y) \\ &= \hbar\,J_+ \\ [J_z, J_-] &= [J_z, J_x - i\,J_y] = [J_z, J_x] - i\,[J_z, J_y] = i\hbar\,J_y - i\,(-i\hbar\,J_x) \\ &= -\hbar\,(J_x - i\,J_y) \\ &= -\hbar\,J_-\end{aligned}$$

Note that since J^2 commutes with each component of angular momentum, it also commutes with the ladder operators:

$$[J^2, J_+] = [J^2, J_x + i\,J_y] = [J^2, J_x] + i\,[J^2, J_y] = 0$$

Similarly, we find that

$$[J^2, J_-] = 0$$

Now consider

$$\begin{aligned}[J_+, J_-] &= [J_x + i\,J_y, J_x - i\,J_y] = [J_x, J_x] - i\,[J_x, J_y] + i\,[J_y, J_x] + [J_y, J_y] \\ &= -i\,[J_x, J_y] + i\,[J_y, J_x] \\ &= -i\,[J_x, J_y] - i\,[J_x, J_y] \\ &= -2i\,[J_x, J_y] \\ &= 2\hbar\,J_z\end{aligned}$$

It is often helpful to write J^2 in terms of the ladder operators

$$J^2 = J_- J_+ + J_z^2 + \hbar\,J_z$$

$$J^2 = J_z^2 + \frac{1}{2}\,(J_+ J_- + J_- J_+)$$

We now consider the eigenstates of angular momentum.

SOLVING THE EIGENVALUE PROBLEM FOR ANGULAR MOMENTUM

States of angular momentum are characterized by two numbers. For orbital angular momentum, these are:

l, the orbital quantum number

m, the azimuthal or magnetic quantum number

When considering general angular momentum **J**, we characterize the states by j, m. The role of these quantum numbers is as follows. The eigenvalues of J^2 are given the label j while the eigenvalues of J_z are given the label m. If one chooses to work with a component other than J_z, the azimuthal quantum number can be labeled with a subscript such as m_x, m_y, m_z.

The eigenvalue equations assume the following form:

$$J^2 |j, m\rangle = \hbar^2 (j)(j+1) |j, m\rangle$$

$$J_z |j, m\rangle = m\hbar |j, m\rangle$$

If we speak of "the angular momentum," this means the quantity $\hbar\sqrt{j(j+1)}$. The value of j fixes the values that m can assume. These are given by

$$m = -j, -j+1, \ldots, 0, \ldots, j-1, j$$

In the case of orbital angular momentum, both l and m are integers. The system is quantized—so only specific values of angular momentum are ever measured. Table10-1 shows a few values of l and m:

Table 10-1

Orbital Quantum Number l	Angular Momentum	Possible Values of m
0	0	0
1	$\hbar\sqrt{2}$	$-1, 0, 1$
2	$\hbar\sqrt{6}$	$-2, -1, 0, 1, 2$
3	$\hbar\sqrt{12}$	$-3, -2, -1, 0, 1, 2, 3$

Orbital angular momentum can never assume a value that lies in between one of these possible measurement results. For example, while angular momentum could be $\hbar\sqrt{2}$ or $\hbar\sqrt{12}$, it could never be $\hbar\sqrt{5}$ or $3\hbar$.

Once the orbital angular momentum number has been fixed, then the values m can assume are fixed as well within the range specified in Table 10-1. For example, if a measurement finds the angular momentum of a particle to be $\hbar\sqrt{2}$, then a measurement of m will find $-\hbar$, 0, or $+\hbar$, but never anything else. For example, $3\hbar$ or $4\hbar$ could never be found for m if l was found to be $\hbar\sqrt{2}$.

While possible values of orbital angular momentum are restricted to the whole integers, in general half-odd integral values of angular momentum are possible. For example, the spin of the electron is $s = 1/2$. In cases where j is half-odd integral, m is also half-odd integral and the value $m = 0$ is not allowed. As an example, consider $j = 3/2$. Possible values for a measurement of m are

$$m = -\frac{3}{2}, -\frac{1}{2}, \frac{1}{2}, \frac{3}{2}$$

If $j = 1/2$, then

$$m = -\frac{1}{2}, +\frac{1}{2}$$

To move up and down in states of angular momentum, we use the ladder operators. They act in the following way:

$$J_+ |j, m\rangle = \hbar\sqrt{j(j+1) - m(m+1)}\, |j, m+1\rangle$$

$$J_- |j, m\rangle = \hbar\sqrt{j(j+1) - m(m-1)}\, |j, m-1\rangle$$

In other words, J_+ is the "raising operator" because it steps up a state from m to $m + 1$, while J_- is the lowering operator because it steps down a state from m to $m - 1$.

Since the value of j fixes the maximum and minimum values that m can assume, we cannot use the ladder operators to move past these maximum and minimum values. These are given by

$$\max(m) = +j$$

$$\min(m) = -j$$

Therefore we require that

$$J_+ |j, m = j\rangle = 0$$

$$J_- |j, m = -j\rangle = 0$$

The $|j, m\rangle$ states are a complete and orthonormal basis for angular momentum, and they satisfy the following orthonormality relation:

$$\langle j_1, m_1 | j_2, m_2 \rangle = \delta_{j_1, j_2} \delta_{m_1, m_2}$$

Completeness is expressed by

$$\sum_{j=0}^{\infty} \sum_{m=-j}^{j} |j, m\rangle \langle j, m| = I$$

EXAMPLE 10.2

For a state of angular momentum $|j, m\rangle$

(a) Show that $\Delta J_z = 0$.

(b) Find $\langle J^2 \rangle$, $\langle J_x \rangle$, $\langle J_y \rangle$, $\langle J_x^2 \rangle$.

SOLUTION

(a) Since $|j, m\rangle$ is an eigenstate of J_z, it is easy to see that

$$\langle J_z \rangle = \langle j, m| J_z |j, m\rangle = m\hbar \langle j, m | j, m \rangle = m\hbar$$

We also obtain

$$\langle J_z^2 \rangle = \langle j, m| J_z (J_z |j, m\rangle) = m\hbar \langle j, m| J_z |j, m\rangle = m^2\hbar^2 \langle j, m| j, m\rangle = m^2\hbar^2$$

and so we have

$$\Delta J_z = \sqrt{\langle J_z^2 \rangle - \langle J_z \rangle^2} = \sqrt{m^2\hbar^2 - (m\hbar)^2} = 0$$

(b) We also know that $|j, m\rangle$ is an eigenstate of J^2 and so

$$\langle J^2 \rangle = \langle j, m| J^2 |j, m\rangle = \hbar^2 j(j+1) \langle j, m | j, m \rangle = \hbar^2 j(j+1)$$

To find the mean values of J_x and J_y, write them in terms of the ladder operators. Considering J_x, we find

$$\begin{aligned}\langle J_x \rangle &= \langle j, m| J_x |j, m\rangle = \frac{1}{2} \langle j, m| J_+ + J_- |j, m\rangle \\ &= \frac{1}{2} \langle j, m| J_+ |j, m\rangle + \frac{1}{2} \langle j, m| J_- |j, m\rangle \\ &= \frac{1}{2}\left(\frac{\hbar}{2}\sqrt{j(j+1) - m(m+1)}\right) \langle j, m| j, m+1\rangle \\ &\quad + \frac{1}{2}\left(\frac{\hbar}{2}\sqrt{j(j+1) - m(m-1)}\right) \langle j, m| j, m-1\rangle = 0\end{aligned}$$

For J_y we obtain

$$\begin{aligned}\langle J_y \rangle &= \langle j, m| J_y |j, m\rangle = \frac{1}{2i} \langle j, m| J_+ - J_- |j, m\rangle = \frac{1}{2i} \langle j, m| J_+ |j, m\rangle \\ &- \frac{1}{2i} \langle j, m| J_- |j, m\rangle = \frac{1}{2i}\left(\frac{\hbar}{2}\sqrt{j(j+1) - m(m+1)}\right) \langle j, m| j, m+1\rangle \\ &- \frac{1}{2i}\left(\frac{\hbar}{2}\sqrt{j(j+1) - m(m-1)}\right) \langle j, m| j, m-1\rangle = 0\end{aligned}$$

To compute the mean of J_x^2, first note that

$$J_x^2 = \left(\frac{J_+ + J_-}{2}\right)^2 = \frac{1}{4}\left(J_+^2 + J_+J_- + J_-J_+ + J_-^2\right)$$

We can disregard the first and last terms, because of the orthonormality of the states. This is because $\langle j, m| J_+^2 |j, m\rangle \propto \langle j, m | j, m+2\rangle = 0$ and $\langle j, m| J_-^2 |j, m\rangle \propto \langle j, m | j, m-2\rangle = 0$. And so we calculate

$$\begin{aligned}
\langle J_x^2\rangle &= \frac{1}{4}\langle j, m| J_+J_- + J_-J_+ |j, m\rangle = \frac{1}{4}\langle j, m| J_+J_- |j, m\rangle + \frac{1}{4}\langle j, m| J_-J_+ |j, m\rangle \\
&= \frac{\hbar\sqrt{j(j+1) - m(m-1)}}{4}\langle j, m| J_+ |j, m-1\rangle \\
&\quad + \frac{\hbar\sqrt{j(j+1) - m(m+1)}}{4}\langle j, m| J_- |j, m+1\rangle \\
&= \left(\frac{\hbar\sqrt{j(j+1) - m(m-1)}}{4}\right)\left(\hbar\sqrt{j(j+1) - (m-1)m}\right)\langle j, m | j, m\rangle \\
&\quad + \left(\frac{\hbar\sqrt{j(j+1) - m(m+1)}}{4}\right)\left(\hbar\sqrt{j(j+1) - m(m+1)}\right)\langle j, m | j, m\rangle \\
&= \frac{\hbar^2}{4}\left[j(j+1) - m(m-1)\right] + \frac{\hbar^2}{4}\left[j(j+1) - m(m+1)\right] \\
&= \frac{\hbar^2}{2}\left[j(j+1) - m^2\right]
\end{aligned}$$

EXAMPLE 10.3

e.g.

Consider orbital angular momentum. A system with $l = 1$ is in the state

$$|\psi\rangle = \frac{1}{\sqrt{2}}|1\rangle - \frac{1}{2}|0\rangle + \frac{1}{2}|-1\rangle$$

Find $\langle L_y\rangle$.

SOLUTION

✔

When working with the x, y components of angular momentum it is easiest to work with the ladder operators. So we write

$$L_y = \frac{L_+ - L_-}{2i}$$

Now we let this operator act on the state:

$$L_y|\psi\rangle = \frac{1}{\sqrt{2}}\left(L_y|1\rangle\right) - \frac{1}{2}\left(L_y|0\rangle\right) + \frac{1}{2}\left(L_y|-1\rangle\right)$$

We consider each of these terms in turn, using the ladder operators. First we have

$$L_y|1\rangle = \frac{1}{2i}\left(L_+|1\rangle - L_-|1\rangle\right)$$

Since $l = 1$, $L_+|1\rangle = 0$ and this term becomes

$$\begin{aligned} L_y|1\rangle &= -\frac{1}{2i}L_-|1\rangle = -\frac{1}{2i}\hbar\sqrt{1(1+1) - 1(1-1)}\,|0\rangle = -\frac{1}{2i}\left(\hbar\sqrt{2}\right)|0\rangle \\ &= i\frac{\hbar}{\sqrt{2}}|0\rangle \end{aligned}$$

where we used $1/i = -i$ in the last step. Next we consider $L_y|0\rangle$:

$$\begin{aligned} L_y|0\rangle &= \frac{1}{2i}\left[L_+|0\rangle - L_-|0\rangle\right] = \frac{1}{2i}\hbar\sqrt{1(1+1)}\,|1\rangle - \frac{1}{2i}\hbar\sqrt{1(1+1)}\,|-1\rangle \\ &= -i\frac{\hbar}{\sqrt{2}}|1\rangle + i\frac{\hbar}{\sqrt{2}}|-1\rangle \end{aligned}$$

When considering the last term, with $l = 1$ we have $L_-|-1\rangle = 0$. Therefore

$$\begin{aligned} L_y|-1\rangle &= \frac{1}{2i}\left(L_+|-1\rangle - L_-|-1\rangle\right) = \frac{1}{2i}L_+|-1\rangle \\ &= \frac{1}{2i}\hbar\sqrt{1(1+1) + 1(-1+1)}\,|0\rangle \\ &= -i\frac{\hbar}{\sqrt{2}}|0\rangle \end{aligned}$$

Putting these results together, the action of the operator on the state is

$$\begin{aligned} L_y|\psi\rangle &= \frac{1}{\sqrt{2}}\left(L_y|1\rangle\right) - \frac{1}{2}\left(L_y|0\rangle\right) + \frac{1}{2}\left(L_y|-1\rangle\right) \\ &= \frac{1}{\sqrt{2}}\left(i\frac{\hbar}{\sqrt{2}}|0\rangle\right) - \frac{1}{2}\left(-i\frac{\hbar}{\sqrt{2}}|1\rangle + i\frac{\hbar}{\sqrt{2}}|-1\rangle\right) + \frac{1}{2}\left(-i\frac{\hbar}{\sqrt{2}}|0\rangle\right) \\ &= i\frac{\hbar}{2\sqrt{2}}\left(|1\rangle - |-1\rangle\right) + i\hbar\left(\frac{\sqrt{2}-1}{2\sqrt{2}}\right)|0\rangle \end{aligned}$$

Therefore the mean value is

$$\begin{aligned}
\langle L_y \rangle &= \langle \psi | \, L_y \, | \psi \rangle \\
&= \left(\frac{1}{\sqrt{2}} \langle 1| - \frac{1}{2} \langle 0| + \frac{1}{2} \langle -1| \right) \left(i \frac{\hbar}{2\sqrt{2}} (|1\rangle - |-1\rangle) + i\hbar \left(\frac{\sqrt{2}-1}{2\sqrt{2}} \right) |0\rangle \right) \\
&= \left(\frac{1}{\sqrt{2}} \right) \left(i \frac{\hbar}{2\sqrt{2}} \right) \langle 1 \, | 1 \rangle + \left(-\frac{1}{2} \right) i\hbar \left(\frac{\sqrt{2}-1}{2\sqrt{2}} \right) \langle 0 \, | 0 \rangle + \left(\frac{1}{2} \right) \left(-i \frac{\hbar}{2\sqrt{2}} \right) \langle -1 \, | -1 \rangle \\
&= \left(\frac{1}{\sqrt{2}} \right) \left(i \frac{\hbar}{2\sqrt{2}} \right) + \left(-\frac{1}{2} \right) i\hbar \left(\frac{\sqrt{2}-1}{2\sqrt{2}} \right) + \left(\frac{1}{2} \right) \left(-i \frac{\hbar}{2\sqrt{2}} \right) \\
&= 0
\end{aligned}$$

EXAMPLE 10.4

Let $j = 3/2$.

(a) What are the possible values of m that can be measured?

(b) Using outer product notation, find a representation of the operators $J_z, \ J_+, \ J_-$.

(c) Construct J_x from J_+ and J_-.

SOLUTION 

(a) Since j is half-odd integral, we first note that $m = 0$ is not permitted. The possible values of m are

$$m = -\frac{3}{2}, \ -\frac{1}{2}, \ \frac{1}{2}, \ \frac{3}{2}$$

(b) The basis states are $|j, m\rangle$. Since we know $j = 3/2$, we simply list the value of m. The states are

$$\left| +\frac{3}{2} \right\rangle, \ \left| +\frac{1}{2} \right\rangle, \ \left| -\frac{1}{2} \right\rangle, \ \left| -\frac{3}{2} \right\rangle$$

Therefore the operator J_z acts on these states in the following way

$$J_z \, |m\rangle = m\hbar \, |m\rangle \, , \quad \Rightarrow$$

$$J_z \left| +\frac{3}{2} \right\rangle = \frac{3\hbar}{2} \left| +\frac{3}{2} \right\rangle$$

$$J_z \left|+\frac{1}{2}\right\rangle = \frac{\hbar}{2}\left|+\frac{1}{2}\right\rangle$$

$$J_z \left|-\frac{1}{2}\right\rangle = -\frac{\hbar}{2}\left|-\frac{1}{2}\right\rangle$$

$$J_z \left|-\frac{3}{2}\right\rangle = -\frac{3\hbar}{2}\left|-\frac{3}{2}\right\rangle$$

We can write the operator in this way

$$J_z = \sum_m \hbar m \,|m\rangle\,\langle m| = \frac{3\hbar}{2}\left|+\frac{3}{2}\right\rangle\left\langle+\frac{3}{2}\right| + \frac{\hbar}{2}\left|+\frac{1}{2}\right\rangle\left\langle+\frac{1}{2}\right| - \frac{\hbar}{2}\left|-\frac{1}{2}\right\rangle\left\langle-\frac{1}{2}\right| - \frac{3\hbar}{2}\left|-\frac{3}{2}\right\rangle\left\langle-\frac{3}{2}\right|$$

To find the expression for J_+, we also consider how it acts on the basis states. Remember that for each state, J_+ will take it to

$$|\ \textit{old state}\ \rangle \rightarrow |\ \textit{new state}\ \rangle$$

$$\left|+\frac{3}{2}\right\rangle \rightarrow 0$$

$$\left|+\frac{1}{2}\right\rangle \rightarrow \left|+\frac{3}{2}\right\rangle$$

$$\left|-\frac{1}{2}\right\rangle \rightarrow \left|+\frac{1}{2}\right\rangle$$

$$\left|-\frac{3}{2}\right\rangle \rightarrow \left|-\frac{1}{2}\right\rangle$$

The form of the operator will be a summation over terms that have the form

$$|\ \textit{new state}\ \rangle\ \langle\ \textit{old state}\ |$$

So we write the operator in this form with undetermined constants

$$J_+ = \alpha\left|+\frac{3}{2}\right\rangle\left\langle+\frac{1}{2}\right| + \beta\left|+\frac{1}{2}\right\rangle\left\langle-\frac{1}{2}\right| + \gamma\left|-\frac{1}{2}\right\rangle\left\langle-\frac{3}{2}\right|$$

From

$$J_+\,|j, m\rangle = \hbar\sqrt{j\,(j+1) - m\,(m+1)}\,|j, m+1\rangle$$

we see that the values of these constants are

$$\alpha = \hbar\sqrt{\frac{3}{2}\left(\frac{3}{2}+1\right) - \frac{1}{2}\left(\frac{1}{2}+1\right)} = \hbar\sqrt{3}$$

$$\beta = 2\hbar$$

$$\gamma = \hbar\sqrt{3}$$

And so we obtain

$$J_+ = \sqrt{3}\hbar\left|+\frac{3}{2}\right\rangle\left\langle+\frac{1}{2}\right| + 2\hbar\left|+\frac{1}{2}\right\rangle\left\langle-\frac{1}{2}\right| + \sqrt{3}\hbar\left|-\frac{1}{2}\right\rangle\left\langle-\frac{3}{2}\right|$$

For J_-, we proceed in a similar fashion. Its action on the basis states is

$$\left|+\frac{3}{2}\right\rangle \to \left|+\frac{1}{2}\right\rangle$$

$$\left|+\frac{1}{2}\right\rangle \to \left|-\frac{1}{2}\right\rangle$$

$$\left|-\frac{1}{2}\right\rangle \to \left|-\frac{3}{2}\right\rangle$$

$$\left|-\frac{3}{2}\right\rangle \to 0$$

So the operator must have the form

$$J_- = \alpha\left|+\frac{1}{2}\right\rangle\left\langle+\frac{3}{2}\right| + \beta\left|-\frac{1}{2}\right\rangle\left\langle+\frac{1}{2}\right| + \gamma\left|-\frac{3}{2}\right\rangle\left\langle-\frac{1}{2}\right|$$

The constants can be found from the formula

$$J_-\left|j, m\right\rangle = \hbar\sqrt{j\left(j+1\right) - m\left(m-1\right)}\left|j, m-1\right\rangle$$

Therefore we find

$$\alpha = \hbar\sqrt{\frac{3}{2}\left(\frac{3}{2}+1\right) - \frac{3}{2}\left(\frac{3}{2}-1\right)} = \sqrt{3}\hbar$$

$$\beta = 2\hbar$$

$$\gamma = \sqrt{3}\hbar$$

And so we obtain for the lowering operator

$$J_- = \sqrt{3}\hbar \left|+\frac{1}{2}\right\rangle\left\langle+\frac{3}{2}\right| + 2\hbar \left|-\frac{1}{2}\right\rangle\left\langle+\frac{1}{2}\right| + \sqrt{3}\hbar \left|-\frac{3}{2}\right\rangle\left\langle-\frac{1}{2}\right|$$

(c) We find J_x from the ladder operators.

$$\begin{aligned} J_x &= \frac{1}{2}\left(J_+ + J_-\right) \\ &= \frac{1}{2}\left[\begin{array}{l} \left(\sqrt{3}\hbar \left|+\tfrac{3}{2}\right\rangle\left\langle+\tfrac{1}{2}\right| + 2\hbar \left|+\tfrac{1}{2}\right\rangle\left\langle-\tfrac{1}{2}\right| + \sqrt{3}\hbar \left|-\tfrac{1}{2}\right\rangle\left\langle-\tfrac{3}{2}\right|\right) \\ \quad + \left(\sqrt{3}\hbar \left|+\tfrac{1}{2}\right\rangle\left\langle+\tfrac{3}{2}\right| + 2\hbar \left|-\tfrac{1}{2}\right\rangle\left\langle+\tfrac{1}{2}\right| + \sqrt{3}\hbar \left|-\tfrac{3}{2}\right\rangle\left\langle-\tfrac{1}{2}\right|\right) \end{array}\right] \\ &= \frac{\sqrt{3}\hbar}{2}\left[\left|+\frac{3}{2}\right\rangle\left\langle+\frac{1}{2}\right| + \left|+\frac{1}{2}\right\rangle\left\langle+\frac{3}{2}\right| + \left|-\frac{1}{2}\right\rangle\left\langle-\frac{3}{2}\right| + \left|-\frac{3}{2}\right\rangle\left\langle-\frac{1}{2}\right|\right] \\ &\quad + 2\hbar\left[\left|+\frac{1}{2}\right\rangle\left\langle-\frac{1}{2}\right| + \left|-\frac{1}{2}\right\rangle\left\langle+\frac{1}{2}\right|\right] \end{aligned}$$

Matrix Representations of Angular Momentum

We now proceed to find matrix representations for the angular momentum operators, find the eigenvalues, and see how to work with states represented as column vectors. As usual, to come up with a matrix representation we must select a basis. We continue to work in the basis $|j, m\rangle$, which are the eigenstates of J^2 and J_z. If the value of j is understood then we can label the states by $|m\rangle$ alone. To illustrate the process, we consider $j = 1$. In this case

$$J^2 |m\rangle = \hbar^2 (1)(1+1)|m\rangle = 2\hbar^2 |m\rangle$$

for all values of m. The permissible values that m can assume are

$$m = -1, 0, 1$$

And so the possible states of the system are

$$|-1\rangle,\ |0\rangle,\ |+1\rangle$$

These states form an orthonormal basis. This means we can write any arbitrary state of the system as a linear combination of them

$$|\psi\rangle = \alpha|-1\rangle + \beta|0\rangle + \gamma|+1\rangle$$

where α, β, γ are complex numbers. If the state is normalized then

$$|\alpha|^2 + |\beta|^2 + |\gamma|^2 = 1$$

In addition, we know that since the states are orthnormal, then

$$\langle -1 | -1 \rangle = \langle 0 | 0 \rangle = \langle +1 | +1 \rangle = 1$$

$$\langle -1 | 0 \rangle = \langle -1 | +1 \rangle = \langle +1 | 0 \rangle = 0$$

Using $J^2 |m\rangle = 2\hbar^2 |m\rangle$, we find the matrix representation of this operator to be

$$\begin{aligned}
[J^2] &= \begin{pmatrix} \langle +1| J^2 |+1\rangle & \langle +1| J^2 |0\rangle & \langle +1| J^2 |-1\rangle \\ \langle 0| J^2 |+1\rangle & \langle 0| J^2 |0\rangle & \langle 0| J^2 |-1\rangle \\ \langle -1| J^2 |+1\rangle & \langle -1| J^2 |0\rangle & \langle -1| J^2 |-1\rangle \end{pmatrix} \\
&= 2\hbar^2 \begin{pmatrix} \langle +1| +1\rangle & \langle +1|0\rangle & \langle +1| -1\rangle \\ \langle 0| +1\rangle & \langle 0|0\rangle & \langle 0| -1\rangle \\ \langle -1| +1\rangle & \langle -1|0\rangle & \langle -1| -1\rangle \end{pmatrix} \\
&= 2\hbar^2 \begin{pmatrix} 1 & 0 & 0 \\ 0 & 1 & 0 \\ 0 & 0 & 1 \end{pmatrix}
\end{aligned}$$

Now we derive the matrix representation of J_z, recalling that

$$\begin{aligned}
& J_z |m\rangle = m\hbar |m\rangle, \quad \Rightarrow \\
& J_z |+1\rangle = \hbar |+1\rangle \\
& J_z |0\rangle = 0 \\
& J_z |-1\rangle = -\hbar |-1\rangle
\end{aligned}$$

And so the matrix representation is found to be

$$\begin{aligned}
[J_z] &= \begin{pmatrix} \langle +1| J_z |+1\rangle & \langle +1| J_z |0\rangle & \langle +1| J_z |-1\rangle \\ \langle 0| J_z |+1\rangle & \langle 0| J_z |0\rangle & \langle 0| J_z |-1\rangle \\ \langle -1| J_z |+1\rangle & \langle -1| J_z |0\rangle & \langle -1| J_z |-1\rangle \end{pmatrix} \\
&= \begin{pmatrix} \hbar \langle +1| +1\rangle & 0 & -\hbar \langle +1| -1\rangle \\ \hbar \langle 0| +1\rangle & 0 & -\hbar \langle 0| -1\rangle \\ \hbar \langle -1| +1\rangle & 0 & -\hbar \langle -1| -1\rangle \end{pmatrix} \\
&= \hbar \begin{pmatrix} 1 & 0 & 0 \\ 0 & 0 & 0 \\ 0 & 0 & -1 \end{pmatrix}
\end{aligned}$$

From this matrix, which is diagonal, it is easy to find the column vector representations of the eigenvectors, which turn out to be

$$|+1\rangle = \begin{pmatrix} 1 \\ 0 \\ 0 \end{pmatrix}, \quad |0\rangle = \begin{pmatrix} 0 \\ 1 \\ 0 \end{pmatrix}, \quad |-1\rangle = \begin{pmatrix} 0 \\ 0 \\ 1 \end{pmatrix}$$

corresponding to eigenvalues $+\hbar, 0, -\hbar$, respectively. Now let's find the matrix representations of the ladder operators. They act on the states in the following way:

$$J_+ \left|+1\right\rangle = 0$$
$$J_+ \left|0\right\rangle = \sqrt{2}\hbar \left|+1\right\rangle$$
$$J_+ \left|-1\right\rangle = \sqrt{2}\hbar \left|0\right\rangle$$

$$J_- \left|+1\right\rangle = \sqrt{2}\hbar \left|0\right\rangle$$
$$J_- \left|0\right\rangle = \sqrt{2}\hbar \left|-1\right\rangle$$
$$J_- \left|-1\right\rangle = 0$$

Using these relationships along with the orthonormality of the basis states, we find

$$\left[J_+\right] = \begin{pmatrix} \langle +1|\, J_+ \,|+1\rangle & \langle +1|\, J_+ \,|0\rangle & \langle +1|\, J_+ \,|-1\rangle \\ \langle 0|\, J_+ \,|+1\rangle & \langle 0|\, J_+ \,|0\rangle & \langle 0|\, J_+ \,|-1\rangle \\ \langle -1|\, J_+ \,|+1\rangle & \langle -1|\, J_+ \,|0\rangle & \langle -1|\, J_+ \,|-1\rangle \end{pmatrix}$$
$$= \sqrt{2}\hbar \begin{pmatrix} 0 & \langle +1\,|+1\rangle & \langle +1\,|0\rangle \\ 0 & \langle 0\,|+1\rangle & \langle 0\,|0\rangle \\ 0 & \langle -1\,|+1\rangle & \langle -1\,|0\rangle \end{pmatrix}$$
$$\sqrt{2}\hbar \begin{pmatrix} 0 & 1 & 0 \\ 0 & 0 & 1 \\ 0 & 0 & 0 \end{pmatrix}$$

and for J_- we find

$$\left[J_-\right] = \begin{pmatrix} \langle +1|\, J_- \,|+1\rangle & \langle +1|\, J_- \,|0\rangle & \langle +1|\, J_- \,|-1\rangle \\ \langle 0|\, J_- \,|+1\rangle & \langle 0|\, J_- \,|0\rangle & \langle 0|\, J_- \,|-1\rangle \\ \langle -1|\, J_- \,|+1\rangle & \langle -1|\, J_- \,|0\rangle & \langle -1|\, J_- \,|-1\rangle \end{pmatrix}$$
$$= \sqrt{2}\hbar \begin{pmatrix} \langle +1\,|0\rangle & \langle +1\,|-1\rangle & 0 \\ \langle 0\,|0\rangle & \langle 0\,|-1\rangle & 0 \\ \langle -1\,|0\rangle & \langle -1\,|-1\rangle & 0 \end{pmatrix}$$
$$= \sqrt{2}\hbar \begin{pmatrix} 0 & 0 & 0 \\ 1 & 0 & 0 \\ 0 & 1 & 0 \end{pmatrix}$$

With the matrix representations of the ladder operators in hand, the matrix representations of $J_{x,y}$ are easy to find (exercise).

FINDING THE POSSIBLE RESULTS OF MEASUREMENT OF ANGULAR MOMENTUM

Suppose that we are given a state that has been specified in a component of angular momentum such as L_z. We are then asked what the possible results of measurement are for another component such as L_x. Let's say that the eigenfunctions of each of these respective operators are ϕ_x and ϕ_z.

If the state of the system is given to us in terms of L_z states and we are asked to find possible L_x measurement results, the first item of business is to expand the ϕ_z states in terms of ϕ_x states. They will be some linear combination

$$\phi_z = \sum_i \alpha_i \phi_{x,i}$$

The possible results of measurement are then the eigenvalues that correspond to each ϕ_x eigenfunction that appears in the expansion, and the probability of each measurement is given by $|\alpha_i|^2$. Since the components of angular momentum do not commute, if the system is in a definite state of L_z and then another component such as L_x is measured, all previous information known about L_z is lost. The angular momentum in the z direction would be completely uncertain. When a measurement of L_x is made and the system is found to be in some definite state of L_x that we label ϕ_x, we can expand this state in terms of ϕ_z states:

$$\phi_x = \sum_i \beta_i \phi_{z,i}$$

So when we make a measurement of L_x and find the system in some definite state ϕ_x, then the system is now in a superposition of states of L_z.

EXAMPLE 10.5

e.g.

A particle is in the $j = 1$ state. J_z is measured and the result $+\hbar$ is found. Now J_x is measured. What values can be found and with what probabilities?

SOLUTION

✓

First we write the matrix representation of J_x:

$$J_x = \frac{J_+ + J_-}{2} = \frac{\hbar\sqrt{2}}{2}\begin{pmatrix} 0 & 1 & 0 \\ 0 & 0 & 1 \\ 0 & 0 & 0 \end{pmatrix} + \frac{\hbar\sqrt{2}}{2}\begin{pmatrix} 0 & 0 & 0 \\ 1 & 0 & 0 \\ 0 & 1 & 0 \end{pmatrix} = \frac{\hbar}{\sqrt{2}}\begin{pmatrix} 0 & 1 & 0 \\ 1 & 0 & 1 \\ 0 & 1 & 0 \end{pmatrix}$$

The eigenvalues of this matrix are $\{-\hbar, 0, +\hbar\}$, the same as the eigenvalues for J_z. Let us label the eigenvectors of J_x as $\{|1_x\rangle, |0_x\rangle, |-1_x\rangle\}$. The eigenvector corresponding to $\hbar m_x = +\hbar$ is found from

$$\frac{\hbar}{\sqrt{2}}\begin{pmatrix} 0 & 1 & 0 \\ 1 & 0 & 1 \\ 0 & 1 & 0 \end{pmatrix}\begin{pmatrix} a \\ b \\ c \end{pmatrix} = \hbar \begin{pmatrix} a \\ b \\ c \end{pmatrix}$$

$$\Rightarrow b = \sqrt{2}\, a, \quad c = a$$

Normalizing,

$$1 = \begin{pmatrix} a^* & \sqrt{2}a^* & a^* \end{pmatrix}\begin{pmatrix} a \\ \sqrt{2}\, a \\ a \end{pmatrix} = |a|^2 + 2\,|a|^2 + |a|^2 = 4\,|a|^2$$

$$\Rightarrow a = \frac{1}{2}$$

and so we can write

$$|1_x\rangle = \frac{1}{2}\begin{pmatrix} 1 \\ \sqrt{2} \\ 1 \end{pmatrix}$$

Likewise we find the eigenvector corresponding to $m_x = 0$ is

$$|0_x\rangle = \frac{1}{\sqrt{2}}\begin{pmatrix} 1 \\ 0 \\ -1 \end{pmatrix}$$

and the eigenvector corresponding to $m_x = -1$ is

$$|-1_x\rangle = \frac{1}{2}\begin{pmatrix} 1 \\ -\sqrt{2} \\ 1 \end{pmatrix}$$

We are told that J_z is measured and is found to be $+\hbar$. Looking at the eigenstates of J_z, the state of the system after measurement is

$$|\psi\rangle = |1_z\rangle = \begin{pmatrix} 1 \\ 0 \\ 0 \end{pmatrix}$$

We use some algebra to rewrite the state in terms of the basis states of J_x. First notice that

$$|1_x\rangle + |-1_x\rangle = \frac{1}{2}\begin{pmatrix} 1 \\ \sqrt{2} \\ 1 \end{pmatrix} + \frac{1}{2}\begin{pmatrix} 1 \\ -\sqrt{2} \\ 1 \end{pmatrix} = \begin{pmatrix} 1 \\ 0 \\ 1 \end{pmatrix}$$

Now see what happens if we form the sum

$$|1_x\rangle + |-1_x\rangle + \sqrt{2}\,|0_x\rangle = \begin{pmatrix}1\\0\\1\end{pmatrix} + \begin{pmatrix}1\\0\\-1\end{pmatrix} = \begin{pmatrix}2\\0\\0\end{pmatrix} = 2\begin{pmatrix}1\\0\\0\end{pmatrix} = 2\,|1_z\rangle$$

So we have found a way to express the state of the system in basis states of J_x:

$$|\psi\rangle = |1_z\rangle = \frac{1}{2}\left(|1_x\rangle + |-1_x\rangle + \sqrt{2}\,|0_x\rangle\right) = \frac{1}{2}\,|1_x\rangle + \frac{1}{\sqrt{2}}\,|0_x\rangle + \frac{1}{2}\,|-1_x\rangle$$

From this expansion, we see the possible results of measurement and their probabilities are

$$+\hbar \qquad prob = \left(\frac{1}{2}\right)^2 = \frac{1}{4}$$

$$0 \qquad prob = \left(\frac{1}{\sqrt{2}}\right)^2 = \frac{1}{2}$$

$$-\hbar \qquad prob = \left(\frac{1}{2}\right)^2 = \frac{1}{4}$$

EXAMPLE 10.6

e.g.

Find a matrix U that diagonalizes J_x.

SOLUTION

✔

In the previous example we found the matrix representation of J_x in the basis of J_z:

$$[J_x] = \frac{\hbar}{\sqrt{2}}\begin{pmatrix}0 & 1 & 0\\1 & 0 & 1\\0 & 1 & 0\end{pmatrix}$$

We seek a matrix U such that

$$J_x' = U^\dagger J_x U$$

where J_x' is a diagonal matrix, the elements along the diagonal being the eigenvalues of the J_x operator, $\{+\hbar, 0, -\hbar\}$. We already have everything we need to form U; the columns of the matrix are the eigenvectors $\{|1_x\rangle\,, |0_x\rangle\,, |-1_x\rangle\}$. So we take the matrix to be

$$U = \begin{pmatrix}\frac{1}{2} & \frac{1}{\sqrt{2}} & \frac{1}{2}\\ \frac{1}{\sqrt{2}} & 0 & -\frac{1}{\sqrt{2}}\\ \frac{1}{2} & -\frac{1}{\sqrt{2}} & \frac{1}{2}\end{pmatrix}$$

It is easy to show this matrix is unitary, $UU^{\dagger} = I$. Now we show that it diagonalizes the matrix representation of J_x

$$U^{\dagger} J_x U = \begin{pmatrix} \frac{1}{2} & \frac{1}{\sqrt{2}} & \frac{1}{2} \\ \frac{1}{\sqrt{2}} & 0 & -\frac{1}{\sqrt{2}} \\ \frac{1}{2} & -\frac{1}{\sqrt{2}} & \frac{1}{2} \end{pmatrix} \frac{\hbar}{\sqrt{2}} \begin{pmatrix} 0 & 1 & 0 \\ 1 & 0 & 1 \\ 0 & 1 & 0 \end{pmatrix} \begin{pmatrix} \frac{1}{2} & \frac{1}{\sqrt{2}} & \frac{1}{2} \\ \frac{1}{\sqrt{2}} & 0 & -\frac{1}{\sqrt{2}} \\ \frac{1}{2} & -\frac{1}{\sqrt{2}} & \frac{1}{2} \end{pmatrix}$$

$$= \frac{\hbar}{\sqrt{2}} \begin{pmatrix} \frac{1}{2} & \frac{1}{\sqrt{2}} & \frac{1}{2} \\ \frac{1}{\sqrt{2}} & 0 & -\frac{1}{\sqrt{2}} \\ \frac{1}{2} & -\frac{1}{\sqrt{2}} & \frac{1}{2} \end{pmatrix} \begin{pmatrix} \frac{1}{\sqrt{2}} & 0 & -\frac{1}{\sqrt{2}} \\ 1 & 0 & 1 \\ \frac{1}{\sqrt{2}} & 0 & -\frac{1}{\sqrt{2}} \end{pmatrix}$$

$$= \frac{\hbar}{\sqrt{2}} \begin{pmatrix} \sqrt{2} & 0 & 0 \\ 0 & 0 & 0 \\ 0 & 0 & -\sqrt{2} \end{pmatrix} = \hbar \begin{pmatrix} 1 & 0 & 0 \\ 0 & 0 & 0 \\ 0 & 0 & -1 \end{pmatrix}$$

---

e.g. EXAMPLE 10.7

Let $j = 1$. Proceeding in the J_z basis,

(a) Find the matrix representation of J_y.

(b) Find the eigenvectors of J_y.

(c) If the system is found in the $|1_y\rangle$ state, what is the probability of finding each projection of angular momentum along the z axis?

✔ SOLUTION

(a) Using the ladder operators, we find

$$J_y = \frac{J_+ - J_-}{2i} = \frac{\hbar}{\sqrt{2}i} \begin{pmatrix} 0 & 1 & 0 \\ 0 & 0 & 1 \\ 0 & 0 & 0 \end{pmatrix} - \frac{\hbar}{\sqrt{2}i} \begin{pmatrix} 0 & 0 & 0 \\ 1 & 0 & 0 \\ 0 & 1 & 0 \end{pmatrix} = \frac{\hbar}{\sqrt{2}} \begin{pmatrix} 0 & -i & 0 \\ i & 0 & -i \\ 0 & i & 0 \end{pmatrix}$$

Note that this is the representation of J_y in the z-basis.

(b) Solving the characteristic equation $\det \left| J_y - \lambda I \right| = 0$ leads to the eigenvalues $\{+\hbar, 0, -\hbar\}$. Solving for the eigenvectors

$$\frac{\hbar}{\sqrt{2}} \begin{pmatrix} 0 & -i & 0 \\ i & 0 & -i \\ 0 & i & 0 \end{pmatrix} \begin{pmatrix} a \\ b \\ c \end{pmatrix} = \hbar \begin{pmatrix} a \\ b \\ c \end{pmatrix}$$

This leads to the relations

$$b = i\sqrt{2}a, \;\; c = -a$$

and so we can write the eigenvector as

$$|1_y\rangle = \begin{pmatrix} a \\ i\sqrt{2}a \\ -a \end{pmatrix}$$

Normalizing to find a,

$$1 = \langle 1_y | 1_y \rangle = \begin{pmatrix} a^* & -i\sqrt{2}a^* & -a^* \end{pmatrix} \begin{pmatrix} a \\ i\sqrt{2}a \\ -a \end{pmatrix} = |a|^2 + 2|a|^2 + |a|^2 = 4|a|^2$$

$$\Rightarrow a = \frac{1}{2}$$

Therefore we have

$$|1_y\rangle = \frac{1}{2} \begin{pmatrix} 1 \\ i\sqrt{2} \\ -1 \end{pmatrix}$$

Next, we consider $\lambda = 0$. The eigenvalue equation is

$$\frac{\hbar}{\sqrt{2}} \begin{pmatrix} 0 & -i & 0 \\ i & 0 & -i \\ 0 & i & 0 \end{pmatrix} \begin{pmatrix} a \\ b \\ c \end{pmatrix} = 0$$

This leads to

$$b = 0, \; c = a$$

$$\Rightarrow |0_y\rangle = \begin{pmatrix} a \\ 0 \\ a \end{pmatrix}$$

Normalizing, we find

$$1 = \langle 0_y | 0_y \rangle = \begin{pmatrix} a^* & 0 & a^* \end{pmatrix} \begin{pmatrix} a \\ 0 \\ a \end{pmatrix} = 2|a|^2, \; \Rightarrow a = \frac{1}{\sqrt{2}}$$

and so we have

$$|0_y\rangle = \frac{1}{\sqrt{2}} \begin{pmatrix} 1 \\ 0 \\ 1 \end{pmatrix}$$

The final eigenvalue equation for $\lambda = -\hbar$ gives

$$\frac{\hbar}{\sqrt{2}} \begin{pmatrix} 0 & -i & 0 \\ i & 0 & -i \\ 0 & i & 0 \end{pmatrix} \begin{pmatrix} a \\ b \\ c \end{pmatrix} = -\hbar \begin{pmatrix} a \\ b \\ c \end{pmatrix}$$

$$\Rightarrow b = -i\sqrt{2}a, \; c = -a$$

This leads to

$$|-1_y\rangle = \begin{pmatrix} a \\ -i\sqrt{2}a \\ -a \end{pmatrix}$$

Normalizing to solve for a leads to

$$|-1_y\rangle = \frac{1}{2}\begin{pmatrix} 1 \\ -i\sqrt{2} \\ -1 \end{pmatrix}$$

(c) We can find the probability of each projection by constructing the squares of each inner product. First we recall the basis states of J_z:

$$|1_z\rangle = \begin{pmatrix} 1 \\ 0 \\ 0 \end{pmatrix}, \quad |0_z\rangle = \begin{pmatrix} 0 \\ 1 \\ 0 \end{pmatrix}, \quad |-1_z\rangle = \begin{pmatrix} 0 \\ 0 \\ 1 \end{pmatrix}$$

If the system is in the state

$$|1_y\rangle = \frac{1}{2}\begin{pmatrix} 1 \\ i\sqrt{2} \\ -1 \end{pmatrix}$$

then the probability that a measurement of J_z is $+\hbar$ is

$$\Pr(+\hbar) = \left|\langle 1_z | 1_y \rangle\right|^2,$$

$$\langle 1_z | 1_y \rangle = (1 \quad 0 \quad 0)\begin{pmatrix} \frac{1}{2} \\ i\frac{\sqrt{2}}{2} \\ -\frac{1}{2} \end{pmatrix} = \frac{1}{2}$$

$$\Rightarrow \Pr(+\hbar) = \left(\frac{1}{2}\right)^2 = \frac{1}{4}$$

For the other states, we find

$$\Pr(0) = \left|\langle 0_z | 1_y \rangle\right|^2,$$

$$\langle 0_z | 1_y \rangle = (0 \quad 1 \quad 0)\begin{pmatrix} \frac{1}{2} \\ i\frac{\sqrt{2}}{2} \\ -\frac{1}{2} \end{pmatrix} = i\frac{\sqrt{2}}{2}$$

$$\Rightarrow \Pr(0) = \left|i\frac{\sqrt{2}}{2}\right|^2 = \frac{1}{2}$$

$$\Pr(-\hbar) = \left|\langle -1_z \,|1_y\rangle\right|^2,$$

$$\langle -1_z \,|1_y\rangle = (0 \quad 0 \quad 1)\begin{pmatrix} \frac{1}{2} \\ i\frac{\sqrt{2}}{2} \\ -\frac{1}{2} \end{pmatrix} = -\frac{1}{2}$$

$$\Rightarrow \Pr(-\hbar) = \left(-\frac{1}{2}\right)^2 = \frac{1}{4}$$

EXAMPLE 10.8

e.g.

A particle is the angular momentum state

$$|\psi\rangle = \frac{1}{\sqrt{13}}\begin{pmatrix} 2 \\ 0 \\ 3 \end{pmatrix}$$

If L_x is measured, what values can be found and with what probabilities?

SOLUTION

The eigenstates of L_x are

$$|1_x\rangle = \frac{1}{2}\begin{pmatrix} 1 \\ \sqrt{2} \\ 1 \end{pmatrix}, \quad |0_x\rangle = \frac{1}{\sqrt{2}}\begin{pmatrix} 1 \\ 0 \\ -1 \end{pmatrix}, \quad |-1_x\rangle = \frac{1}{2}\begin{pmatrix} 1 \\ -\sqrt{2} \\ 1 \end{pmatrix}$$

corresponding to the measurements $+\hbar$, 0, $-\hbar$, respectively. The inner products of these states with $|\psi\rangle$ are

$$\langle 1_x \,|\psi\,\rangle = \frac{1}{2}\left(1 \quad \sqrt{2} \quad 1\right)\begin{pmatrix} 2 \\ 0 \\ 3 \end{pmatrix}\frac{1}{\sqrt{13}} = \frac{1}{2\sqrt{13}}\left(1 * 2 + \sqrt{2} * 0 + 1 * 3\right) = \frac{5}{2\sqrt{13}}$$

and so the probability of finding $+\hbar$ upon measurement is

$$|\langle 1_x \,|\psi\,\rangle|^2 = \left(\frac{5}{2\sqrt{13}}\right)^2 = \frac{25}{52}$$

Now, for the next state we find

$$\langle 0_x \,|\psi\,\rangle = \frac{1}{\sqrt{2}}(1 \quad 0 \quad -1)\begin{pmatrix} 2 \\ 0 \\ 3 \end{pmatrix}\frac{1}{\sqrt{13}} = \frac{1}{\sqrt{26}}(1 * 2 - 1 * 3) = -\frac{1}{\sqrt{26}}$$

so the probability of obtaining 0 upon measurement is

$$|\langle 0_x|\psi\rangle|^2 = \left(-\frac{1}{\sqrt{26}}\right)^2 = \frac{1}{26}$$

Finally, we have

$$\langle -1_x|\psi\rangle = \frac{1}{2}\begin{pmatrix}1 & -\sqrt{2} & 1\end{pmatrix}\begin{pmatrix}2\\0\\3\end{pmatrix}\frac{1}{\sqrt{13}}$$

$$= \frac{1}{2\sqrt{13}}\left(1*2-\sqrt{2}*0+1*3\right) = \frac{5}{2\sqrt{13}}$$

and so the probability of finding $-\hbar$ is

$$|\langle -1_x|\psi\rangle|^2 = \left(\frac{5}{2\sqrt{13}}\right)^2 = \frac{25}{52}$$

(It is a good idea to verify that the probabilities sum to one).

e.g. EXAMPLE 10.9

Suppose that a system is in the state

$$|\psi\rangle = \frac{1}{\sqrt{17}}\begin{pmatrix}2\\-3\\2\end{pmatrix}$$

(a) Write the state in ket notation using the eigenstates of J_z.

(b) If J_z is measured, what results can be found and with what probabilities?

(c) Suppose J_z is measured and the result is 0. What is the state of the particle immediately after measurement? If J_x is now measured, what are the possible results of measurement and with what probabilities?

SOLUTION

(a)

$$|\psi\rangle = \frac{1}{\sqrt{17}}\begin{pmatrix}2\\-3\\2\end{pmatrix} = \frac{1}{\sqrt{17}}\left[\begin{pmatrix}2\\0\\0\end{pmatrix}-\begin{pmatrix}0\\3\\0\end{pmatrix}+\begin{pmatrix}0\\0\\2\end{pmatrix}\right]$$

$$= \frac{2}{\sqrt{17}}\begin{pmatrix}1\\0\\0\end{pmatrix}-\frac{3}{\sqrt{17}}\begin{pmatrix}0\\1\\0\end{pmatrix}+\frac{2}{\sqrt{17}}\begin{pmatrix}0\\0\\1\end{pmatrix}$$

$$= \frac{2}{\sqrt{17}}|1_z\rangle - \frac{3}{\sqrt{17}}|0_z\rangle + \frac{2}{\sqrt{17}}|-1_z\rangle$$

(b) Since we have expanded the state in terms of the eigenstates of J_z, we can simply read off the possible measurement results and their probabilities. These are shown in Table 10-2.

Table 10-2

Eigenvector	Measurement Result	Probability
$\lvert 1_z\rangle$	$\hbar$	$\left(\frac{2}{\sqrt{17}}\right)^2 = \frac{4}{17}$
$\lvert 0_z\rangle$	0	$\left(-\frac{3}{\sqrt{17}}\right)^2 = \frac{9}{17}$
$\lvert -1_z\rangle$	$-\hbar$	$\left(\frac{2}{\sqrt{17}}\right)^2 = \frac{4}{17}$

(c) If J_z is measured and 0 is the measurement result, the state of the particle is

$$|\psi'\rangle = |0_z\rangle = \begin{pmatrix} 0 \\ 1 \\ 0 \end{pmatrix}$$

We follow a different method than we did in the last example. This time we expand the state in terms of the eigenstates of J_x. Some algebra shows that

$$\begin{aligned} |0_z\rangle &= \begin{pmatrix} 0 \\ 1 \\ 0 \end{pmatrix} = \frac{1}{\sqrt{2}}\begin{pmatrix} 0 \\ \sqrt{2} \\ 0 \end{pmatrix} = \frac{1}{\sqrt{2}}\left(\frac{1}{2}\right)\begin{pmatrix} 1-1 \\ \sqrt{2}+\sqrt{2} \\ 1-1 \end{pmatrix} \\ &= \frac{1}{\sqrt{2}}\left(\frac{1}{2}\right)\begin{pmatrix} 1 \\ \sqrt{2} \\ 1 \end{pmatrix} - \frac{1}{\sqrt{2}}\left(\frac{1}{2}\right)\begin{pmatrix} 1 \\ -\sqrt{2} \\ 1 \end{pmatrix} \\ &= \frac{1}{\sqrt{2}}\left(|1_x\rangle - |-1_x\rangle\right) \end{aligned}$$

We see that a measurement of J_x can result in $+\hbar$ with a probability $(1/\sqrt{2})^2 = 1/2$ and a measurement can result in $-\hbar$ with a probability of $(-1/\sqrt{2})^2 = 1/2$.

Coordinate Representation of Orbital Angular Momentum and the Spherical Harmonics

We now turn to the problem of representing the states of orbital angular momentum in terms of coordinate functions. In this case it is convenient to work with spherical coordinates. We recall that these are related to Cartesian coordinates in the following way:

$$x = r\sin\theta\cos\phi, \quad y = r\sin\theta\sin\phi, \quad z = r\cos\theta$$

With this coordinate transformation we can write the components of angular momentum as

$$L_x = i\hbar\left(\sin\phi\frac{\partial}{\partial\theta} + \frac{\cos\phi}{\tan\theta}\frac{\partial}{\partial\phi}\right), \quad L_y = i\hbar\left(-\cos\phi\frac{\partial}{\partial\theta} + \frac{\sin\phi}{\tan\theta}\frac{\partial}{\partial\phi}\right)$$

$$L_z = -i\hbar\frac{\partial}{\partial\phi}$$

We also have

$$L^2 = -\hbar^2\left(\frac{\partial^2}{\partial\theta^2} + \frac{1}{\tan\theta}\frac{\partial}{\partial\theta} + \frac{1}{\sin^2\theta}\frac{\partial^2}{\partial\phi^2}\right),$$

$$L_+ = \hbar e^{i\phi}\left(\frac{\partial}{\partial\theta} + i\cot\theta\frac{\partial}{\partial\phi}\right), \quad L_- = \hbar e^{-i\phi}\left(-\frac{\partial}{\partial\theta} + i\cot\theta\frac{\partial}{\partial\phi}\right)$$

Since these operators depend only on the angular variables, the eigenfunctions of angular momentum will also. These are the spherical harmonics

$$\langle\theta, \phi\,|l, m\rangle = Y_l^m(\theta, \phi)$$

The eigenfunctions of L^2 and L_z are separable functions of the angular variables

$$Y_l^m(\theta, \phi) = f(\phi)\,\Theta(\theta)$$

Since L_z depends only on the first derivative with respect to the ϕ coordinate, solving its eigenvalue equation is a simple matter. We already know the eigenvalue is labeled by m, and so we have

$$L_z f = -i\hbar\frac{\partial f}{\partial\phi} = m\hbar f$$

This leads to

$$\frac{df}{f} = im\,d\phi, \Rightarrow \ln(f) = im\phi$$

ignoring the constant of integration. In other words,

$$f(\phi) = e^{im\phi}$$

The normalization of the Y_l^m functions is calculated by integration over the surface of the unit sphere:

$$\int \left|Y_l^m\right|^2 \sin\theta \, d\theta \, d\phi = 1$$

More generally, they satisfy the following orthonormality relationship:

$$\int_0^{2\pi} \int_0^{\pi} \left(Y_{l'}^{m'}\right)^* Y_l^m \sin\theta \, d\theta \, d\phi = \delta_{l,l'} \delta_{m,m'}$$

We won't derive the $\Theta(\theta)$ part of the wavefunction here, which is lengthy and can be found in most quantum mechanics textbooks. Here we simply state the results. These functions are written in terms of the *Legendre polynomials* and *associated Legendre functions*. The Legendre polynomials are given by the formula

$$P_l(x) = \frac{(-1)^l}{2^l l!} \frac{d^l}{dx^l} \left(1 - x^2\right)^l$$

and the associated Legendre functions are

$$P_l^m(x) = \sqrt{\left(1 - x^2\right)^m} \frac{d^m}{dx^m} P_l(x)$$

The spherical harmonics can then be written in the following horrific manner:

$$Y_l^m(\theta, \phi) = (-1)^m \sqrt{\frac{(2l+1)(l-m)!}{4\pi (l+m)!}} P_l^m(\cos\theta) e^{im\phi}$$

(for $m > 0$). If $m < 0$, then the formula is

$$Y_l^m(\theta, \phi) = (-1)^{|m|} \sqrt{\frac{(2l+1)(l-|m|)!}{4\pi (l+m)!}} P_l^{|m|}(\cos\theta) e^{im\phi}$$

Here is a listing of the first few spherical harmonic functions:

$$Y_0^0 = \sqrt{\frac{1}{4\pi}}$$

$$Y_1^0 = \sqrt{\frac{3}{4\pi}} \cos\theta = \frac{z}{r}, \quad Y_1^{\pm 1} = \mp\sqrt{\frac{3}{8\pi}} e^{\pm i\phi} \sin\theta = \mp\sqrt{\frac{3}{8\pi}} \frac{x \pm iy}{r}$$

$$Y_2^0 = \sqrt{\frac{5}{16\pi}} \left(3\cos^2\theta - 1\right) = \sqrt{\frac{5}{16\pi}} \frac{3z^2 - r^2}{r^2}$$

$$Y_2^{\pm 1} = \mp\sqrt{\frac{15}{8\pi}}e^{\pm i\phi}\cos\theta\sin\theta = \mp\sqrt{\frac{15}{8\pi}}\frac{(x \pm iy)\,z}{r^2}$$

$$Y_2^{\pm 2} = \sqrt{\frac{15}{32\pi}}e^{\pm i2\phi}\sin^2\theta = \sqrt{\frac{15}{32\pi}}\frac{(x \pm iy)^2}{r^2}$$

e.g. **EXAMPLE 10.10**

A given wavefunction is

$$\Psi = N\sin\theta\cos\phi$$

(a) Write the wavefunction in terms of spherical harmonics.

(b) Find the normalization constant N.

(c) What is the mean value of L^2 for this state?

(d) L_z is measured. What are the possible measurement results? What are the respective probabilities?

✔ **SOLUTION**

(a) A glance at the spherical harmonics shows that terms involving ϕ are written as exponentials. We use Euler's formula to rewrite the state:

$$\Psi = N\sin\theta\cos\phi = N\sin\theta\left(\frac{e^{i\phi} + e^{-i\phi}}{2}\right) = N\left(\frac{\sin\theta}{2}e^{i\phi} + \frac{\sin\theta}{2}e^{-i\phi}\right)$$

From the listing of spherical harmonics we find

$$Y_1^{\pm 1} = \mp\sqrt{\frac{3}{8\pi}}e^{\pm i\phi}\sin\theta$$

so we rewrite the state in this form

$$\begin{aligned}\Psi &= N\left(\frac{\sin\theta}{2}e^{i\phi} + \frac{\sin\theta}{2}e^{-i\phi}\right)\\ &= \frac{N}{2}\left(-\sqrt{\frac{8\pi}{3}}\right)\left(-\sqrt{\frac{3}{8\pi}}\right)\sin\theta e^{i\phi} + \frac{N}{2}\left(\sqrt{\frac{8\pi}{3}}\right)\left(\sqrt{\frac{3}{8\pi}}\right)\sin\theta e^{-i\phi}\\ &= -\frac{N}{2}\sqrt{\frac{8\pi}{3}}Y_1^1 + \frac{N}{2}\sqrt{\frac{8\pi}{3}}Y_1^{-1}\end{aligned}$$

(b) Let's write the state out in ket notation to make it easier to work with:

$$|\Psi\rangle = -\frac{N}{2}\sqrt{\frac{8\pi}{3}}\,|1, 1\rangle + \frac{N}{2}\sqrt{\frac{8\pi}{3}}\,|1, -1\rangle$$

Computing the norm of the state and using $\langle 1, 1 | 1, -1 \rangle = \langle 1, -1 | 1, 1 \rangle = 0$ we find

$$\langle \Psi | \Psi \rangle = \frac{N^2}{4}\left(\frac{8\pi}{3}\right)\langle 1, 1 | 1, 1 \rangle + \frac{N^2}{4}\left(\frac{8\pi}{3}\right)\langle 1, -1 | 1, -1 \rangle$$

$$= \frac{N^2}{2}\left(\frac{8\pi}{3}\right)$$

Setting this equal to unity for normalization, we can solve for N:

$$N^2 = \frac{6}{8\pi}, \quad \Rightarrow N = \sqrt{\frac{3}{4\pi}}$$

And so the normalized state is

$$|\Psi\rangle = -\frac{N}{2}\sqrt{\frac{8\pi}{3}}\,|1, 1\rangle + \frac{N}{2}\sqrt{\frac{8\pi}{3}}\,|1, -1\rangle$$

$$= -\frac{1}{2}\sqrt{\frac{3}{4\pi}}\sqrt{\frac{8\pi}{3}}\,|1, 1\rangle + \frac{1}{2}\sqrt{\frac{3}{4\pi}}\sqrt{\frac{8\pi}{3}}\,|1, -1\rangle$$

$$= -\frac{1}{\sqrt{2}}|1, 1\rangle + \frac{1}{\sqrt{2}}|1, -1\rangle$$

(c) Recalling that $L^2 |l, m\rangle = \hbar^2 l(l+1)\,|l, m\rangle$, we find

$$L^2 |\Psi\rangle = -\frac{1}{\sqrt{2}}L^2 |1, 1\rangle + \frac{1}{\sqrt{2}}L^2 |1, -1\rangle$$

$$= -\frac{1}{\sqrt{2}}\left(2\hbar^2\right)|1, 1\rangle + \frac{1}{\sqrt{2}}\left(2\hbar^2\right)|1, -1\rangle$$

$$= \left(2\hbar^2\right)|\Psi\rangle$$

The expectation value in this state is then

$$\langle L^2 \rangle = \langle \Psi | L^2 | \Psi \rangle = 2\hbar^2 \langle \Psi | \Psi \rangle = 2\hbar^2$$

(d) Since angular momentum states are written in the form $|l, m\rangle$ and in this case the state is found to be

$$|\Psi\rangle = -\frac{1}{\sqrt{2}}|1, 1\rangle + \frac{1}{\sqrt{2}}|1, -1\rangle$$

we see the possible results of measurement are $\pm\hbar$, occurring with respective probabilities

$$\left(-\frac{1}{\sqrt{2}}\right)^2 = \frac{1}{2} \quad \text{and} \quad \left(\frac{1}{\sqrt{2}}\right)^2 = \frac{1}{2}$$

EXAMPLE 10.11

A system is in the state

$$\psi(x, y, z) = N(xy + yz)$$

If L^2 is measured, what value is found? What is the probability that $2\hbar$ is found if we measure L_z? N is a normalization constant and r is fixed.

SOLUTION

We rewrite the state in spherical coordinates

$$\begin{aligned}\psi(x, y, z) &= N(xy + yz) \\ &= N\left[(r\sin\theta\cos\phi)(r\sin\theta\sin\phi) + (r\sin\theta\sin\phi)(r\cos\theta)\right] \\ &= Nr^2\left[\sin^2\theta\cos\phi\sin\phi + \sin\theta\cos\theta\sin\phi\right]\end{aligned}$$

Again, we use Euler's formulas

$$\cos\phi = \frac{e^{i\phi} + e^{-i\phi}}{2}, \quad \sin\phi = \frac{e^{i\phi} - e^{-i\phi}}{2i}$$

allowing us to put the state in the following form:

$$\begin{aligned}\psi &= N\sin^2\theta\left(\frac{e^{i\phi} + e^{-i\phi}}{2}\right)\left(\frac{e^{i\phi} - e^{-i\phi}}{2i}\right) + N\sin\theta\cos\theta\left(\frac{e^{i\phi} - e^{-i\phi}}{2i}\right) \\ &= N\sin^2\theta\left(\frac{e^{i2\phi} - e^{-i2\phi}}{4i}\right) + N\sin\theta\cos\theta\left(\frac{e^{i\phi} - e^{-i\phi}}{2i}\right) \\ &= N\left[\frac{1}{4i}\sin^2\theta e^{i2\phi} - \frac{1}{4i}\sin^2\theta e^{-i2\phi} + \frac{1}{2i}\sin\theta\cos\theta e^{i\phi} - \frac{1}{2i}\sin\theta\cos\theta e^{-i\phi}\right]\end{aligned}$$

Comparing each term to the spherical harmonics, we note that

$$Y_2^{\pm 2} = \sqrt{\frac{15}{32\pi}}e^{\pm i2\phi}\sin^2\theta \quad \text{and} \quad Y_2^{\pm 1} = \mp\sqrt{\frac{15}{8\pi}}e^{\pm i\phi}\cos\theta\sin\theta$$

We rewrite the first term of the wavefunction as follows:

$$\frac{1}{4i}\sin^2\theta e^{i2\phi} = \frac{1}{4i}\sqrt{\frac{32\pi}{15}}\sqrt{\frac{15}{32\pi}}\sin^2\theta e^{i2\phi} = \frac{1}{4i}\sqrt{\frac{32\pi}{15}}Y_2^2$$

The third term can be rewritten as

$$\frac{1}{2i}\sin\theta\cos\theta e^{i\phi} = -\frac{1}{2i}\sqrt{\frac{8\pi}{15}}\left(-\sqrt{\frac{15}{8\pi}}\right)\sin\theta\cos\theta e^{i\phi} = -\frac{1}{2i}\sqrt{\frac{8\pi}{15}}Y_2^1$$

A similar procedure can be used to write the other terms in the wavefunction as spherical harmonics, giving us

$$|\psi\rangle = N\left[\frac{1}{4i}\sqrt{\frac{32\pi}{15}}Y_2^2 - \frac{1}{4i}\sqrt{\frac{32\pi}{15}}Y_2^{-2} + \frac{1}{2i}\sqrt{\frac{8\pi}{15}}Y_2^1 - \frac{1}{2i}\sqrt{\frac{8\pi}{15}}Y_2^{-1}\right]$$

$$= N\left[\frac{1}{4i}\sqrt{\frac{32\pi}{15}}\,|2,2\rangle - \frac{1}{4i}\sqrt{\frac{32\pi}{15}}\,|2,-2\rangle + \frac{1}{2i}\sqrt{\frac{8\pi}{15}}\,|2,1\rangle - +\frac{1}{2i}\sqrt{\frac{8\pi}{15}}\,|2,-1\rangle\right]$$

The state must be normalized to find N

$$1 = \langle\psi\,|\psi\rangle = N^2\left[\left(\frac{1}{16}\right)\left(\frac{32\pi}{15}\right) + \left(\frac{1}{16}\right)\left(\frac{32\pi}{15}\right) + \left(\frac{1}{4}\right)\left(\frac{8\pi}{15}\right) + \left(\frac{1}{4}\right)\left(\frac{8\pi}{15}\right)\right]$$

$$= N^2\frac{8\pi}{15}, \quad \Rightarrow \quad N = \sqrt{\frac{15}{8\pi}}$$

and so the normalized state is

$$|\psi\rangle = \frac{1}{2i}\,|2,2\rangle - \frac{1}{2i}\,|2,-2\rangle + \frac{1}{2i}\,|2,1\rangle - \frac{1}{2i}\,|2,-1\rangle$$

We see that $l = 2$ for all the states, so if L^2 is measured we find $l(l+1)\hbar^2 = 6\hbar^2$. The probability that $2\hbar$ is found for a measurement of L_z is found by squaring the coefficient of $|2,2\rangle$:

$$\text{Prob} = \left|\frac{1}{2i}\right|^2 = \left(\frac{1}{2i}\right)\left(-\frac{1}{2i}\right) = \frac{1}{4}$$

EXAMPLE 10.12 *e.g.*

A particle is in the state

$$\psi(x,y,z) = \sqrt{\frac{3}{8\pi}}\frac{-iy+z}{r}$$

(a) What is the total angular momentum of the particle?

(b) Suppose L_z is measured. What are the possible results? What is the probability of obtaining each result?

(c) Find $\langle L_-\rangle$.

SOLUTION

(a) We use some algebra to write the state in terms of spherical harmonics. Noting that only single powers of y and z appear in the expression, we guess that the state might have terms involving $Y_1^{\pm 1}$, Y_1^0:

$$\sqrt{\frac{3}{8\pi}}\frac{-iy+z}{r} = \sqrt{\frac{3}{8\pi}}\frac{z}{r} - \sqrt{\frac{3}{8\pi}}\frac{iy}{r}$$

$$= \frac{1}{\sqrt{2}}Y_1^0 - -\sqrt{\frac{3}{8\pi}}\frac{iy}{r} + \frac{1}{2}\sqrt{\frac{3}{8\pi}}\frac{x}{r} - \frac{1}{2}\sqrt{\frac{3}{8\pi}}\frac{x}{r}$$

$$= \frac{1}{\sqrt{2}}Y_1^0 - \frac{1}{2}\sqrt{\frac{3}{8\pi}}\frac{x+iy}{r} + \frac{1}{2}\sqrt{\frac{3}{8\pi}}\frac{x-iy}{r}$$

$$= \frac{1}{\sqrt{2}}Y_1^0 - \frac{1}{2}Y_1^1 + \frac{1}{2}Y_1^{-1}$$

Or using ket notation, we can write

$$|\psi\rangle = \frac{1}{\sqrt{2}}|0\rangle - \frac{1}{2}|1\rangle + \frac{1}{2}|-1\rangle$$

For each term, l=1, and so measurement of L^2 is certain to give $1(1+1)\hbar^2 = 2\hbar^2$. The total angular momentum of the particle is

$$\sqrt{\langle\psi|\,L^2\,|\psi\rangle} = \sqrt{2\hbar^2} = \sqrt{2}\hbar$$

(b) Looking at the ket expansion of the state, we see the possible results of measurement of L_z are $\{0, \hbar, -\hbar\}$, with the probabilities given by squaring the respective coefficient of each eigenfunction (check to see the state is normalized). We have for the probability of finding 0:

$$prob(m=0) = \left(\frac{1}{\sqrt{2}}\right)^2 = \frac{1}{2}$$

The probability of finding $m = +1$ is

$$prob(m=1) = \left(-\frac{1}{2}\right)^2 = \frac{1}{4}$$

and

$$prob(m=-1) = \left(\frac{1}{2}\right)^2 = \frac{1}{4}$$

(c) The expectation value of the lowering operator is

$$\langle L_- \rangle = \langle \psi | L_- | \psi \rangle$$

$$L_- |\psi\rangle = \frac{1}{\sqrt{2}} L_- |0\rangle - \frac{1}{2} L_- |1\rangle + \frac{1}{2} L_- |-1\rangle$$

$$= \frac{1}{\sqrt{2}} \hbar\sqrt{2} |-1\rangle - \frac{1}{2}\hbar\sqrt{2}\, |0\rangle$$

$$\Rightarrow \langle L_- \rangle = \langle \psi | L_- | \psi \rangle = \hbar\sqrt{2}\left(\frac{1}{\sqrt{2}} \langle 0| - \frac{1}{2}\langle 1| + \frac{1}{2}\langle -1|\right)\left(\frac{1}{\sqrt{2}}|-1\rangle - \frac{1}{2}|0\rangle\right)$$

$$= -\left(\frac{\hbar}{2}\right)\langle 0 | 0\rangle + \left(\frac{\hbar}{2}\right)\langle -1 | -1\rangle$$

$$= 0$$

EXAMPLE 10.13

Using

$$L_- = \hbar e^{-i\phi}\left(-\frac{\partial}{\partial \theta} + i \cot\theta \frac{\partial}{\partial \phi}\right)$$

show that the operator takes $Y_1^1 \to Y_1^0$.

SOLUTION

First we recall that

$$Y_1^0 = \sqrt{\frac{3}{4\pi}} \cos\theta$$

Applying the lowering operator to Y_1^1 we obtain

$$\hbar e^{-i\phi}\left(-\frac{\partial}{\partial \theta} + i \cot\theta \frac{\partial}{\partial \phi}\right) Y_1^1 = \hbar e^{-i\phi}\left(-\frac{\partial}{\partial \theta} + i \cot\theta \frac{\partial}{\partial \phi}\right)\left(-\sqrt{\frac{3}{8\pi}} \sin\theta e^{i\phi}\right)$$

$$= \hbar e^{-i\phi}\sqrt{\frac{3}{8\pi}} \frac{\partial}{\partial \theta}\left(\sin\theta e^{i\phi}\right) - i\hbar e^{-i\phi}\cot\theta \sqrt{\frac{3}{8\pi}} \frac{\partial}{\partial \phi}\left(\sin\theta e^{i\phi}\right)$$

$$= \hbar\sqrt{\frac{3}{8\pi}} \cos\theta - i\hbar e^{-i\phi}\sqrt{\frac{3}{8\pi}}\left(i \cot\theta \sin\theta\right) e^{i\phi}$$

$$= 2\hbar\sqrt{\frac{3}{8\pi}}\cos\theta$$

$$= \hbar\sqrt{2}\sqrt{\frac{3}{4\pi}}\cos\theta = \hbar\sqrt{2}Y_1^0$$

EXAMPLE 10.14

It is found that the Hamiltonian for a symmetrical top is

$$H = \frac{1}{2I_x}L^2 + \left(\frac{1}{2I_z} - \frac{1}{2I_x}\right)L_z^2$$

What values of energy can be measured? What are the eigenstates of the system?

SOLUTION

I_x, I_z are components of the moment of inertia and are not operators. Since the eigenvalues of L^2 are $\hbar^2 l\,(l+1)$ and the eigenvalues of L_z are $m\hbar$, the eigenvalues of H which are the energy can be written by inspection:

$$E_{lm} = \frac{\hbar^2}{2I_x}l(l+1) + \left(\frac{1}{2I_z} - \frac{1}{2I_x}\right)\hbar^2 m^2$$

The eigenstates of the Hamiltonian are the eigenstates of L^2 and L_z, which are the spherical harmonics.

ANGULAR MOMENTUM AND ROTATIONS

The angular momentum operator L is said to be a generator of rotations. A state vector in a given coordinate system O can be rotated to a new coordinate system O' by a rotation operator

$$|\psi'\rangle = U_R\,|\psi\rangle$$

The infinitesimal rotation operator about an axis defined by unit vector $\hat{n}$ by infinitesimal angle ε is

$$U_R = 1 - \frac{i}{\hbar}\varepsilon\vec{L}\cdot\hat{n}$$

To obtain an operator for a system O', where the rotation is now by an angle θ, we use a limiting process:

$$U_R\left(\theta,\hat{n}\right) = \lim_{N\to\infty}\left(1 - \frac{i}{\hbar}\frac{\varepsilon}{N}\vec{L}\cdot\hat{n}\right)^N = \exp\left(-i\theta\,\hat{n}\cdot\vec{L}/\hbar\right)$$

A rotation about the z-axis by an angle ϕ would be

$$U_R(\phi, z) = \exp(-i\phi L_z/\hbar)$$

The rotation operator must satisfy

$$U_R\left(0, \hat{n}\right) = U_R\left(2\pi, \hat{n}\right) = 1$$

Operators are transformed between systems O and O' by

$$A' = U_R A U_R^{\dagger}, \quad A = U_R^{\dagger} A' U_R$$

Quiz

1. A system is in the state

$$|\psi\rangle = \frac{1}{\sqrt{14}}\begin{pmatrix} 3 \\ 1 \\ 2 \end{pmatrix}$$

Suppose L_y is measured. What results can be found and with what probabilities?

2. A particle is in the state

$$\psi = N\left[\frac{2}{\sqrt{5}}Y_1^0 + \frac{1}{2}Y_1^{-1} - \frac{1}{\sqrt{5}}Y_1^1\right]$$

(a) Find the normalization constant N.
(b) L_z is measured. What are the probabilities of finding $\{+\hbar, 0, -\hbar\}$?
(c) Find the action of the ladder operators on this state.

3. Suppose that $J = 2$. What values can m assume?

4. Write $\sin 2\theta \cos\phi$ in terms of spherical harmonics.

5. The commutator $[L_x, x]$ is equal to
(a) 0
(b) $-i\hbar x$
(c) $i\hbar L_y$
(d) $-i\hbar L_y$

CHAPTER 11

Spin-1/2 Systems

Spin is an intrinsic property of particles that is akin to a type of angular momentum. Following the procedure outlined in the previous chapter, we define a spin operator **S** that plays the role of **J** and use it to construct operators S^2 and S_z. Labeling the eigenstates of these operators by $|s, m\rangle$, the operators act in the usual way:

$$S^2 |s, m\rangle = \hbar^2 s (s + 1) |s, m\rangle, \quad S_z |s, m\rangle = m \hbar |s, m\rangle$$

Unlike orbital angular momentum, spin does not depend on spatial coordinates in any way. Therefore the spherical harmonics are not used to represent spin wavefunctions. In fact we will not look at "wavefunctions" at all when dealing with spin and will work exclusively with kets or column vectors when describing spin states. From the previous chapter we also carry over the ladder operators

$$S_{\pm} |s, m\rangle = \hbar \sqrt{s (s + 1) - m (m \pm 1)} |s, m \pm 1\rangle$$

where $S_+ = 1/2 \left(S_x + S_y\right)$ and $S_- = 1/2i \left(S_x - S_y\right)$. In addition, the components of the spin operator obey the usual commutation relations:

$$\left[S_x, S_y\right] = i\hbar S_z, \quad \left[S_y, S_z\right] = i\hbar S_x, \quad \left[S_z, S_x\right] = i\hbar S_y$$

While the value of orbital angular momentum l can vary, the spin s is fixed for all time for a given species of particle. Think of it as a built-in property like mass or charge. Different types of particles have different values of spin. For example, the force-carrying boson particles have spin $s = 1$. The postulated graviton has spin $s = 2$. In this chapter we focus on spin $s = 1/2$, which is the spin carried by the fundamental particles that make up matter, such as electrons and quarks. For a spin 1/2 particle we have

$$S^2 |1/2, m\rangle = \hbar^2 \left(\frac{1}{2}\right)\left(\frac{1}{2}+1\right)|1/2, m\rangle = \hbar^2 \frac{3}{4}|1/2, m\rangle$$

The rules for the allowed values of m are the same as those described for (J, m) in the previous chapter. Therefore if 1/2, then

$$m = -\frac{1}{2}, +\frac{1}{2}$$

are the only permissible values of m.

The Stern-Gerlach Experiment

We quickly review the experiment that led to the discovery of spin. In the Stern-Gerlach experiment, neutral particles such as silver atoms are passed through an inhomogeneous field. The field is created by two magnetic pole pieces, one with an irregular shape that produces the inhomogeneity in the desired direction (that could be the z-axis, say). Particles travel in a perpendicular direction that we label the x-axis. The force on a particle with a magnetic moment $\vec{\mu}$ is found to be

$$\vec{F} = \vec{\nabla}\left(\vec{\mu} \cdot \vec{B}\right)$$

where B is the magnetic field. The z-component of the magnetic field leads to a force in the z-direction given by

$$F_z = \mu_z \frac{d B_z}{d z}$$

This force will cause particles in the beam to be deflected. It was thought that the deflections would be random based on a random distribution in the values of μ_z. It was not known at the time, but it turns out that the magnetic dipole moment of the particle is proportional to its spin. Specifically, in the case of the electron,

$$\vec{\mu} = \frac{e}{mc}\vec{S}$$

Therefore we have

$$\mu_z = \frac{e}{mc} S_z$$

As we said above, classically, one would expect particles subjected to this force to experience a deflection that varied continuously between $\pm\mu$. However, when the experiment is actually carried out the incoming beam splits exactly in two as it emerges from the apparatus. This is a spectacular demonstration of quantum theory in action. The result is completely quantized—one emerging beam heads in the positive z-direction or "up" while the other heads in the negative z-direction or "down." In Fig. 11-1 we illustrate the experimental setup and label particles that head in the up direction by the state $|+\rangle$ and particles that head in the down direction by $|-\rangle$.

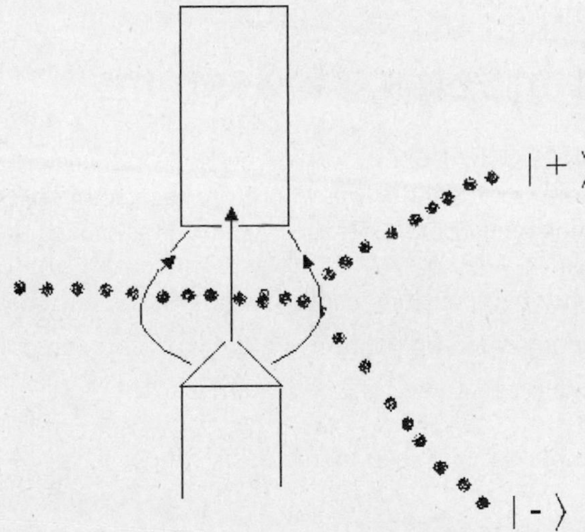

Fig. 11-1

These are the two spin states of $m = +1/2$ and $m = -1/2$. A Stern-Gerlach apparatus can be used to prepare a system in one state or another by filtering the output. For example, we can block the beam of spin-down particles to produce a beam consisting of entirely spin-up particles. We represent this schematically in Fig. 11-2.

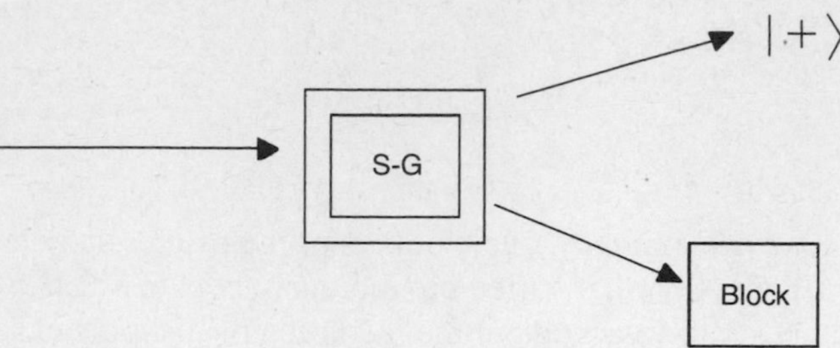

Fig. 11-2

We can position the apparatus to create the magnetic field in any direction we like; therefore we can measure spin in the x and y directions as well by rotating the magnets 90 degrees. We can measure spin in an arbitrary direction defined by a unit vector $\hat{n}$ by rotating the apparatus by the appropriate angle. As we will see later, spin states in any given direction are superpositions of spin states in the other orthogonal directions.

The Basis States for Spin-1/2 Systems

A measurement of the spin of an electron or any spin-1/2 particle can have only two possible results: spin-up or spin-down. Therefore the vector space used to describe spin is a two-dimensional complex Hilbert space. This means the following:

- There are only two possible results of a measurement; these are $\pm\frac{\hbar}{2}$
- States are represented by column vectors with two components
- Operators are represented by 2×2 matrices

As we described in the last section, the basis states of S^2, S_z are labeled in the following way:

$$|+\rangle, \Rightarrow S_z|+\rangle = \frac{\hbar}{2}|+\rangle$$

$$|-\rangle, \Rightarrow S_z|-\rangle = -\frac{\hbar}{2}|-\rangle$$

These states are orthonormal, and so the inner products between them are

$$\langle + | + \rangle = \langle - | - \rangle = 1$$

$$\langle + | - \rangle = \langle - | + \rangle = 0$$

The column vectors representing the states are

$$|+\rangle = \begin{pmatrix} 1 \\ 0 \end{pmatrix}, \quad |-\rangle = \begin{pmatrix} 0 \\ 1 \end{pmatrix}$$

The most general state we can write is a superposition of these basis states

$$|\psi\rangle = \alpha\,|+\rangle + \beta\,|-\rangle$$

If the state is normalized, then

$$|\alpha|^2 + |\beta|^2 = 1$$

where $|\alpha|^2$ gives the probability of finding the system in the spin-up state $|+\rangle$ and $|\beta|^2$ is the probability of finding the system in the spin-down state $|-\rangle$. Using the column vector representations of the basis states, we can write $|\psi\rangle$ in the compact form

$$|\psi\rangle = \alpha\,|+\rangle + \beta\,|-\rangle = \alpha \begin{pmatrix} 1 \\ 0 \end{pmatrix} + \beta \begin{pmatrix} 0 \\ 1 \end{pmatrix} = \begin{pmatrix} \alpha \\ \beta \end{pmatrix}$$

Since $\cos^2\theta + \sin^2\theta = 1$, the normalization condition allows us to write the expansion coefficients in the following way:

$$\alpha = e^{-i\frac{\phi}{2}} \cos\left(\frac{\theta}{2}\right)$$

$$\beta = e^{i\frac{\phi}{2}} \sin\left(\frac{\theta}{2}\right)$$

It is a simple exercise to verify that writing the coefficients this way leads to $|\alpha|^2 + |\beta|^2 = 1$.

Matrix representations of operators in this basis are written in the following way:

$$\left[\hat{O}\right] = \begin{pmatrix} \langle +|\,\hat{O}\,|+\rangle & \langle +|\,\hat{O}\,|-\rangle \\ \langle -|\,\hat{O}\,|+\rangle & \langle -|\,\hat{O}\,|-\rangle \end{pmatrix}$$

Therefore, the matrix representation of S_z is given by

$$[S_z] = \begin{pmatrix} \langle +|\,S_z\,|+\rangle & \langle +|\,S_z\,|-\rangle \\ \langle -|\,S_z\,|+\rangle & \langle -|\,S_z\,|-\rangle \end{pmatrix} = \begin{pmatrix} \langle +|\left(\frac{\hbar}{2}\right)|+\rangle & \langle +|\left(-\frac{\hbar}{2}\right)|-\rangle \\ \langle -|\left(\frac{\hbar}{2}\right)|+\rangle & \langle -|\left(-\frac{\hbar}{2}\right)|-\rangle \end{pmatrix}$$

$$= \frac{\hbar}{2} \begin{pmatrix} \langle +\,|\,+\rangle & -\langle +\,|\,-\rangle \\ \langle -\,|\,+\rangle & -\langle -\,|\,-\rangle \end{pmatrix}$$

$$= \frac{\hbar}{2} \begin{pmatrix} 1 & 0 \\ 0 & -1 \end{pmatrix}$$

Using the Ladder Operators to Construct S_x, S_y

Reminding ourselves that the ladder operators act in the following way:

$$S_{\pm}\left|s, m\right\rangle = \hbar\sqrt{s\left(s+1\right) - m\left(m \pm 1\right)}\left|s, m \pm 1\right\rangle, \quad S_{+}\left|s, s\right\rangle = 0, \; S_{-}\left|s, -s\right\rangle = 0$$

For the spin-1/2 case we have

$$S_{+}\left|+\right\rangle = 0, \quad S_{-}\left|-\right\rangle = 0$$

$$s = 1/2, \qquad m = -1/2 \Rightarrow$$

$$\hbar\sqrt{s\left(s+1\right) - m\left(m+1\right)} = \hbar\sqrt{\frac{1}{2}\left(\frac{1}{2}+1\right) + \frac{1}{2}\left(-\frac{1}{2}+1\right)} = \hbar$$

and so we have

$$S_{+}\left|-\right\rangle = \hbar\left|+\right\rangle$$

Likewise, we find that

$$S_{-}\left|+\right\rangle = \hbar\left|-\right\rangle$$

These results allow us to construct matrix representations of these operators in the $\{\left|+\right\rangle, \left|-\right\rangle\}$ basis. We find that

$$\left[S_{+}\right] = \begin{pmatrix} \langle +|S_{+}|+\rangle & \langle +|S_{+}|-\rangle \\ \langle -|S_{+}|+\rangle & \langle -|S_{+}|-\rangle \end{pmatrix} = \begin{pmatrix} 0 & \hbar\langle + | +\rangle \\ 0 & \hbar\langle - | +\rangle \end{pmatrix} = \hbar\begin{pmatrix} 0 & 1 \\ 0 & 0 \end{pmatrix}$$

$$\left[S_{-}\right] = \begin{pmatrix} \langle +|S_{-}|+\rangle & \langle +|S_{-}|-\rangle \\ \langle -|S_{-}|+\rangle & \langle -|S_{-}|-\rangle \end{pmatrix} = \begin{pmatrix} \hbar\langle + | -\rangle & 0 \\ \hbar\langle - | -\rangle & 0 \end{pmatrix} = \hbar\begin{pmatrix} 0 & 0 \\ 1 & 0 \end{pmatrix}$$

We can obtain the S_x operator from

$$S_x = \frac{S_{+} + S_{-}}{2} = \frac{1}{2}\left[\hbar\begin{pmatrix} 0 & 1 \\ 0 & 0 \end{pmatrix} + \hbar\begin{pmatrix} 0 & 0 \\ 1 & 0 \end{pmatrix}\right] = \frac{\hbar}{2}\begin{pmatrix} 0 & 1 \\ 1 & 0 \end{pmatrix}$$

and for S_y we find

$$S_y = \frac{S_{+} - S_{-}}{2i} = \frac{1}{2i}\left[\hbar\begin{pmatrix} 0 & 1 \\ 0 & 0 \end{pmatrix} - \hbar\begin{pmatrix} 0 & 0 \\ 1 & 0 \end{pmatrix}\right] = \frac{\hbar}{2}\begin{pmatrix} 0 & \frac{1}{i} \\ -\frac{1}{i} & 0 \end{pmatrix} = \frac{\hbar}{2}\begin{pmatrix} 0 & -i \\ i & 0 \end{pmatrix}$$

The actions of these operators on the basis states can be worked out using kets (or very easily using the matrix representation). We have

$$S_x\left|+\right\rangle = \left(\frac{S_{+} + S_{-}}{2}\right)\left|+\right\rangle = \frac{1}{2}S_{+}\left|+\right\rangle + \frac{1}{2}S_{-}\left|+\right\rangle = \frac{1}{2}S_{-}\left|+\right\rangle = \frac{\hbar}{2}\left|-\right\rangle$$

$$S_x\left|-\right\rangle = \left(\frac{S_{+} + S_{-}}{2}\right)\left|-\right\rangle = \frac{1}{2}S_{+}\left|-\right\rangle + \frac{1}{2}S_{-}\left|-\right\rangle = \frac{1}{2}S_{+}\left|-\right\rangle = \frac{\hbar}{2}\left|+\right\rangle$$

e.g.

EXAMPLE 11.1

A spin-1/2 system is in the state

$$|\psi\rangle = \frac{1}{\sqrt{2}}\left(|+\rangle + |-\rangle\right)$$

Show that $\langle S_x S_y \rangle = 0$.

✔

SOLUTION

Writing S_x, S_y in terms of the ladder operators we have

$$S_x S_y = \left(\frac{S_+ + S_-}{2}\right)\left(\frac{S_+ - S_-}{2i}\right) = \frac{S_+^2 + S_- S_+ - S_+ S_- - S_-^2}{4i}$$

In a two-state system, either of the squares of the ladder operators acting twice will result in zero (consider $S_+^2 |-\rangle = \hbar\,(S_+ |+\rangle) = 0$), and so we can drop those terms. Therefore we have

$$S_x S_y |\psi\rangle = \left(\frac{S_+^2 + S_- S_+ - S_+ S_- - S_-^2}{4i}\right)|\psi\rangle = \left(\frac{S_- S_+ - S_+ S_-}{4i}\right)|\psi\rangle$$

Now we have

$$S_- S_+ |+\rangle = 0, \quad S_+ S_- |-\rangle = 0$$

since

$$S_+ |+\rangle = 0, \qquad S_- |-\rangle = 0$$

and so the result is

$$\begin{aligned} S_x S_y |\psi\rangle &= \frac{1}{4i\sqrt{2}}\left(S_- S_+ |-\rangle - S_+ S_- |+\rangle\right) \\ &= \frac{\hbar}{4i\sqrt{2}}\left(S_- |+\rangle - S_+ |-\rangle\right) \\ &= \frac{\hbar^2}{4i\sqrt{2}}\left(|-\rangle - |+\rangle\right) \end{aligned}$$

For the expectation value we obtain

$$\begin{aligned} \langle S_x S_y \rangle &= \langle \psi | S_x S_y |\psi\rangle \\ &= \frac{1}{\sqrt{2}}\left(\langle +| + \langle -|\right)\frac{\hbar^2}{4i\sqrt{2}}\left(|-\rangle - |+\rangle\right) \end{aligned}$$

$$= \frac{\hbar^2}{8i}\left(\left(\langle +| + \langle -|\right)\left(|-\rangle - |+\rangle\right)\right) = \frac{\hbar^2}{8i}\left(\langle + \mid -\rangle - \langle + \mid +\rangle + \langle - \mid -\rangle - \langle - \mid +\rangle\right)$$

$$= \frac{\hbar^2}{8i}\left(-\langle + \mid +\rangle + \langle - \mid -\rangle\right) = 0$$

EXAMPLE 11.2

Find the eigenvectors and eigenvalues of S_y.

SOLUTION

The characteristic equation is found to be

$$\det\left|S_y - \lambda I\right| = \frac{\hbar}{2}\left|\begin{pmatrix} 0 & -i \\ i & 0 \end{pmatrix} - \begin{pmatrix} \lambda & 0 \\ 0 & \lambda \end{pmatrix}\right| = \frac{\hbar}{2}\begin{vmatrix} -\lambda & -i \\ i & -\lambda \end{vmatrix} = 0$$

$$\Rightarrow \frac{\hbar}{2}\left(\lambda^2 - 1\right) = 0 \quad or \quad \lambda_{1,2} = \pm\frac{\hbar}{2}$$

The first eigenvector associated with eigenvalue $+\hbar/2$ is given the label $|+_y\rangle$ for spin-up in the y-direction:

$$S_y\,|+_y\rangle = \frac{\hbar}{2}\,|+_y\rangle$$

Calling the components of the eigenvector α, β we need to solve

$$\begin{pmatrix} 0 & -i \\ i & 0 \end{pmatrix}\begin{pmatrix} \alpha \\ \beta \end{pmatrix} = \begin{pmatrix} \alpha \\ \beta \end{pmatrix}$$

This leads to the two equations

$$-i\beta = \alpha$$
$$i\alpha = \beta$$

Using the second relation, we eliminate β and write the state as

$$|+_y\rangle = \begin{pmatrix} \alpha \\ i\alpha \end{pmatrix}$$

Now we normalize to find α:

$$1 = \langle +_y \mid +_y\rangle = \begin{pmatrix} \alpha^* & -i\alpha^* \end{pmatrix}\begin{pmatrix} \alpha \\ i\alpha \end{pmatrix} = |\alpha|^2 + |\alpha|^2 = 2\,|\alpha|^2$$

$$\Rightarrow \alpha = \frac{1}{\sqrt{2}}$$

Therefore we can write the state as

$$|+_y\rangle = \frac{1}{\sqrt{2}}\begin{pmatrix}1\\ i\end{pmatrix} = \frac{|+\rangle + i\,|-\rangle}{\sqrt{2}}$$

A similar exercise using the eigenvalue $\lambda_2 = -\hbar/2$ leads us to conclude that

$$|-_y\rangle = \frac{1}{\sqrt{2}}\begin{pmatrix}1\\ -i\end{pmatrix} = \frac{|+\rangle - i\,|-\rangle}{\sqrt{2}}$$

Notice that these relationships can be inverted easily, allowing us to write the standard basis in terms of the eigenstates of S_y:

$$|+_y\rangle + |-_y\rangle = \frac{|+\rangle + i\,|-\rangle}{\sqrt{2}} + \frac{|+\rangle - i\,|-\rangle}{\sqrt{2}} = \frac{2}{\sqrt{2}}|+\rangle$$

$$\Rightarrow |+\rangle = \frac{|+_y\rangle + |-_y\rangle}{\sqrt{2}}$$

We also find that we can write

$$|-\rangle = -i\left(\frac{|+_y\rangle - |-_y\rangle}{\sqrt{2}}\right)$$

The ability to express each basis in terms of the other is a useful tool. Consider an arbitrary state written in the basis of S_z. We can re-express the state as follows:

$$|\psi\rangle = \alpha\,|+\rangle + \beta\,|-\rangle = \alpha\left(\frac{|+_y\rangle + |-_y\rangle}{\sqrt{2}}\right) + \beta\left(\frac{-i\,|+_y\rangle + i\,|-_y\rangle}{\sqrt{2}}\right)$$

$$= \left(\frac{\alpha - i\beta}{\sqrt{2}}\right)|+_y\rangle + \left(\frac{\alpha + i\beta}{\sqrt{2}}\right)|-_y\rangle$$

EXAMPLE 11.3 e.g.

Describe the spin operator S_n for an arbitrary direction defined by the unit vector $\hat{n}$, which describes a new axis z', and this to write the spin up state $|+_n\rangle$ as an expansion of the standard basis of eigenfunctions of S_z.

SOLUTION ✔

An arbitrary unit vector can be written in the following form

$$\hat{n} = \cos\phi \sin\theta\,\hat{x} + \sin\phi \sin\theta\,\hat{y} + \cos\theta\,\hat{z}$$

Now we have

$$\vec{S} \cdot \hat{n} = S_n = \cos\phi \sin\theta S_x + \sin\phi \sin\theta S_y + \cos\theta S_z$$

The eigenvectors of S_n are such that

$$S_n \left|+_n\right\rangle = +\frac{\hbar}{2} \left|+_n\right\rangle, \quad S_n \left|-_n\right\rangle = -\frac{\hbar}{2} \left|-_n\right\rangle$$

We can expand these eigenvectors in terms of the standard basis in some way, i.e.

$$\left|+_n\right\rangle = \alpha \left|+\right\rangle + \beta \left|-\right\rangle$$

Using the eigenvalue equation $S_n \left|+_n\right\rangle = +\frac{\hbar}{2} \left|+_n\right\rangle$, we can write

$$\left(\cos\phi \sin\theta S_x + \sin\phi \sin\theta S_y + \cos\theta S_z\right) \left(\alpha \left|+\right\rangle + \beta \left|-\right\rangle\right) = +\frac{\hbar}{2} \left(\alpha \left|+\right\rangle + \beta \left|-\right\rangle\right)$$

Using the action of each of the operators $S_x, \ S_y, \ S_z$ on the standard basis states, the equation $S_n \left|+_n\right\rangle = +\hbar/2 \left|+_n\right\rangle$ gives us the following relationships:

$$\alpha \sin\theta \cos\phi + i\alpha \ \sin\theta \sin\phi - \beta \cos\theta = \beta$$

$$\alpha \ \cos\theta + \beta \sin\theta \cos\phi - i\beta \sin\theta \sin\phi = \alpha$$

Eliminating α we find

$$\alpha = \frac{1 + \cos\theta}{\sin\theta} \ e^{-i\phi} \beta$$

Since $|\alpha|^2 + |\beta|^2 = 1$ we find that

$$|\beta|^2 = \frac{\sin^2\theta}{2\left(1 + \cos\theta\right)} = 4\frac{\sin^2\left(\frac{\theta}{2}\right)\cos^2\left(\frac{\theta}{2}\right)}{4\cos^2\left(\theta/2\right)} = \sin^2\left(\frac{\theta}{2}\right)$$

We can include a phase with this term, and so we choose

$$\beta = e^{i\phi} \sin\frac{\theta}{2}$$

A little algebra shows that

$$\alpha = \cos\frac{\theta}{2}$$

So in conclusion, we can write the state as

$$\left|+_n\right\rangle = \alpha \left|+\right\rangle + \beta \left|-\right\rangle = \cos\frac{\theta}{2} \left|+\right\rangle + e^{i\phi} \sin\frac{\theta}{2} \left|-\right\rangle$$

EXAMPLE 11.4

e.g.

A particle is in the state

$$|\psi\rangle = \frac{1}{\sqrt{5}}\begin{pmatrix} 2 \\ i \end{pmatrix}$$

Find the probabilities of

(a) Measuring spin-up or spin-down in the z direction.
(b) Measuring spin-up or spin-down in the y direction.

SOLUTION

(a) First we expand the state in the standard basis $|\pm\rangle$:

$$|\psi\rangle = \frac{1}{\sqrt{5}}\begin{pmatrix} 2 \\ i \end{pmatrix} = \frac{1}{\sqrt{5}}\begin{pmatrix} 2 \\ 0 \end{pmatrix} + \frac{1}{\sqrt{5}}\begin{pmatrix} 0 \\ i \end{pmatrix} = \frac{2}{\sqrt{5}}\begin{pmatrix} 1 \\ 0 \end{pmatrix} + \frac{i}{\sqrt{5}}\begin{pmatrix} 0 \\ 1 \end{pmatrix} = \frac{2}{\sqrt{5}}|+\rangle + \frac{i}{\sqrt{5}}|-\rangle$$

The Born rule determines the probability of measuring spin-up in the z-direction, which is found from computing $|\langle + \mid \psi\rangle|^2$. In this case we have

$$|\langle + \mid \psi\rangle|^2 = \left|\frac{2}{\sqrt{5}}\right|^2 = \frac{4}{5} = 0.8$$

Application of the Born rule allows us to find the probability of measuring spin-down

$$|\langle - \mid \psi\rangle|^2 = \left|\frac{i}{\sqrt{5}}\right|^2 = \left(\frac{-i}{\sqrt{5}}\right)\left(\frac{i}{\sqrt{5}}\right) = \frac{1}{5} = 0.2$$

Notice that the probabilities sum to one, as they should.

(b) To find the probabilities of finding spin-up/down along the y-axis, we can use the relationship we derived earlier that allows us to express a state written in the $|\pm\rangle$ in the S_y states. We restate this relationship here:

$$|\psi\rangle = \alpha\,|+\rangle + \beta\,|-\rangle = \alpha\left(\frac{|+_y\rangle + |-_y\rangle}{\sqrt{2}}\right) + \beta\left(\frac{-i\,|+_y\rangle + i\,|-_y\rangle}{\sqrt{2}}\right)$$

$$= \left(\frac{\alpha - i\beta}{\sqrt{2}}\right)|+_y\rangle + \left(\frac{\alpha + i\beta}{\sqrt{2}}\right)|-_y\rangle$$

For the state in this problem, we find

$$|\psi\rangle = \frac{2}{\sqrt{5}}|+\rangle + \frac{i}{\sqrt{5}}|-\rangle = \frac{1}{\sqrt{2}}\left(\frac{2}{\sqrt{5}} + \frac{1}{\sqrt{5}}\right)|+_y\rangle + \frac{1}{\sqrt{2}}\left(\frac{2}{\sqrt{5}} - \frac{1}{\sqrt{5}}\right)|-_y\rangle$$

$$= \frac{3}{\sqrt{10}}|+_y\rangle + \frac{1}{\sqrt{10}}|-_y\rangle$$

Therefore the probability of measuring spin-up along the y-direction is

$$|\langle +_y \mid \psi \rangle|^2 = \left(\frac{3}{\sqrt{10}}\right)^2 = \frac{9}{10} = 0.9$$

and the probability of finding spin-down is

$$|\langle -_y \mid \psi \rangle|^2 = \left(\frac{1}{\sqrt{10}}\right)^2 = \frac{1}{10} = 0.1$$

---

e.g. **EXAMPLE 11.5**

A spin-1/2 system is in the state

$$|\psi\rangle = \frac{1+i}{\sqrt{3}}|+\rangle + \frac{1}{\sqrt{3}}|-\rangle$$

(a) If spin is measured in the z-direction, what are the probabilities of finding $\pm\hbar/2$?

(b) If instead, spin is measured in the x-direction, what is the probability of finding spin-up?

(c) Calculate $\langle S_z \rangle$ and $\langle S_x \rangle$ for this state.

SOLUTION

(a) The probability of finding $+\hbar/2$ is found from the Born rule, and so we calculate

$$|\langle + \mid \psi \rangle|^2 = \left|\frac{1+i}{\sqrt{3}}\right|^2 = \left(\frac{1+i}{\sqrt{3}}\right)\left(\frac{1-i}{\sqrt{3}}\right) = \frac{2}{3}$$

The probability of finding $-\hbar/2$ is given by

$$|\langle - \mid \psi \rangle|^2 = \left|\frac{1}{\sqrt{3}}\right|^2 = \frac{1}{3}$$

(b) In the chapter quiz, you will show that

$$|+_x\rangle = \frac{|+\rangle + |-\rangle}{\sqrt{2}}$$

From the Born rule, the probability of finding spin up in the x-direction is $|\langle +_x \mid \psi\rangle|^2$. Now

$$\langle +_x \mid \psi\rangle = \left(\frac{\langle +| + \langle -|}{\sqrt{2}}\right)\left(\frac{1+i}{\sqrt{3}}\,|+\rangle + \frac{1}{\sqrt{3}}\,|-\rangle\right)$$
$$= \left(\frac{1}{\sqrt{2}}\right)\left(\frac{1+i}{\sqrt{3}}\right)\langle + \mid +\rangle + \left(\frac{1}{\sqrt{2}}\right)\left(\frac{1}{\sqrt{3}}\right)\langle - \mid -\rangle$$
$$= \frac{2+i}{6}$$

Therefore the probability is

$$|\langle +_x \mid \psi\rangle|^2 = \left(\frac{2-i}{6}\right)\left(\frac{2+i}{6}\right) = \frac{5}{6}$$

(Exercise: Calculate $|\langle -_x \mid \psi\rangle|^2$ and verify the probabilities sum to one.)

(c) The expectation values are given by

$$S_z\,|\psi\rangle = \left(\frac{1+i}{\sqrt{3}}\right) S_z\,|+\rangle + \frac{1}{\sqrt{3}} S_z\,|-\rangle = \frac{\hbar}{2}\left[\left(\frac{1+i}{\sqrt{3}}\right)|+\rangle - \frac{1}{\sqrt{3}}\,|-\rangle\right]$$

$\Rightarrow$

$$\langle S_z\rangle = \langle\psi|\,S_z\,|\psi\rangle = \frac{\hbar}{2}\left[\left(\frac{1-i}{\sqrt{3}}\right)\langle +| + \frac{1}{\sqrt{3}}\langle -|\right]\left[\left(\frac{1+i}{\sqrt{3}}\right)|+\rangle - \frac{1}{\sqrt{3}}\,|-\rangle\right]$$
$$= \frac{\hbar}{2}\left[\left(\frac{1-i}{\sqrt{3}}\right)\left(\frac{1+i}{\sqrt{3}}\right)\langle + \mid +\rangle + \left(\frac{1}{\sqrt{3}}\right)\left(-\frac{1}{\sqrt{3}}\right)\langle - \mid -\rangle\right]$$
$$= \frac{\hbar}{2}\left(\frac{2}{3} - \frac{1}{3}\right) = \frac{\hbar}{6}$$

For S_x, recalling that it flips the states (i.e. $S_x\,|\pm\rangle = \hbar/2\,|\mp\rangle$), we have

$$S_x\,|\psi\rangle = \left(\frac{1+i}{\sqrt{3}}\right) S_x\,|+\rangle + \frac{1}{\sqrt{3}} S_x\,|-\rangle = \frac{\hbar}{2}\left[\left(\frac{1+i}{\sqrt{3}}\right)|-\rangle + \frac{1}{\sqrt{3}}\,|+\rangle\right]$$

and so the expectation value is

$$\langle S_x\rangle = \langle\psi|\,S_x\,|\psi\rangle = \frac{\hbar}{2}\left[\left(\frac{1-i}{\sqrt{3}}\right)\langle +| + \frac{1}{\sqrt{3}}\langle -|\right]\left[\left(\frac{1+i}{\sqrt{3}}\right)|-\rangle + \frac{1}{\sqrt{3}}\,|+\rangle\right]$$
$$= \frac{\hbar}{2}\left[\left(\frac{1-i}{\sqrt{3}}\right)\frac{1}{\sqrt{3}}\langle + \mid +\rangle + \left(\frac{1+i}{\sqrt{3}}\right)\frac{1}{\sqrt{3}}\langle - \mid -\rangle\right]$$

$$= \frac{\hbar}{2}\left[\frac{1}{3} + \frac{1}{3}\right] = \frac{\hbar}{3}$$

Unitary Transformations for Spin-1/2 Systems

We now consider the unitary matrix that can be used to transform from the S_z basis to the S_x basis. The matrix is constructed as follows

$$U = \begin{pmatrix} \langle +_x \mid + \rangle & \langle +_x \mid - \rangle \\ \langle -_x \mid + \rangle & \langle -_x \mid - \rangle \end{pmatrix}$$

In the chapter quiz you will show that in terms of the standard basis of S_z, the eigenvectors of S_x can be written as

$$|+_x\rangle = \frac{1}{\sqrt{2}}\left(|+\rangle + |-\rangle\right), \quad |-_x\rangle = \frac{1}{\sqrt{2}}\left(|+\rangle - |-\rangle\right)$$

Calculating each term in the matrix, we find

$$\langle +_x \mid + \rangle = \frac{1}{\sqrt{2}}\left(\langle +| + \langle -|\right)|+\rangle = \frac{1}{\sqrt{2}}\langle + \mid + \rangle + \frac{1}{\sqrt{2}}\langle - \mid + \rangle = \frac{1}{\sqrt{2}}\langle + \mid + \rangle = \frac{1}{\sqrt{2}}$$

$$\langle +_x \mid - \rangle = \frac{1}{\sqrt{2}}\left(\langle +| + \langle -|\right)|-\rangle = \frac{1}{\sqrt{2}}\langle + \mid - \rangle + \frac{1}{\sqrt{2}}\langle - \mid - \rangle = \frac{1}{\sqrt{2}}\langle - \mid - \rangle = \frac{1}{\sqrt{2}}$$

$$\langle -_x \mid + \rangle = \frac{1}{\sqrt{2}}\left(\langle +| - \langle -|\right)|+\rangle = \frac{1}{\sqrt{2}}\langle + \mid + \rangle - \frac{1}{\sqrt{2}}\langle - \mid + \rangle = \frac{1}{\sqrt{2}}\langle + \mid + \rangle = \frac{1}{\sqrt{2}}$$

$$\langle -_x \mid - \rangle = \frac{1}{\sqrt{2}}\left(\langle +| - \langle -|\right)|-\rangle = \frac{1}{\sqrt{2}}\langle + \mid + \rangle - \frac{1}{\sqrt{2}}\langle - \mid - \rangle = -\frac{1}{\sqrt{2}}\langle - \mid - \rangle = -\frac{1}{\sqrt{2}}$$

and so we find the unitary matrix of transformation is

$$U = \begin{pmatrix} \langle +_x \mid + \rangle & \langle +_x \mid - \rangle \\ \langle -_x \mid + \rangle & \langle -_x \mid - \rangle \end{pmatrix} = \frac{1}{\sqrt{2}}\begin{pmatrix} 1 & 1 \\ 1 & -1 \end{pmatrix}$$

We can use this matrix to transform any state written in terms of the S_z basis in terms of the S_x basis. Application of this matrix to an arbitrary state gives

$$|\psi\rangle = \alpha\,|+\rangle + \beta\,|-\rangle = \begin{pmatrix} \alpha \\ \beta \end{pmatrix}, \quad \Rightarrow$$

$$U\,|\psi\rangle = \frac{1}{\sqrt{2}}\begin{pmatrix} 1 & 1 \\ 1 & -1 \end{pmatrix}\begin{pmatrix} \alpha \\ \beta \end{pmatrix} = \frac{1}{\sqrt{2}}\begin{pmatrix} \alpha + \beta \\ \alpha - \beta \end{pmatrix}$$

It is clear that U is Hermitian. Is this matrix unitary? We check

$$UU^{\dagger} = \frac{1}{\sqrt{2}}\begin{pmatrix} 1 & 1 \\ 1 & -1 \end{pmatrix}\frac{1}{\sqrt{2}}\begin{pmatrix} 1 & 1 \\ 1 & -1 \end{pmatrix}$$

$$= \frac{1}{2}\begin{pmatrix} 1*1+1*1 & 1*1+1*-1 \\ 1*1+-1*1 & 1*1+(-1)*(-1) \end{pmatrix} = \frac{1}{2}\begin{pmatrix} 2 & 0 \\ 0 & 2 \end{pmatrix} = \begin{pmatrix} 1 & 0 \\ 0 & 1 \end{pmatrix} = I$$

Since $U = U^{\dagger}$ we have shown that $U^{\dagger} = U^{-1}$ and the matrix is unitary. This matrix can be used to diagonalize S_x

$$US_xU^{\dagger} = \frac{1}{\sqrt{2}}\begin{pmatrix} 1 & 1 \\ 1 & -1 \end{pmatrix}\left(\frac{\hbar}{2}\right)\begin{pmatrix} 0 & 1 \\ 1 & 0 \end{pmatrix}\frac{1}{\sqrt{2}}\begin{pmatrix} 1 & 1 \\ 1 & -1 \end{pmatrix}$$

$$= \frac{\hbar}{4}\begin{pmatrix} 1 & 1 \\ 1 & -1 \end{pmatrix}\begin{pmatrix} 1 & -1 \\ 1 & 1 \end{pmatrix} = \frac{\hbar}{4}\begin{pmatrix} 2 & 0 \\ 0 & -2 \end{pmatrix} = \frac{\hbar}{2}\begin{pmatrix} 1 & 0 \\ 0 & -1 \end{pmatrix}$$

EXAMPLE 11.6

In the S_z basis a system is in the state

$$|\phi\rangle = \frac{1}{\sqrt{5}}\begin{pmatrix} i \\ 2 \end{pmatrix}$$

If S_x is measured, what are the probabilities of finding spin-up/down?

SOLUTION

We apply the transformation matrix U to the state to express it in the basis of S_x:

$$U\,|\phi\rangle = \frac{1}{\sqrt{2}}\begin{pmatrix} 1 & 1 \\ 1 & -1 \end{pmatrix}\frac{1}{\sqrt{5}}\begin{pmatrix} i \\ 2 \end{pmatrix} = \frac{1}{\sqrt{10}}\begin{pmatrix} 2+i \\ -2+i \end{pmatrix}$$

Now that we have transformed to the S_x basis, the column vector representations of the states are

$$|+_x\rangle = \begin{pmatrix} 1 \\ 0 \end{pmatrix}, \quad |-_x\rangle = \begin{pmatrix} 0 \\ 1 \end{pmatrix}$$

and so we can write

$$\frac{1}{\sqrt{10}}\begin{pmatrix} 2+i \\ -2+i \end{pmatrix} = \frac{1}{\sqrt{10}}\begin{pmatrix} 2+i \\ 0 \end{pmatrix} + \frac{1}{\sqrt{10}}\begin{pmatrix} 0 \\ -2+i \end{pmatrix} = \frac{2+i}{\sqrt{10}}\begin{pmatrix} 1 \\ 0 \end{pmatrix} + \frac{-2+i}{\sqrt{10}}\begin{pmatrix} 0 \\ 1 \end{pmatrix}$$

$$= \frac{2+i}{\sqrt{10}}|+_x\rangle + \frac{-2+i}{\sqrt{10}}|-_x\rangle$$

We apply the Born rule to find the probability that a measurement of S_x results in $+\hbar/2$

$$|\langle +_x \mid \phi\rangle|^2 = \left|\frac{2+i}{\sqrt{10}}\right|^2 = \left(\frac{2+i}{\sqrt{10}}\right)\left(\frac{2-i}{\sqrt{10}}\right) = \frac{5}{10} = \frac{1}{2}$$

Likewise we find the probability of measuring $-\hbar/2$ is

$$|\langle -_x \mid \phi\rangle|^2 = \left|\frac{2-i}{\sqrt{10}}\right|^2 = \left(\frac{2+i}{\sqrt{10}}\right)\left(\frac{2-i}{\sqrt{10}}\right) = \frac{5}{10} = \frac{1}{2}$$

Exercise: Show that in this state, the probability of finding spin-up in the z-*direction* is 1/5 and the probability of finding spin-down in the z-direction is 4/5.

The Outer Product Representation of the Spin Operators

From the matrix representations of the spin operators

$$S_x = \frac{\hbar}{2}\begin{pmatrix} 0 & 1 \\ 1 & 0 \end{pmatrix}, \quad S_y = \frac{\hbar}{2}\begin{pmatrix} 0 & -i \\ i & 0 \end{pmatrix}, \quad S_z = \frac{\hbar}{2}\begin{pmatrix} 1 & 0 \\ 0 & -1 \end{pmatrix}$$

Using the matrix representation of an operator in this basis

$$\left[\hat{O}\right] = \begin{pmatrix} \langle + | \hat{O} | + \rangle & \langle + | \hat{O} | - \rangle \\ \langle - | \hat{O} | + \rangle & \langle - | \hat{O} | - \rangle \end{pmatrix}$$

we deduce that

$$S_x = \frac{\hbar}{2}\left(|+\rangle\langle -| + |-\rangle\langle +|\right),$$

$$S_y = \frac{\hbar}{2}\left(-i\,|+\rangle\langle -| + i\,|-\rangle\langle +|\right),$$

$$S_z = \frac{\hbar}{2}\left(|+\rangle\langle +| - |-\rangle\langle -|\right)$$

Now consider the projection operators

$$P_+ = |+\rangle\langle +|, \quad P_- = |-\rangle\langle -|$$

The matrix representation of these operators is

$$[P_+] = \begin{pmatrix} \langle + \mid + \rangle \langle + \mid + \rangle & \langle + \mid + \rangle \langle + \mid - \rangle \\ \langle - \mid + \rangle \langle + \mid + \rangle & \langle - \mid + \rangle \langle + \mid - \rangle \end{pmatrix} = \begin{pmatrix} 1 & 0 \\ 0 & 0 \end{pmatrix}$$

$$[P_-] = \begin{pmatrix} \langle + \mid - \rangle \langle - \mid + \rangle & \langle + \mid - \rangle \langle - \mid - \rangle \\ \langle - \mid - \rangle \langle - \mid + \rangle & \langle - \mid - \rangle \langle - \mid - \rangle \end{pmatrix} = \begin{pmatrix} 0 & 0 \\ 0 & 1 \end{pmatrix}$$

Notice that if we add these operators together, we obtain the identity matrix

$$P_+ + P_- = \begin{pmatrix} 1 & 0 \\ 0 & 0 \end{pmatrix} + \begin{pmatrix} 0 & 0 \\ 0 & 1 \end{pmatrix} = \begin{pmatrix} 1 & 0 \\ 0 & 1 \end{pmatrix}$$

This tells us that we can express the identity operator in the following way:

$$P_+ + P_- = |+\rangle \langle +| + |-\rangle \langle -| = I$$

EXAMPLE 11.7

A spin-1/2 system is in the state

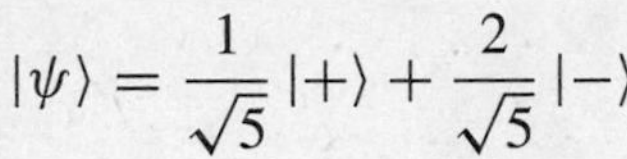

$$|\psi\rangle = \frac{1}{\sqrt{5}} |+\rangle + \frac{2}{\sqrt{5}} |-\rangle$$

Find ΔS_x for this state.

SOLUTION

Using the outer product representation of the operator

$$S_x = \frac{\hbar}{2} \left(|+\rangle \langle -| + |-\rangle \langle +| \right)$$

we find that

$$S_x |\psi\rangle = \frac{\hbar}{2} \left(|+\rangle \langle -| + |-\rangle \langle +| \right) \left(\frac{1}{\sqrt{5}} |+\rangle + \frac{2}{\sqrt{5}} |-\rangle \right)$$

$$= \frac{\hbar}{2} \left(\frac{1}{\sqrt{5}} |+\rangle \langle -| +\rangle + \frac{2}{\sqrt{5}} |+\rangle \langle -| -\rangle + \frac{1}{\sqrt{5}} |-\rangle \langle +| +\rangle + \frac{2}{\sqrt{5}} |-\rangle \langle +| -\rangle \right)$$

$$= \frac{\hbar}{2} \left(\frac{2}{\sqrt{5}} |+\rangle \langle -| -\rangle + \frac{1}{\sqrt{5}} |-\rangle \langle +| +\rangle \right) = \frac{\hbar}{2\sqrt{5}} \left(2 |+\rangle + |-\rangle \right)$$

Using this result we find the expectation value to be

$$\langle S_x \rangle = \langle \psi | S_x | \psi \rangle = \left[\frac{1}{\sqrt{5}} \langle +| + \frac{2}{\sqrt{5}} \langle -| \right] \left[\frac{\hbar}{2\sqrt{5}} \left(2 |+\rangle + |-\rangle \right) \right]$$

$$= \frac{\hbar}{10} \left(\left(\langle +| + 2 \langle -| \right) \left(2 |+\rangle + |-\rangle \right) \right) = \frac{\hbar}{10} \left(2 \langle + \mid + \rangle + 2 \langle - \mid - \rangle \right)$$

$$= \frac{\hbar}{10} (2 + 2) = \frac{4\hbar}{10} = \frac{2\hbar}{5}$$

Now

$$S_x^2 \left|\psi\right\rangle = S_x \frac{\hbar}{2\sqrt{5}} (2\left|+\right\rangle + \left|-\right\rangle) = \frac{\hbar^2}{4}\left(\frac{1}{\sqrt{5}}\left|+\right\rangle + \frac{2}{\sqrt{5}}\left|-\right\rangle\right) = \frac{\hbar^2}{4}\left|\psi\right\rangle$$

$$\Rightarrow \left\langle S_x^2 \right\rangle = \left\langle\psi\right| S_x \left|\psi\right\rangle = \frac{\hbar^2}{4}\left\langle \psi \mid \psi \right\rangle = \frac{\hbar^2}{4}$$

Putting these results together we obtain

$$\Delta S_x = \sqrt{\left\langle S_x^2 \right\rangle - \left\langle S_x \right\rangle^2} = \sqrt{\frac{\hbar^2}{4} - \left(\frac{2\hbar}{5}\right)^2} = \hbar\sqrt{\frac{25}{100} - \frac{16}{100}} = \frac{3\hbar}{10}$$

The Pauli Matrices

Using the standard basis of S_z we set

$$S_x = \frac{\hbar}{2}\sigma_x, \quad S_y = \frac{\hbar}{2}\sigma_y, \quad S_z = \frac{\hbar}{2}\sigma_z$$

These are the famous Pauli matrices:

$$\sigma_x = \begin{pmatrix} 0 & 1 \\ 1 & 0 \end{pmatrix}, \quad \sigma_y = \begin{pmatrix} 0 & -i \\ i & 0 \end{pmatrix}, \quad \sigma_z = \begin{pmatrix} 1 & 0 \\ 0 & -1 \end{pmatrix}$$

The commutation relations obeyed by these matrices follow immediately from the standard commutation relations for angular momentum:

$$\left[\sigma_x, \sigma_y\right] = 2i\,\sigma_z, \quad \left[\sigma_y, \sigma_z\right] = 2i\,\sigma_x, \quad \left[\sigma_z, \sigma_x\right] = 2i\,\sigma_y$$

If we make the following definition:

$$\varepsilon_{ijk} = \begin{cases} +1 & \text{for cyclic permutations of } ijk \\ -1 & \text{for anti-cyclic permutations} \\ 0 & \text{otherwise} \end{cases}$$

(for example, $\varepsilon_{123} = \varepsilon_{312} = \varepsilon_{231} = +1, \ \varepsilon_{213} = -1, \ \varepsilon_{223} = 0$). We can use this to write the commutation relations in the compact form

$$\left[\sigma_i, \ \sigma_j\right] = 2i\,\varepsilon_{ijk}\,\sigma_k$$

The Pauli matrices satisfy the following anti-commutation relation:

$$\left\{\sigma_i, \ \sigma_j\right\} = \sigma_i\sigma_j + \sigma_j\sigma_i = 2\delta_{ij}$$

As you can see by inspection, the trace of any Pauli matrix (the sum of the diagonal elements) vanishes:

$$Tr(\sigma_i) = 0$$

EXAMPLE 11.8

Show that

$$Tr(\sigma_i \sigma_j) = 2\delta_{ij}$$

SOLUTION

First we write out the commutation relation

$$\left[\sigma_i, \sigma_j\right] = \sigma_i \sigma_j - \sigma_j \sigma_i = 2i\varepsilon_{ijk}\sigma_k$$

and the anti-commutation relation

$$\left\{\sigma_i, \ \sigma_j\right\} = \sigma_i \sigma_j + \sigma_j \sigma_i = 2\delta_{ij} I$$

Adding the two together, we obtain

$$\left[\sigma_i, \ \sigma_j\right] + \left\{\sigma_i, \ \sigma_j\right\} = \left(\sigma_i \sigma_j - \sigma_j \sigma_i\right) + \left(\sigma_i \sigma_j + \sigma_j \sigma_i\right) = 2\sigma_i \sigma_j$$

while adding the terms on the right side gives

$$\left[\sigma_i, \ \sigma_j\right] + \left\{\sigma_i, \ \sigma_j\right\} = 2i\varepsilon_{ljk}\sigma_k + 2\delta_{ij} I$$

Putting these two results together gives

$$\sigma_l \sigma_j = i \ \varepsilon_{ijk}\sigma_k + \delta_{ij} I$$

This allows us to compute the required trace:

$$\begin{aligned} Tr(\sigma_i \sigma_j) &= Tr(i\varepsilon_{ijk}\sigma_k + \delta_{ij} I) = Tr(i\varepsilon_{ijk}\sigma_k) + Tr(\delta_{ij} I) \\ &= i\varepsilon_{ijk} Tr(\sigma_k) + \delta_{ij} Tr(I) \end{aligned}$$

The Pauli matrices are traceless, so the first term is $Tr(\sigma_k) = 0$. The trace of the identity matrix is

$$Tr(I) = Tr\begin{pmatrix} 1 & 0 \\ 0 & 1 \end{pmatrix} = 1 + 1 = 2$$

Therefore we conclude $Tr(\sigma_i \sigma_j) = 2\delta_{ij}$.

Now we consider some other properties of the Pauli matrices. The determinant of any Pauli matrix, which is the product of its eigenvalues, is

$$\det |\sigma_i| = -1$$

For example,

$$\det |\sigma_x| = \begin{vmatrix} 0 & 1 \\ 1 & 0 \end{vmatrix} = 0 - 1 = -1$$

Since the eigenvalues of S_i are $\pm\hbar/2$, and $S_i = \hbar/2\sigma_i$, it follows that the eigenvalues of the Pauli matrices are ± 1. Since the Pauli matrices differ from the spin operators by a constant, it follows that they also have the same eigenvectors. As an example, we note that the eigenvectors of σ_z are

$$\sigma_z |+\rangle = |+\rangle$$

$$\sigma_z |-\rangle = -|-\rangle$$

The square of any Pauli matrix gives the identity, i.e.

$$\sigma_x^2 = \sigma_y^2 = \sigma_z^2 = I$$

We can also define the ladder operators $\sigma_\pm = 1/2(\sigma_x \pm i\sigma_y)$ such that

$$\sigma_+ |+\rangle = 0, \quad \sigma_+ |-\rangle = |+\rangle$$

$$\sigma_- |+\rangle = |-\rangle, \quad \sigma_- |-\rangle = 0$$

The matrix representations of these operators are

$$\sigma_+ = \begin{pmatrix} 0 & 1 \\ 0 & 0 \end{pmatrix}, \quad \sigma_- = \begin{pmatrix} 0 & 0 \\ 1 & 0 \end{pmatrix}$$

EXAMPLE 11.9

Find the commutators $[\sigma_+, \sigma_-]$ and $[\sigma_z, \sigma_\pm]$.

SOLUTION

$$[\sigma_+, \sigma_-] = \left[\sigma_+, \frac{1}{2}\left(\sigma_x - i\sigma_y\right)\right] = \left[\frac{1}{2}\left(\sigma_x + i\sigma_y\right), \frac{1}{2}\left(\sigma_x - i\sigma_y\right)\right]$$

$$= \frac{1}{4}\left\{[\sigma_x, \sigma_x] - i\left[\sigma_x, \sigma_y\right] + i\left[\sigma_y, \sigma_x\right] + \left[\sigma_y, \sigma_y\right]\right\}$$

Now since every operator commutes with itself $[A, A] = 0$ and $[A, B] = -[B, A]$, we can write this as

$$[\sigma_+, \sigma_-] = \frac{1}{4}\left\{-i\left[\sigma_x, \sigma_y\right] + i\left[\sigma_y, \sigma_x\right]\right\}$$

$$= \frac{1}{4}\left\{-i\left[\sigma_x, \sigma_y\right] - i\left[\sigma_x, \sigma_y\right]\right\}$$

$$= -\frac{i}{2}\left[\sigma_x, \sigma_y\right]$$

$$= -\frac{i}{2}(2i\sigma_z) = \sigma_z$$

For the other commutator, we find

$$\left[\sigma_z, \sigma_\pm\right] = \left[\sigma_z, \frac{1}{2}\left(\sigma_x \pm i\sigma_y\right)\right]$$

$$= \frac{1}{2}\left[\sigma_z, \sigma_x\right] \pm \frac{i}{2}\left[\sigma_z, \sigma_y\right]$$

Looking at the commutation relations for the Pauli matrices, we find

$$\left[\sigma_z, \sigma_\pm\right] = \frac{1}{2}\left(2i\sigma_y\right) \pm \frac{i}{2}\left(-2i\sigma_x\right) = i\sigma_y \pm \sigma_x = \pm 2\left(\sigma_x \pm i\sigma_y\right) = \pm 2\sigma_\pm$$

EXAMPLE 11.10 e.g.

Show that any operator in a two-dimensional complex Hilbert space can be written in terms of the Pauli matrices as

$$A = \frac{1}{2}\left(a_0 I + \vec{a}\cdot\vec{\sigma}\right)$$

where $a_0 = Tr(A), \vec{a} = Tr(A\vec{\sigma})$.

SOLUTION 

An operator in a two-dimensional Hilbert space is represented by a 2×2 matrix. We write

$$A = \begin{pmatrix} a & b \\ c & d \end{pmatrix}$$

then

$$a_0 = Tr\,(A) = a + d$$

The vector $\vec{\sigma} = \sigma_x\hat{x} + \sigma_y\hat{y} + \sigma_z\hat{z}$ and the components of $A\vec{\sigma}$ are

$$A\sigma_x = \begin{pmatrix} a & b \\ c & d \end{pmatrix}\begin{pmatrix} 0 & 1 \\ 1 & 0 \end{pmatrix} = \begin{pmatrix} b & a \\ d & c \end{pmatrix}$$

$$A\sigma_y = \begin{pmatrix} a & b \\ c & d \end{pmatrix}\begin{pmatrix} 0 & -i \\ i & 0 \end{pmatrix} = \begin{pmatrix} ib & -ia \\ id & -ic \end{pmatrix}$$

$$A\sigma_z = \begin{pmatrix} a & b \\ c & d \end{pmatrix}\begin{pmatrix} 1 & 0 \\ 0 & -1 \end{pmatrix} = \begin{pmatrix} a & -b \\ c & -d \end{pmatrix}$$

Therefore we find the following

$$Tr(A\sigma_x) = b + c$$

$$Tr(A\sigma_y) = ib - ic$$

$$Tr(A\sigma_z) = a - d$$

and so we have

$$\vec{a} = Tr(A\vec{\sigma}) = Tr(A\sigma_x\hat{x} + A\sigma_y\hat{y} + A\sigma_z\hat{z}) = (b+c)\hat{x} + (ib-ic)\hat{y} + (a-d)\hat{z}$$

Forming the dot product

$$\begin{aligned}
\vec{a}\cdot\vec{\sigma} &= (b+c)\sigma_x + (ib-ic)\sigma_y + (a-d)\sigma_z \\
&= (b+c)\begin{pmatrix} 0 & 1 \\ 1 & 0 \end{pmatrix} + (ib-ic)\begin{pmatrix} 0 & -i \\ i & 0 \end{pmatrix} + (a-d)\begin{pmatrix} 1 & 0 \\ 0 & -1 \end{pmatrix} \\
&= \begin{pmatrix} 0 & b+c \\ b+c & 0 \end{pmatrix} + \begin{pmatrix} 0 & b-c \\ -b+c & 0 \end{pmatrix} + \begin{pmatrix} a-d & 0 \\ 0 & -a+d \end{pmatrix} \\
&= \begin{pmatrix} a-d & 2b \\ 2c & -a+d \end{pmatrix}
\end{aligned}$$

Using this with $a_0 = Tr(A) = a + d$ we obtain

$$\begin{aligned}
A &= \frac{1}{2}\left(a_0 I + \vec{a}\cdot\vec{\sigma}\right) = \frac{1}{2}(a+d)\begin{pmatrix} 1 & 0 \\ 0 & 1 \end{pmatrix} + \frac{1}{2}\begin{pmatrix} a-d & 2b \\ 2c & -a+d \end{pmatrix} \\
&= \frac{1}{2}\begin{pmatrix} a+d & 0 \\ 0 & a+d \end{pmatrix} + \frac{1}{2}\begin{pmatrix} a-d & 2b \\ 2c & -a+d \end{pmatrix} \\
&= \frac{1}{2}\begin{pmatrix} 2a & 2b \\ 2c & 2d \end{pmatrix} = \begin{pmatrix} a & b \\ c & d \end{pmatrix}
\end{aligned}$$

Using this result, it can be shown that the projection operators can be written

$$P_+ = \frac{1}{2}\left(I + \sigma_z\right), \quad P_- = \frac{1}{2}\left(I - \sigma_z\right)$$

EXAMPLE 11.11

Show that

$$e^{i\theta\sigma_x} = I\cos\theta + i\sigma_x\sin\theta$$

SOLUTION

First recall the series expansion of the exponential

$$e^x = 1 + x + \frac{x^2}{2!} + \frac{x^3}{3!} + \cdots$$

Also recall that we can write

$$\sin x = x - \frac{x^3}{3!} + \frac{x^5}{5!} - \cdots$$

$$\cos x = 1 - \frac{x^2}{2!} + \frac{x^4}{4!} - \cdots.$$

Now using the fact that $\sigma_x^2 = I$ we find that

$$\begin{aligned}
e^{i\theta\sigma_x} &= I + i\theta\sigma_x + \frac{(i\theta\sigma_x)^2}{2!} + \frac{(i\theta\sigma_x)^3}{3!} + \frac{(i\theta\sigma_x)^4}{4!} + \frac{(i\theta\sigma_x)^5}{5!} + \cdots \\
&= I + \frac{(i\theta\sigma_x)^2}{2!} + \frac{(i\theta\sigma_x)^4}{4!} + \cdots + i\theta\sigma_x + \frac{(i\theta\sigma_x)^3}{3!} + \frac{(i\theta\sigma_x)^5}{5!} + \cdots \\
&= I - \frac{\theta^2}{2!} + \frac{\theta^4}{4!} + \cdots + i\theta\sigma_x + \frac{(i\theta\sigma_x)^3}{3!} + \frac{(i\theta\sigma_x)^5}{5!} + \cdots
\end{aligned}$$

On the last line we used $\sigma_x^{2n} = I$. We do some manipulation on the second series and use this fact again:

$$\begin{aligned}
e^{i\theta\sigma_x} &= I - \frac{\theta^2}{2!} + \frac{\theta^4}{4!} + \cdots + i\theta\sigma_x + \frac{(i\theta)^3\,\sigma_x\sigma_x^2}{3!} + \frac{(i\theta)^5\,\sigma_x\sigma_x^4}{5!} + \cdots \\
&= I - \frac{\theta^2}{2!} + \frac{\theta^4}{4!} + \cdots + i\theta\sigma_x + \frac{(i\theta)^3\,\sigma_x}{3!} + \frac{(i\theta)^5\,\sigma_x}{5!} + \cdots \\
&= I\left(1 - \frac{\theta^2}{2!} + \frac{\theta^4}{4!} + \cdots\right) + i\sigma_x\left(\theta + \frac{(\theta)^3}{3!} + \frac{(\theta)^5}{5!} + \cdots\right) \\
&= I\cos\theta + i\sigma_x \sin\theta
\end{aligned}$$

The Time Evolution of Spin-1/2 States

The Hamiltonian for a particle of spin **S** in a magnetic field **B** is given by

$$H = -\vec{\mu} \cdot \vec{B}$$

We have seen that the magnetic moment is proportional to the spin:

$$\vec{\mu} = \frac{e}{mc}\vec{S}$$

The *Bohr magneton* is defined as $\mu_B = e\hbar/2mc$. If we call the gyromagnetic ratio $\gamma = -2\mu_B/\hbar$, we can write the Hamiltonian as

$$H = -\gamma \vec{S} \cdot \vec{B}$$

The factor of 2 is a relativistic correction and is known as the "g-factor." In Larmor precession, we consider a constant magnetic field in the z-direction, and so the Hamiltonian is

$$H = -\gamma B_z S_z$$

B_z and γ are just numbers, and so the eigenstates of the Hamiltonian are the eigenstates of S_z.

$$H|+\rangle = -\gamma B_z S_z |+\rangle = -\gamma B_z \frac{\hbar}{2} |+\rangle = E_+ |+\rangle$$

$$H|-\rangle = -\gamma B_z S_z |-\rangle = \gamma B_z \frac{\hbar}{2} |-\rangle = E_- |-\rangle$$

To find the time evolution of spin states, we refer to the Schrödinger equation:

$$i\hbar \frac{d|\psi\rangle}{dt} = H|\psi\rangle$$

The Hamiltonian is time-independent, so the solution can be found using separation of variables in the usual way. Therefore the general state

$$|\psi\rangle = \alpha|+\rangle + \beta|-\rangle$$

will evolve in time under this Hamiltonian as

$$|\psi(t)\rangle = \alpha \exp(-iE_+t/\hbar)|+\rangle + \beta \exp(-iE_-t/\hbar)|-\rangle$$

e.g. **EXAMPLE 11.12**

A system subject to the Hamiltonian $H = -\gamma B_z S_z$ is initially in the state $|+_x\rangle$ at time $t = 0$. Write the state at some later time t. If we measure S_x at time t, what are the probabilities of finding $\pm\hbar/2$? What are the probabilities of finding $\pm\hbar/2$ if we measure S_z at time t?

✔ **SOLUTION**

The initial state of the system is

$$|\psi(0)\rangle = |+_x\rangle = \frac{1}{\sqrt{2}}|+\rangle + \frac{1}{\sqrt{2}}|-\rangle$$

At the initial time $t = 0$, note that if we measure S_x we obtain $+\hbar/2$ with certainty. We see from the expansion of the state that at $t = 0$ if we measure S_z there is a 50% probability of finding $+\hbar/2$ and a 50% probability of finding $-\hbar/2$. At a later time t, we can write the state as

$$|\psi(t)\rangle = \frac{1}{\sqrt{2}} \exp(i\gamma B_z t/2)\,|+\rangle \;+\; \frac{1}{\sqrt{2}} \exp(-i\gamma B_z t/2)\,|-\rangle$$

Now consider a measurement of S_x at time t. The probability of finding $+\hbar/2$ is

$$|\langle +_x \mid \psi(t)\rangle|^2$$

Notice that the inner product of the state at time t with the basis states (in z) is

$$\begin{aligned}\langle + \mid \psi(t)\rangle &= \frac{1}{\sqrt{2}} \exp(i\gamma B_z t/2)\,\langle + \mid +\rangle + \frac{1}{\sqrt{2}} \exp(-i\gamma B_z t/2)\,\langle + \mid -\rangle \\ &= \frac{1}{\sqrt{2}} \exp(i\gamma B_z t/2)\end{aligned}$$

For the other basis state we have

$$\begin{aligned}\langle - \mid \psi(t)\rangle &= \frac{1}{\sqrt{2}} \exp(i\gamma B_z t/2)\,\langle - \mid +\rangle + \frac{1}{\sqrt{2}} \exp(-i\gamma B_z t/2)\,\langle - \mid -\rangle \\ &= \frac{1}{\sqrt{2}} \exp(-i\gamma B_z t/2)\end{aligned}$$

And so, to calculate $|\langle +_x \mid \psi(t)\rangle|^2$, we have

$$\begin{aligned}\langle +_x \mid \psi(t)\rangle &= \frac{1}{\sqrt{2}} \left(\langle +| + \langle -|\right)|\psi(t)\rangle = \frac{1}{\sqrt{2}} \langle + \mid \psi(t)\rangle + \frac{1}{\sqrt{2}} \langle - \mid \psi(t)\rangle \\ &= \frac{1}{2} \exp(i\gamma B_z t/2) + \frac{1}{2} \exp(-i\gamma B_z t/2) \\ &= \cos\left(\frac{\gamma B_z t}{2}\right)\end{aligned}$$

Therefore the probability is

$$|\langle +_x \mid \psi(t)\rangle|^2 = \cos^2\left(\frac{\gamma B_z t}{2}\right)$$

The spin-down case can be worked out in a similar manner, where one should find that

$$\langle -_x \mid \psi(t)\rangle = i \sin\left(\frac{\gamma B_z t}{2}\right),$$

$$\Rightarrow |\langle -_x \mid \psi(t)\rangle|^2 = \sin^2\left(\frac{\gamma B_z t}{2}\right)$$

So we arrive at the interesting result where the probabilities are oscillating in time. For a measurement of S_z, we found earlier that

$$\langle + | \psi(t) \rangle = \frac{1}{\sqrt{2}} \exp(i\gamma B_z t/2)$$

$$\langle - | \psi(t) \rangle = \frac{1}{\sqrt{2}} \exp(-i\gamma B_z t/2)$$

Therefore we find that the probabilities of finding $\pm\hbar/2$ for a measurement of S_z at time t are

$$|\langle + | \psi(t) \rangle|^2 = \left(\frac{1}{\sqrt{2}} \exp(i\gamma B_z t/2)\right)\left(\frac{1}{\sqrt{2}} \exp(-i\gamma B_z t/2)\right) = \frac{1}{2}$$

$$|\langle - | \psi(t) \rangle|^2 = \left(\frac{1}{\sqrt{2}} \exp(-i\gamma B_z t/2)\right)\left(\frac{1}{\sqrt{2}} \exp(i\gamma B_z t/2)\right) = \frac{1}{2}$$

---

e.g. **EXAMPLE 11.13**

For the state in the previous example, show that $\langle S_z \rangle = 0$ and that $\langle S_x \rangle$ oscillates in time.

✔ **SOLUTION**

Since we have calculated the probabilities of obtaining the various measurement results, we can find the expectation values by calculating

$$\langle A \rangle = \sum p_i a_i$$

where a_i is a possible measurement result and p_i is the respective probability. And so we find

$$\langle S_z \rangle = prob\left(+\frac{\hbar}{2}\right)\left(+\frac{\hbar}{2}\right) + prob\left(-\frac{\hbar}{2}\right)\left(-\frac{\hbar}{2}\right)$$

$$= \frac{1}{2}\left(+\frac{\hbar}{2}\right) + \frac{1}{2}\left(-\frac{\hbar}{2}\right) = 0$$

For $\langle S_x \rangle$ we obtain

$$\begin{aligned}\langle S_x \rangle &= prob\left(+\frac{\hbar}{2}\right)\left(\frac{\hbar}{2}\right) + prob\left(-\frac{\hbar}{2}\right)\left(-\frac{\hbar}{2}\right)\\ &= \cos^2\left(\frac{\gamma B_z t}{2}\right)\left(\frac{\hbar}{2}\right) + \sin^2\left(\frac{\gamma B_z t}{2}\right)\left(-\frac{\hbar}{2}\right)\\ &= \left(\frac{\hbar}{2}\right)\left[\cos^2\left(\frac{\gamma B_z t}{2}\right) - \sin^2\left(\frac{\gamma B_z t}{2}\right)\right]\\ &= \left(\frac{\hbar}{2}\right)\cos(\gamma B_z t)\end{aligned}$$

$\langle S_x \rangle$ oscillates in time with frequency $\omega = \gamma B_z$ (see Fig. 11-3).

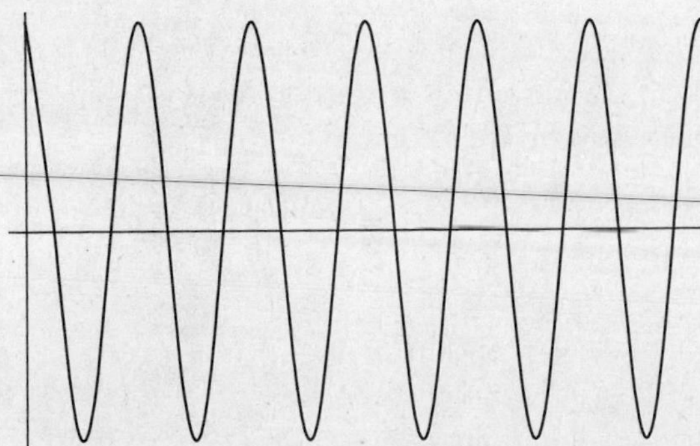

Fig. 11-3

EXAMPLE 11.14

e.g.

A particle is under the influence of a magnetic field $\vec{B} = B_0 \sin(\omega t)\, \hat{z}$.

(a) Write down the state of the system at time t and find the Hamiltonian.

(b) If $|\psi(0)\rangle = |+_y\rangle$, find the probability that a measurement of S_y at time t results in $+\hbar/2$

SOLUTION

(a) The Hamiltonian is given by

$$H = -\gamma \vec{B} \cdot \vec{S} = -\gamma B_0 \sin(\omega t) S_z$$

Since the Hamiltonian is expressed only in terms of S_z, the basis states of S_z are the eigenstates of the Hamiltonian. Therefore we write the state of the system at time t in terms of these states:

$$|\psi(t)\rangle = \alpha(t)|+\rangle + \beta(t)|-\rangle$$

where we must have

$$|\alpha(t)|^2 + |\beta(t)|^2 = 1$$

to conserve probability. To solve for the coefficients, we write the state in column vector form

$$|\psi(t)\rangle = \begin{pmatrix} \alpha(t) \\ \beta(t) \end{pmatrix}$$

The time evolution of the state is found by solving

$$i\hbar \frac{d}{dt}|\psi\rangle = H|\psi\rangle$$

In matrix form, representing time derivatives using "dot" notation (i.e. $df/dt = \dot{f}$), we can write this equation as

$$i\hbar \begin{pmatrix} \dot{\alpha}(t) \\ \dot{\beta}(t) \end{pmatrix} = -\gamma B_0 \sin(\omega t) \left(\frac{\hbar}{2}\right) \begin{pmatrix} 1 & 0 \\ 0 & -1 \end{pmatrix} \begin{pmatrix} \alpha(t) \\ \beta(t) \end{pmatrix}$$

$$= -\gamma B_0 \sin(\omega t) \left(\frac{\hbar}{2}\right) \begin{pmatrix} \alpha(t) \\ -\beta(t) \end{pmatrix}$$

This leads to the two equations

$$i\hbar\dot{\alpha}(t) = -\frac{\gamma\hbar B_0}{2} \sin(\omega t)\, \alpha(t)$$

$$i\hbar\dot{\beta}(t) = \frac{\gamma\hbar B_0}{2} \sin(\omega t)\, \beta(t)$$

We solve the first equation explicitly. Dividing through by $i\hbar$ and using $1/i = -i$ we have

$$\frac{d\alpha}{dt} = i\frac{\gamma B_0}{2} \sin(\omega t)\, \alpha(t)$$

$$\Rightarrow \frac{d\alpha}{\alpha} = i\frac{\gamma B_0}{2} \sin(\omega t)\, dt$$

Integrating we obtain

$$\ln(\alpha) = -i\frac{\gamma B_0}{2\omega}\cos(\omega t) + C,$$

$$\Rightarrow \alpha(t) = \alpha(0)\exp\left[-i\frac{\gamma B_0}{2\omega}\cos(\omega t)\right]$$

Using Euler's formula, we can write this as

$$\alpha(t) = \alpha(0)\cos\left[\frac{\gamma B_0}{2\omega}\cos(\omega t)\right] - i\alpha(0)\sin\left[\frac{\gamma B_0}{2\omega}\cos(\omega t)\right]$$

A similar procedure shows that

$$\beta(t) = \beta(0)\cos\left[\frac{\gamma B_0}{2\omega}\cos(\omega t)\right] + i\beta(0)\sin\left[\frac{\gamma B_0}{2\omega}\cos(\omega t)\right]$$

and so the wavefunction at time t is

$$|\psi\rangle = \begin{pmatrix} \alpha(0)\cos\left[\frac{\gamma B_0}{2\omega}\cos(\omega t)\right] - i\alpha(0)\sin\left[\frac{\gamma B_0}{2\omega}\cos(\omega t)\right] \\ \beta(0)\cos\left[\frac{\gamma B_0}{2\omega}\cos(\omega t)\right] + i\beta(0)\sin\left[\frac{\gamma B_0}{2\omega}\cos(\omega t)\right] \end{pmatrix}$$

(b) We start by writing $|\psi(0)\rangle = |+_y\rangle$ as a column vector:

$$|\psi(0)\rangle = \frac{1}{\sqrt{2}}\begin{pmatrix} 1 \\ i \end{pmatrix}$$

$$\Rightarrow \alpha(0) = \frac{1}{\sqrt{2}}, \ \beta(0) = \frac{i}{\sqrt{2}}$$

This means the state of the system at time t is

$$|\psi\rangle = \begin{pmatrix} \frac{1}{\sqrt{2}}\cos\left[\frac{\gamma B_0}{2\omega}\cos(\omega t)\right] - \frac{i}{\sqrt{2}}\sin\left[\frac{\gamma B_0}{2\omega}\cos(\omega t)\right] \\ \frac{i}{\sqrt{2}}\cos\left[\frac{\gamma B_0}{2\omega}\cos(\omega t)\right] - \frac{1}{\sqrt{2}}\sin\left[\frac{\gamma B_0}{2\omega}\cos(\omega t)\right] \end{pmatrix}$$

$$= \begin{pmatrix} \frac{1}{\sqrt{2}}\cos\left[\frac{\gamma B_0}{2\omega}\cos(\omega t)\right] \\ \frac{i}{\sqrt{2}}\cos\left[\frac{\gamma B_0}{2\omega}\cos(\omega t)\right] \end{pmatrix} + \begin{pmatrix} -\frac{i}{\sqrt{2}}\sin\left[\frac{\gamma B_0}{2\omega}\cos(\omega t)\right] \\ -\frac{1}{\sqrt{2}}\sin\left[\frac{\gamma B_0}{2\omega}\cos(\omega t)\right] \end{pmatrix}$$

$$= \frac{1}{\sqrt{2}}\cos\left[\frac{\gamma B_0}{2\omega}\cos(\omega t)\right]\begin{pmatrix} 1 \\ i \end{pmatrix} - i\frac{1}{\sqrt{2}}\sin\left[\frac{\gamma B_0}{2\omega}\cos(\omega t)\right]\begin{pmatrix} 1 \\ -i \end{pmatrix}$$

$$= \cos\left[\frac{\gamma B_0}{2\omega}\cos(\omega t)\right]|+_y\rangle - i\sin\left[\frac{\gamma B_0}{2\omega}\cos(\omega t)\right]|-_y\rangle$$

Now at time t if S_y is measured, the probability of finding $+\hbar/2$ is found by calculating

$$\left|\langle +_y \mid \psi(t)\rangle\right|^2 = \left(\cos\left[\frac{\gamma B_0}{2\omega}\cos(\omega t)\right]\right)^2$$

This function oscillates as shown in Fig. 11-4.

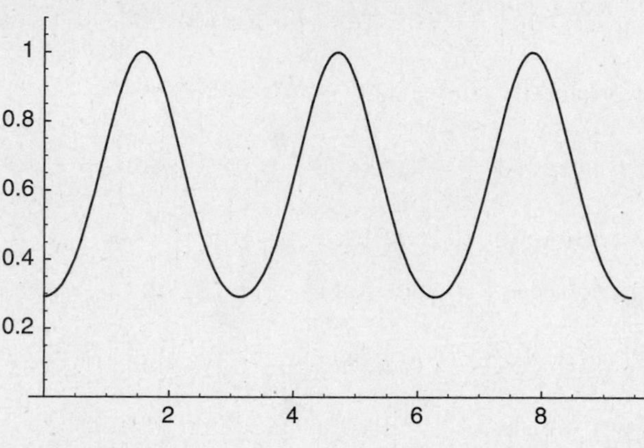

Fig. 11-4

The probability of finding $-\hbar/2$ oscillates as shown in Fig. 11-5.

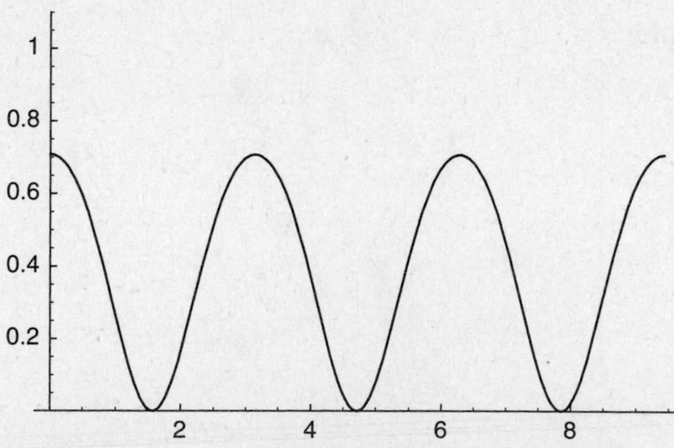

Fig. 11-5

EXAMPLE 11.15

e.g.

A particle is under the influence of the magnetic field

$$\vec{B} = B_0 \cos(\omega t)\,\hat{x} + B_0 \sin(\omega t)\,\hat{y}$$

At time $t = 0$, the system is in the state $|\psi(0)\rangle = |+\rangle = \begin{pmatrix} 1 \\ 0 \end{pmatrix}$.

(a) Find the Hamiltonian of the system.

(b) Find a transformation that would allow one to write a time-independent Hamiltonian.

(c) Find the probability that at time t, the system is in the state $|-\rangle$. Find the first time $t > 0$ when the system is in the state $|-\rangle$ with certainty.

SOLUTION

(a) The Hamiltonian is found from $H = -\gamma \vec{B} \cdot \vec{S}$. If we let $\omega_o = -\gamma B_0$, then we have

$$H = \omega_o \cos(\omega t)\, S_x + \omega_o \sin(\omega t)\, S_y$$

$$= \omega_o \cos(\omega t) \left(\frac{\hbar}{2}\right) \begin{pmatrix} 0 & 1 \\ 1 & 0 \end{pmatrix} + \omega_o \sin(\omega t) \left(\frac{\hbar}{2}\right) \begin{pmatrix} 0 & -i \\ i & 0 \end{pmatrix}$$

Adding these matrices together and using Euler's formula we find

$$H = \frac{\hbar \omega_o}{2} \begin{pmatrix} 0 & e^{-i\omega t} \\ e^{i\omega t} & 0 \end{pmatrix}$$

(b) The system evolves in time according to

$$i\hbar \frac{d}{dt} |\psi(t)\rangle = H\, |\psi(t)\rangle$$

If we write $|\psi(t)\rangle = \alpha(t)\,|+\rangle + \beta(t)\,|-\rangle$, then we have

$$i\hbar \begin{pmatrix} \dot{\alpha} \\ \dot{\beta} \end{pmatrix} = \frac{\hbar \omega_o}{2} \begin{pmatrix} 0 & e^{-i\omega t} \\ e^{i\omega t} & 0 \end{pmatrix} \begin{pmatrix} \alpha \\ \beta \end{pmatrix}$$

This leads to the coupled equations

$$i \frac{d\alpha}{dt} = \frac{\omega_o}{2} e^{-i\omega t} \beta$$

$$i \frac{d\beta}{dt} = \frac{\omega_o}{2} e^{i\omega t} \alpha$$

Now consider the following transformation. We set

$$c_+ = e^{\frac{i\omega t}{2}}\alpha, \quad c_- = e^{-\frac{i\omega t}{2}}\beta$$

This results in the following

$$\alpha = e^{-i\frac{\omega t}{2}}c_+, \Rightarrow \dot{\alpha} = -\frac{i\omega}{2}e^{-i\frac{\omega t}{2}}c_+ + e^{-i\frac{\omega t}{2}}\dot{c}_+$$

$$\beta = e^{i\frac{\omega t}{2}}c_-, \Rightarrow \dot{\beta} = \frac{i\omega}{2}e^{i\frac{\omega t}{2}}c_- + e^{i\frac{\omega t}{2}}\dot{c}_-$$

Substitution into the original set of equations derived from the Hamiltonian results in this new set of equations:

$$i\dot{c}_+ = -\frac{\omega}{2}c_+ + \frac{\omega_o}{2}c_-$$

$$i\dot{c}_- = \frac{\omega}{2}c_- + \frac{\omega_o}{2}c_+$$

We have redefined the system giving us the time-independent Hamiltonian

$$\hat{H} = \frac{\hbar}{2}\begin{pmatrix} -\omega & \omega_o \\ \omega_o & \omega \end{pmatrix}$$

(c) To find the probability that the system is in the state $|-\rangle$ at time t, we solve for $c_\pm$. First we differentiate the equations for these variables a second time:

$$i\dot{c}_+ = -\frac{\omega}{2}c_+ + \frac{\omega_o}{2}c_-,$$

$$\Rightarrow i\ddot{c}_+ = -\frac{\omega}{2}\dot{c}_+ + \frac{\omega_o}{2}\dot{c}_-$$

Now insert $\dot{c}_+$ back into this equation in terms of $c_\pm$:

$$i\ddot{c}_+ = -\frac{\omega}{2}\dot{c}_+ + \frac{\omega_o}{2}\dot{c}_- = -\frac{\omega}{2}\left(i\frac{\omega}{2}c_+ - i\frac{\omega_o}{2}c_-\right) + \frac{\omega_o}{2}\dot{c}_-$$

Now we use $i\dot{c}_- = \omega/2c_- + \omega_o/2c_+$ to eliminate $\dot{c}_-$; this gives

$$\begin{aligned} i\ddot{c}_+ &= -\frac{\omega}{2}\left(i\frac{\omega}{2}c_+ - i\frac{\omega_o}{2}c_-\right) + \frac{\omega_o}{2}\dot{c}_- \\ &= -\frac{\omega}{2}\left(i\frac{\omega}{2}c_+ - i\frac{\omega_o}{2}c_-\right) + \frac{\omega_o}{2}\left(-i\frac{\omega}{2}c_- - i\frac{\omega_o}{2}c_+\right) \\ &= -i\frac{\omega^2}{4}c_+ + i\frac{\omega\omega_o}{4}c_- - i\frac{\omega\omega_o}{4}c_- - i\frac{\omega_o^2}{4}c_+ \\ &= -i\frac{\omega^2 + \omega_o^2}{4}c_+ \end{aligned}$$

Dividing both sides by i and moving everything to one side, we obtain the equation

$$\ddot{c}_+ + \frac{\omega^2 + \omega_o^2}{4} c_+ = 0$$

with solution

$$c_+ (t) = A \cos (\Gamma t) + B \sin (\Gamma t) ,$$

$$\Gamma = \sqrt{\frac{\omega^2 + \omega_o^2}{4}}$$

Returning to the statement of the problem, we are told that at time $t = 0$, the system is in the state $|+\rangle$. We also recall that

$$\alpha (t) = e^{-i\omega t/2} c_+ (t) , \quad \Rightarrow \ c_+ (0) = \alpha (0)$$

For this requirement to be satisfied, we must have

$$c_+ (0) = 1 = A \cos(0) + B \sin(0) = A$$

So we take

$$c_+ (t) = \cos (\Gamma t)$$

and so we have

$$\alpha (t) = e^{-i\omega t/2} c_+ = e^{-i\omega t/2} \cos (\Gamma t)$$

The normalization condition

$$|\alpha (t)|^2 + |\beta (t)|^2 = 1$$

tells us that we must have

$$c_- (t) = \sin (\Gamma t) , \ \Rightarrow \ \beta (t) = e^{i\omega t/2} \sin (\Gamma t)$$

Therefore we can write the state of the system at time t as

$$|\psi (t)\rangle = \begin{pmatrix} e^{-i\omega t/2} \cos (\Gamma t) \\ e^{i\omega t/2} \sin (\Gamma t) \end{pmatrix} = e^{-i\omega t/2} \cos (\Gamma t) |+\rangle + e^{i\omega t/2} \sin (\Gamma t) |-\rangle$$

And so the probability that a measurement of S_z finds the system in the spin-down state is

$$|\langle - | \psi (t)\rangle|^2 = \left| e^{i\omega t/2} \sin (\Gamma t) \right|^2 = \sin^2 (\Gamma t)$$

The first time t at which this is certainty occurs when

$$\Gamma t = \sqrt{\frac{\omega^2 + \omega_o^2}{4}}t = \frac{\pi}{2},$$

$$\Rightarrow \quad t = \pi \frac{4}{\sqrt{\omega^2 + \omega_o^2}}$$

At this time the spin, which was prepared in the spin-up state, has flipped to spin-down.

The Density Operator for Spin-1/2 Systems

In Chapter 8 we introduced the density operator. We now briefly explore the density operator for a spin-1/2 system. We compare two situations—the pure state given by

$$\rho_1 = |+_y\rangle\langle +_y|$$

and a completely mixed state made of half spin-up states and half-spin down:

$$\rho_2 = \frac{1}{2}|+\rangle\langle +| + \frac{1}{2}|-\rangle\langle -|$$

Using $|+_y\rangle = 1/\sqrt{2}(|+\rangle + i|-\rangle)$ we find

$$\begin{aligned}\rho_1 &= \frac{1}{\sqrt{2}}(|+\rangle + i|-\rangle)\frac{1}{\sqrt{2}}(\langle +| - i\langle -|)\\ &= \frac{1}{2}|+\rangle\langle +| - i\frac{1}{2}|+\rangle\langle -| + i\frac{1}{2}|-\rangle\langle +| + \frac{1}{2}|-\rangle\langle -|\end{aligned}$$

and so the matrix representations of the two density operators are

$$\rho_1 = \begin{pmatrix} \frac{1}{2} & -\frac{i}{2} \\ \frac{i}{2} & \frac{1}{2} \end{pmatrix}, \quad \rho_2 = \begin{pmatrix} \frac{1}{2} & 0 \\ 0 & \frac{1}{2} \end{pmatrix}$$

Notice that, as required for a density operator, both matrices are Hermitian and have unit trace. For a pure state, $Tr\rho^2 = 1$. For these matrices we have

$$Tr(\rho_1^2) = Tr\left[\begin{pmatrix} \frac{1}{2} & -\frac{i}{2} \\ \frac{i}{2} & \frac{1}{2} \end{pmatrix}\begin{pmatrix} \frac{1}{2} & -\frac{i}{2} \\ \frac{i}{2} & \frac{1}{2} \end{pmatrix}\right] = Tr\begin{pmatrix} \frac{1}{2} & -\frac{i}{4} \\ \frac{i}{4} & \frac{1}{2} \end{pmatrix} = 1$$

and for the other density operator we have

$$Tr(\rho_2^2) = Tr\left[\begin{pmatrix} \frac{1}{2} & 0 \\ 0 & \frac{1}{2} \end{pmatrix}\begin{pmatrix} \frac{1}{2} & 0 \\ 0 & \frac{1}{2} \end{pmatrix}\right] = Tr\begin{pmatrix} \frac{1}{4} & 0 \\ 0 & \frac{1}{4} \end{pmatrix} = \frac{1}{2} < 1$$

as expected for a mixed state. If we measure S_z, what is the probability of finding the result $+\hbar/2$? The projection operator for spin-up is $|+\rangle\langle+|$, using the results of Chapter 8 we find

$$Tr(\rho_1 |+\rangle \langle+|) = \langle+| \rho_1 |+\rangle$$

$$= (1 \quad 0)\begin{pmatrix} \frac{1}{2} & -\frac{i}{2} \\ \frac{i}{2} & \frac{1}{2} \end{pmatrix}\begin{pmatrix} 1 \\ 0 \end{pmatrix} = (1 \quad 0)\begin{pmatrix} \frac{1}{2} \\ \frac{i}{2} \end{pmatrix} = \frac{1}{2}$$

The probability of finding spin-up in the mixed state is

$$Tr(\rho_2 |+\rangle \langle+|) = \langle+| \rho_2 |+\rangle$$

$$= (1 \quad 0)\begin{pmatrix} \frac{1}{2} & 0 \\ 0 & \frac{1}{2} \end{pmatrix}\begin{pmatrix} 1 \\ 0 \end{pmatrix} = (1 \quad 0)\begin{pmatrix} \frac{1}{2} \\ 0 \end{pmatrix} = \frac{1}{2}$$

There seems to be no difference in the physical predictions of the two states. However, consider a measurement of S_y instead. What is the probability of obtaining $+\hbar/2$? For the pure state we find

$$Tr(\rho_1|+_y\rangle\langle+_y|) = \langle+_y|\rho_1|+_y\rangle = 1$$

For the mixed state, we obtain

$$Tr(\rho_2|+_y\rangle\langle+_y|) = \langle+_y|\rho_2|+_y\rangle$$

$$= \left(\frac{1}{2}\right)(1 \quad -i)\begin{pmatrix} \frac{1}{2} & 0 \\ 0 & \frac{1}{2} \end{pmatrix}\begin{pmatrix} 1 \\ i \end{pmatrix} = \left(\frac{1}{2}\right)(1 \quad -i)\begin{pmatrix} \frac{1}{2} \\ \frac{i}{2} \end{pmatrix} = \left(\frac{1}{2}\right)\left(\frac{1}{2} + \frac{1}{2}\right) = \left(\frac{1}{2}\right)$$

Quiz

1. Consider the matrix

$$S_x = \frac{\hbar}{2}\begin{pmatrix} 0 & 1 \\ 1 & 0 \end{pmatrix}$$

The eigenvalues of the matrix are

(a) $\{0, 1\}$

(b) $\{-1, 1\}$

(c) $\{-\hbar/2, \hbar/2\}$

(d) $\{-\hbar, \hbar\}$

2. Using the matrix of the previous problem, what are its normalized eigenvectors?

(a) $|+_x\rangle = \begin{pmatrix} 1 \\ i \end{pmatrix}, |-_x\rangle = \begin{pmatrix} 1 \\ -i \end{pmatrix}$

(b) $|+_x\rangle = \frac{1}{\sqrt{2}}\begin{pmatrix} 1 \\ i \end{pmatrix}, |-_x\rangle = \frac{1}{\sqrt{2}}\begin{pmatrix} 1 \\ -i \end{pmatrix}$

(c) $|+_x\rangle = \frac{1}{\sqrt{2}}\begin{pmatrix} -1 \\ i \end{pmatrix}, |-_x\rangle = \frac{1}{\sqrt{2}}\begin{pmatrix} -1 \\ -i \end{pmatrix}$
(d) $|+_x\rangle = \frac{1}{\sqrt{2}}\begin{pmatrix} 1 \\ 1 \end{pmatrix}, |-_x\rangle = \frac{1}{\sqrt{2}}\begin{pmatrix} 1 \\ -1 \end{pmatrix}$

3. A spin-1/2 system is in the state

$$|\psi\rangle = \frac{1}{\sqrt{8}}\begin{pmatrix} \sqrt{7} \\ 1 \end{pmatrix}$$

A measurement of spin is made along the z-direction. The probability of finding spin-up is
(a) 1/2
(b) 1/4
(c) 7/8
(d) 1/8

4. Using the same state as in the previous problem, now suppose a measurement of spin is made in the y-direction. The probability of finding $+\hbar/2$ is
(a) 3/5
(b) 1/2
(c) 3/4
(d) 1/8
(e) 7/8

5. At time $t = 0$ a spin-1/2 system is in the state

$$|\psi(0)\rangle = |+_y\rangle$$

The system evolves with time under a magnetic field $\vec{B} = B_o\hat{x}$. At some later time t, S_y is measured. The probability of finding $+\hbar/2$ is
(a) 1/2
(b) $\frac{1}{2}\sin\left(\frac{B_o}{2}t\right)$
(c) $\cos^2\left(\frac{\gamma B_o}{2}t\right)$

6. Consider the same system used in problem 5. If instead s_x is measured, the probability of finding $+\hbar/2$ is
(a) 1/2
(b) $\sin^2\left(\frac{\gamma B_o}{2}t\right)$
(c) $2\cos^2\left(\frac{\gamma B_o}{2}t\right)$
(d) $\cos^2(\gamma B_o t)$

CHAPTER 12

Quantum Mechanics in Three Dimensions

The examples we have considered so far have been one-dimensional. The real world, of course, consists of three spatial dimensions. While the one-dimensional situations we have considered are often very useful in real situations, to study atomic or molecular physics or chemistry it will be important to work with all three coordinates. Perhaps the most important case is the hydrogen atom, which we study in this chapter. To acclimate ourselves to doing quantum mechanics in multiple dimensions, we consider two simple examples. First we generalize the infinite square well to two dimensions in Cartesian coordinates. We will see that by adding an extra degree of freedom, degeneracies creep in.

In the second example, we consider a particle trapped inside a cylinder. This will give us a chance to look at the momentum operator in a different coordinate system and solve a more complicated case before moving on to the hydrogen atom. Not surprisingly, we will find that the solutions in the cylindrical well are Bessel functions.

A general rule of thumb when dealing with multiple dimensions involves keeping an eye on the number of degrees of freedom available to a system. If a system has n

degrees of freedom, it will be necessary to carry out n measurements to completely characterize the state. Another way of saying this is to say the system will have n "quantum numbers." To see how this works, think back to the infinite square well, where we found that the wavefunction for energy level m could be written as

$$\Psi(x) = \sqrt{\frac{2}{a}} \sin\left(\frac{m\pi x}{a}\right)$$

There is only one degree of freedom for the particle-motion in the x-direction—and so we can specify the state of the particle with one quantum number, which in the above wavefunction is labeled m.

As we'll see in a moment, when a particle is trapped in a two-dimensional well, the wavefunction in Cartesian coordinates is found to be

$$\Psi(x, y) = \sqrt{\frac{2}{a}} \sin\left(\frac{m\pi x}{a}\right) \sin\left(\frac{n\pi y}{a}\right)$$

The sytem has two degrees of freedom—it can move in the x and y directions. This means that it requires two quantum numbers to specify the state, which we have labeled m and n. To know the wavefunction *and* the energy of the system, we must specify both numbers. But as we will soon find out, we could exchange the values of m and n but arrive at the same energy. This means that there is degeneracy—two different quantum states have the same energy. We now explore this case in some detail.

The 2-D Square Well

The first step in the transition to multiple dimensions is to write the momentum operator as a vector. This is done by making the transition

$$p \to -i\hbar\nabla$$

In two dimensions the momentum p is a vector with x and y components:

$$\vec{p} = p_x\hat{x} + p_y\hat{y}$$

The components of momentum are defined in the usual way:

$$p_x = -i\hbar\frac{\partial}{\partial x}, \quad p_y = -i\hbar\frac{\partial}{\partial y}$$

Using these definitions, the time-independent Schrödinger equation becomes

$$-\frac{\hbar^2}{2m}\nabla^2\Psi(x, y) + V(x, y)\Psi(x, y) = E\Psi(x, y)$$

As a specific example, we consider the 2-D square well. The inside of the well is defined by

$$0 \le x \le a$$
$$0 \le y \le a$$

where the potential $V = 0$. Outside the well, $V \to \infty$. Inside the well, the Schrödinger equation is

$$-\frac{\hbar^2}{2m}\nabla^2\Psi = -\frac{\hbar^2}{2m}\left(\frac{\partial^2}{\partial x^2} + \frac{\partial^2}{\partial y^2}\right)\Psi = E\,\Psi$$

We assume a separable solution of the form

$$\Psi\,(x, y) = f\,(x)\,g\,(y)$$

Inserting this into the Schrödinger equation, we obtain

$$-\frac{\hbar^2}{2m}\left(g\,(y)\,\frac{d^2 f}{dx^2} + f\,(x)\,\frac{d^2 g}{dy^2}\right) = E\,f\,(x)\,g\,(y)$$

We divide through by $\Psi\,(x, y) = f\,(x)\,g\,(y)$ giving us

$$-\frac{\hbar^2}{2m}\frac{d^2 f}{dx^2} - \frac{\hbar^2}{2m}\frac{d^2 g}{dy^2} = E$$

This equation tells us that each term is separately equal to a constant. We separate the energy E into the energy of the particle in the x-direction, and the energy of the particle in the y-direction:

$$E = E_x + E_y$$

This allows us to set up the two equations

$$-\frac{\hbar^2}{2m}\frac{d^2 f}{dx^2} = E_x$$
$$-\frac{\hbar^2}{2m}\frac{d^2 g}{dy^2} = E_y$$

If we define the wavenumbers

$$k_x = \frac{2mE_x}{\hbar^2}, \quad k_y = \frac{2mE_y}{\hbar^2}$$

then we obtain solutions of the form

$$f(x) = \sin(k_x x), \quad g(y) = \sin(k_y y)$$

The wavenumbers are fixed by the boundary conditions of the square well in the usual way. The wavefunctions must go to zero at $x = a$, $y = a$, and this leads to the conditions

$$k_x = \frac{n_x \pi}{a}, \quad k_y = \frac{n_y \pi}{a}$$
$$n_x = 0, 1, 2, \ldots$$
$$n_y = 0, 1, 2, \ldots$$

Notice that since the problem is two-dimensional, we must specify two quantum numbers to characterize the state. The total energy is then the sum of the energies in the x- and y-directions

$$E = \frac{\hbar^2}{2m}\left(k_x^2 + k_y^2\right)$$

where we have

$$E_x = \frac{\hbar^2 n_x^2 \pi^2}{2ma^2}, \quad E_y = \frac{\hbar^2 n_y^2 \pi^2}{2ma^2}$$

Therefore we can write the total energy as

$$E = E_x + E_y = \frac{\hbar^2 n_x^2 \pi^2}{2ma^2} + \frac{\hbar^2 n_y^2 \pi^2}{2ma^2} = \frac{\hbar^2 \pi^2}{2ma^2}\left(n_x^2 + n_y^2\right)$$

This situation leads to degeneracies. Suppose, for example that we have $n_x = 1, n_y = 2$. The total energy of the particle is

$$E = E_x + E_y = \frac{\hbar^2 \pi^2}{2ma^2}(1 + 4) = \frac{5\hbar^2 \pi^2}{2ma^2}$$

This energy corresponds to the state

$$\Psi \propto \sin\left(\frac{\pi x}{a}\right) \sin\left(\frac{2\pi y}{a}\right)$$

(we are ignoring the normalization constant for now). Suppose that instead we have the energy state with $n_x = 2,\ n_y = 1$. The energy is

$$E = E_x + E_y = \frac{\hbar^2 \pi^2}{2ma^2}(4 + 1) = \frac{5\hbar^2 \pi^2}{2ma^2}$$

but the state is

$$\Psi \propto \sin\left(\frac{2\pi x}{a}\right)\sin\left(\frac{\pi y}{a}\right)$$

showing that the energy is degenerate. You are going to find degenerate energy states in multiple-dimensional systems.

Normalization proceeds as follows. Not surprisingly, we must integrate over both variables:

$$1 = \int\int \left|\Psi(x, y)\right|^2 dx\, dy$$

In this case, calling the normalization constant N, we find

$$N^2 \int_0^a \int_0^a \Psi^*(x, y)\,\Psi(x, y)\; dx\; dy$$

$$= N^2 \int_0^a \int_0^a \sin^2\left(\frac{n_x \pi x}{a}\right)\sin^2\left(\frac{n_y \pi y}{a}\right) dx\; dy = N^2 \frac{a^2}{4}$$

and so we find

$$N = \frac{2}{a}$$

The normalized wavefunction is

$$\Psi(x, y) = \left(\frac{2}{a}\right)\sin\left(\frac{n_x \pi x}{a}\right)\sin\left(\frac{n_y \pi y}{a}\right)$$

As an example, we consider $a = 3$, $n_x = 2$, $n_y = 3$, for which the state is

$$\Psi(x, y) = \frac{2}{3}\sin\left(\frac{2\pi x}{3}\right)\sin(\pi y)$$

Fig. 12-1 shows a plot of the wavefunction.

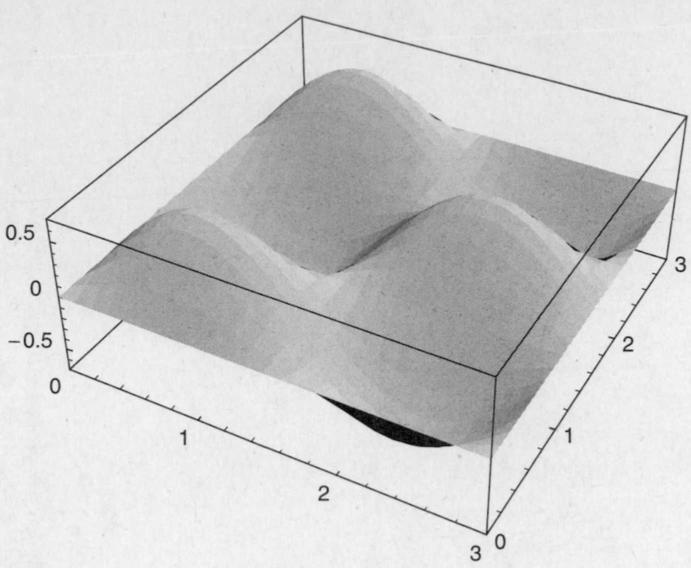

Fig. 12-1

As usual, the square of the wavefunction is a probability density, telling us where we are most likely to find the particle (see Fig. 12-2).

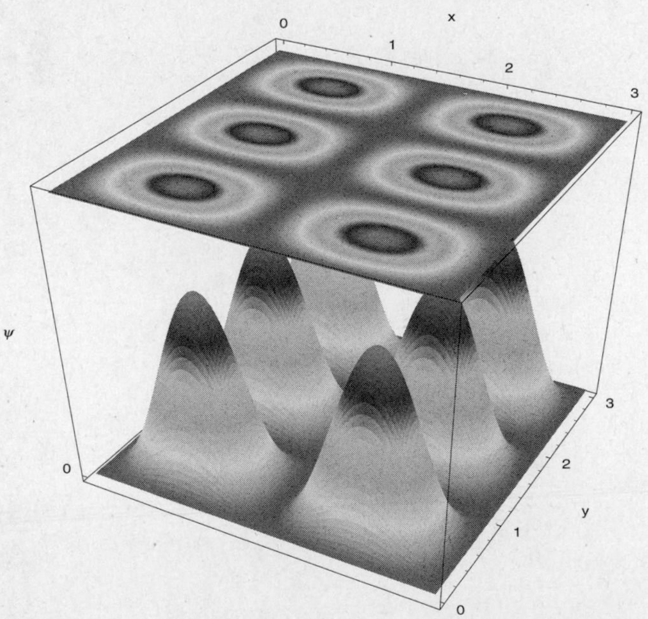

Fig. 12-2

EXAMPLE 12.1

A particle is trapped in a cylindrical well, for which the potential is

$$V = \begin{cases} 0 & \text{for } 0 < r < a,\ 0 < z < h \\ \infty & \text{otherwise} \end{cases}$$

Find the allowed energy levels.

SOLUTION

We use cylindrical coordinates and assume a separable solution. Therefore we write the wavefunction as

$$\Psi(r, \phi, z) = R(r)\Phi(\phi)Z(z)$$

In cylindrical coordinates, the Laplacian is written as

$$\nabla^2\Psi = \frac{1}{r}\frac{\partial}{\partial r}\left(r\frac{\partial\Psi}{\partial r}\right) + \frac{1}{r^2}\frac{\partial^2\Psi}{\partial\phi^2} + \frac{\partial^2\Psi}{\partial z^2}$$

Inside the well the potential is zero, and so we have

$$-\frac{\hbar^2}{2m}\nabla^2\Psi = E\,\Psi$$

Using the separable solution, we have

$$\frac{\partial\Psi}{\partial r} = \Phi(\phi)Z(z)\frac{dR}{dr}, \quad \frac{\partial^2\Psi}{\partial\phi^2} = R(r)Z(z)\frac{d^2\Phi}{d\phi^2}, \quad \frac{\partial^2\Psi}{\partial z^2} = R(r)\Phi(\phi)\frac{d^2Z}{dz^2}$$

This gives us

$$\nabla^2\Psi = \frac{\Phi Z}{r}\left(\frac{dR}{dr} + r\frac{d^2R}{dr^2}\right) + \frac{RZ}{r^2}\frac{d^2\Phi}{d\phi^2} + R\Phi\frac{d^2Z}{dz^2}$$

We insert this into the Schrödinger equation, divide through by $\Psi = R\Phi Z$ and $-\hbar^2/2m$, giving

$$\frac{1}{rR(r)}\left(\frac{dR}{dr} + r\frac{d^2R}{dr^2}\right) + \frac{1}{r^2\Phi(\phi)}\frac{d^2\Phi}{d\phi^2} + \frac{1}{Z(z)}\frac{d^2Z}{dz^2} = -\frac{2mE}{\hbar^2} = -k^2$$

Each part of the equation, which depends on one variable only, must be separately constant. For the z-equation we have

$$\frac{1}{Z(z)}\frac{d^2Z}{dz^2} = -k_z^2$$

An equation with the usual solution (consider the boundary condition at 0):

$$Z(z) = \sin(k_z z)$$

In the usual way, from the boundary condition $Z(h) = 0$, we find

$$k_z h = n_z \pi,$$
$$n_z = 1, 2, 3, \ldots.$$

Now we consider the angular part of the wavefunction. Calling the constant m_ϕ, we have

$$\frac{1}{\Phi(\phi)} \frac{d^2\Phi}{d\phi^2} = -m_\phi^2$$

The solution is an exponential:

$$\Phi(\phi) = e^{im_\phi \phi}$$

This part of the wavefunction is subject to periodic boundary conditions. It must satisfy

$$\Phi(\phi) = \Phi(\phi + 2\pi)$$

This condition forces us to take

$$m_\phi = 0, \ \pm 1, \ \pm 2, \ \ldots$$

With these results in hand, we can write the equation for R in the following form:

$$\frac{1}{rR(r)} \left(\frac{dR}{dr} + r\frac{d^2R}{dr^2} \right) - \frac{m_\phi^2}{r^2} - k_z^2 = -k^2$$

We move all terms to one side, and define $\lambda^2 = k^2 - k_z^2$, giving

$$\frac{1}{rR(r)} \left(\frac{dR}{dr} + r\frac{d^2R}{dr^2} \right) - \frac{m_\phi^2}{r^2} + \lambda^2 = 0$$

We now make a change of variables, setting $\lambda r = \rho$. This defines the following relationship:

$$\frac{\partial}{\partial \rho} = \frac{\partial r}{\partial \rho} \frac{\partial}{\partial r} = \frac{1}{\lambda} \frac{\partial}{\partial r}$$

This allows us to rewrite derivatives in the following way:

$$\frac{dR}{dr} = \lambda \frac{dR}{d\rho}$$

Using this change of variables, the equation for R becomes (exercise):

$$\frac{\lambda^2}{\rho R}\left(\frac{dR}{d\rho}+\rho\frac{d^2R}{d\rho^2}\right)+\lambda^2\left(1-\frac{m_\phi^2}{\rho^2}\right)=0$$

Dividing through by λ^2 gives Bessel's equation:

$$\frac{d^2R}{d\rho^2}+\frac{1}{\rho}\frac{dR}{d\rho}+\left(1-\frac{m_\phi^2}{\rho^2}\right)R=0$$

Solutions to this equation are given by Bessel functions, which have the series representation:

$$J_{m_\phi}(r)=\sum_{k=0}^{\infty}(-1)^k\frac{(r/2)^{m_\phi+2k}}{k!(m_\phi+k)!}$$

(*Neuman functions* are also solutions to this equation, but must be rejected because they blow up at the origin.) To get an idea of the form of these wavefunctions, see Figs. 12-3 and 12-4 for plots of the first three Bessel functions.

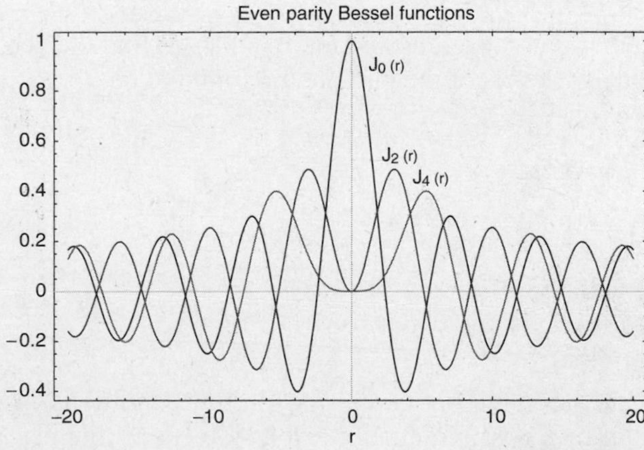

Fig. 12-3

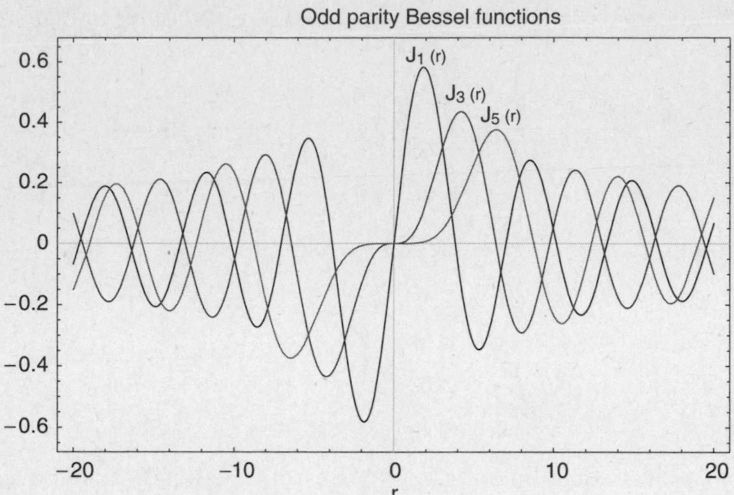

Fig. 12-4

Now, the wavefunction must vanish at $r = a$. To satisfy the boundary condition, using $\lambda r = \rho$ we solve

$$J_{m_\phi}(a/\lambda) = 0$$

We solve this numerically (or you can look in tables to find the zeros of the Bessel function). The energy levels of the particle are found to be

$$E = \frac{\hbar^2 k^2}{2m} = \frac{\hbar^2}{2m}\left(\lambda^2 + k_z^2\right)$$

Earlier we found the conditions on k_z. The limiting conditions on λ, and therefore the conditions limiting possible energy levels, are set by the zeros of the Bessel function.

We now consider a particle in a central potential.

An Overview of a Particle in a Central Potential

We now consider a particle trapped in a central potential—that is, a potential V that depends only on the radial coordinate r. In this case it is convenient to work in spherical coordinates where the Laplacian is:

$$\nabla^2 = \frac{1}{r^2}\frac{\partial}{\partial r}\left(r^2\frac{\partial}{\partial r}\right) + \frac{1}{r^2 \sin\theta}\frac{\partial}{\partial\theta}\left(\sin\theta\frac{\partial}{\partial\theta}\right) + \frac{1}{r^2\sin^2\theta}\frac{\partial^2}{\partial\phi^2}$$

In spherical polar coordinates, the angular momentum operator L^2 is given by

$$L^2 = -\hbar^2\left[\frac{1}{\sin\theta}\frac{\partial}{\partial\theta}\left(\sin\theta\frac{\partial}{\partial\theta}\right) + \frac{1}{\sin^2\theta}\frac{\partial^2}{\partial\phi^2}\right]$$

This allows us to write the Hamiltonian in the following way:

$$H = -\frac{\hbar^2}{2m}\frac{1}{r^2}\frac{\partial}{\partial r}\left(r^2\frac{\partial}{\partial r}\right) + \frac{1}{2mr^2}L^2 + V(r)$$

Recalling that the eigenstates of L^2, are also the common eigenstates of L_z, this Hamiltonian leads to three equations:

$$H\,\Psi(r,\theta,\phi) = E\,\Psi(r,\theta,\phi)$$
$$L^2\,\Psi(r,\theta,\phi) = \hbar^2(l)(l+1)\Psi(r,\theta,\phi)$$
$$L_z\,\Psi(r,\theta,\phi) = m\,\hbar\,\Psi(r,\theta,\phi)$$

In this problem there are three quantum numbers:

n – energy, from H

l – angular momentum, from L^2

m – from L_z

Again we assume a separable solution and set $\Psi = R(r)\,\Theta(\theta)\,\Phi(\phi)$. We have already seen that the angular part of the equation in spherical coordinates is solved by the spherical harmonics $Y_l^m(\theta,\phi)$, and so the problem at hand is actually simplified to finding a solution to the radial equation. The radial function R depends on the quantum numbers n and l. The equation is

$$-\frac{\hbar^2}{2mr}\frac{d^2}{dr^2}(r\,R_{nl}(r)) + \left[\frac{l(l+1)\hbar^2}{2mr^2} + V(r)\right]R_{nl}(r) = E\,R_{nl}(r)$$

It is helpful to make the following definition:

$$R_{nl}(r) = \frac{1}{r} U_{nl}(r)$$

The radial equation is then simplified somewhat to

$$-\frac{\hbar^2}{2m}\frac{d^2}{dr^2}(r\, U_{nl}(r)) + \left[\frac{l(l+1)\hbar^2}{2mr^2} + V(r)\right] U_{nl}(r) = E\, U_{nl}(r)$$

The *effective potential* is defined as

$$V_{eff} = V(r) + \frac{l(l+1)\hbar^2}{2mr^2}$$

This allows us to further simplify the radial equation to

$$-\frac{\hbar^2}{2m}\frac{d^2}{dr^2}(r\, U_{nl}(r)) + V_{eff} U_{nl}(r) = E\, U_{nl}(r)$$

To obtain actual solutions, we must consider a specific form of the potential. We will examine the hydrogen atom.

An Overview of the Hydrogen Atom

In this section we summarize the results of the solution for the hydrogen atom. A hydrogen atom is a bound system consisting of a proton and neutron. The potential is given by the electrostatic Coulomb potential

$$V(r) = -\frac{1}{4\pi\varepsilon_o}\frac{q^2}{r}$$

where q is the charge on the electron.

To find a solution in the case of the hydrogen atom, since the problem involves only two particles, the proton and electron, it is convenient to define a reduced mass μ, given by

$$\mu = \frac{m_e m_p}{m_e + m_p}$$

Recalling that the mass of the proton is much larger than the mass of the electron, we can write the reduced mass in approximate form as

$$\mu \approx m_e\left(1 - \frac{m_e}{m_p}\right)$$

The center of mass of a two-particle system is

$$r_{cm} = \frac{m_1 r_1 + m_2 r_2}{m_1 + m_2}$$

Since the proton mass is so much larger than the mass of the electron (about 2000 times as large), we can take the center of mass to be the proton.

The size of the lowest-energy orbit of the electron is given by the Bohr radius

$$a_o = 0.52 \ \mathring{A}$$

With the Coulomb potential, the radial equation becomes

$$-\frac{\hbar^2}{2m}\frac{d^2}{dr^2}\left(r\, U_{nl}\left(r\right)\right) + \left[\frac{l\left(l+1\right)\hbar^2}{2mr^2} - \frac{1}{4\pi\varepsilon_o}\frac{q^2}{r}\right] U_{nl}\left(r\right) = E\, U_{nl}\left(r\right)$$

The radial equation can be simplified in the following way. We let

$$\rho = \frac{r}{a_o}$$

We also define

$$\lambda_{kl} = \sqrt{-E_{kl}/E_1}$$

where the ground state energy is given by

$$E_1 = \frac{\mu\, q^4}{2\hbar^2}$$

The radial equation can be written in the new variables as

$$\frac{d^2}{d\rho^2}U_{kl}\left(\rho\right) - \frac{l\left(l+1\right)}{\rho^2}U_{kl}\left(\rho\right) + \frac{2}{\rho}U_{kl}\left(\rho\right) - \lambda_{kl}^2 U_{kl}\left(\rho\right) = 0$$

To obtain a solution to this equation, we first consider the case where ρ is very large. The equation in this case becomes

$$\frac{d^2}{d\rho^2}U_{kl}\left(\rho\right) - \lambda_{kl}^2 U_{kl}\left(\rho\right) = 0$$

This equation is solved by

$$U_{kl}\left(\rho\right) \sim e^{\lambda_{kl}\rho} + e^{-\lambda_{kl}\rho}$$

modulo constants. We immediately reject the $e^{\lambda_{kl}\rho}$ term, because it blows up as ρ gets large. The radial function must vanish as $\rho \to \infty$. Therefore we take

$$U_{kl}(\rho) \sim e^{-\lambda_{kl}\rho}$$

The general solution to the radial equation is a series solution that we call $\xi_{kl}(\rho)$. The complete solution is the product of the series solution and the asymptotic solution

$$U_{kl}(\rho) = e^{-\lambda_{kl}\rho}\xi_{kl}(\rho)$$

where

$$\xi_{kl}(\rho) = \rho^s \sum_{j=0}^{\infty} C_j \rho^j$$

This series must terminate to keep the solution from blowing up, and we determine the constant C_o from normalization. The series solution is given in terms of *Laguerre polynomials.* Simply stating the result whose derivation can be found for example in Griffiths, the radial part of the wavefunction depends on two quantum numbers, n and l, and is found to be

$$R_{nl}(r) = -\sqrt{\left(\frac{2}{n\,a_o}\right)^3 \frac{(n-l-1)!}{2n\,[(n-1)!]^2}}\, e^{-\rho/2}\rho^l L_{n+1}^{2l+1}(\rho)$$

$L_{n+1}^{2l+1}(\rho)$ are the associated Laguerre polynomials. The Laguerre polynomials are defined as

$$L_n(r) = e^r \left(\frac{d}{dr}\right)^n \left(e^{-r} r^n\right)$$

and so

$$L_{n-l}^{l}(r) = (-1)^l \left(\frac{d}{dr}\right)^l L_n(r)$$

We normalize the radial part of the wavefunction in this way:

$$\int_0^\infty r^2 \,|R(r)|^2\, dr = 1$$

The expectation value of r^k is given by

$$\langle r^k \rangle = \int_0^\infty r^{2+k}\,|R(r)|^2\, dr = 1$$

The complete wavefunction is a product of the radial wavefunctions and the spherical harmonics

$$\Psi_{nlm}(r,\theta,\phi) = \sqrt{\left(\frac{2}{n\,a_o}\right)^3 \frac{(n-l-1)!}{2n\,[(n-1)!]^2}}\, e^{-\rho/2}\rho^l L_{n+1}^{2l+1}(\rho)\, Y_l^m(\theta,\phi)$$

As we stated earlier, n identifies the energy of the state. The energy levels in the hydrogen atom depend only on n and are given by

$$E_{kl} = -\frac{E_1}{(k+l)^2} = -\frac{E_1}{n^2}$$

Here $E_1 = -13.6\,eV$ is the ground state energy. The values that l can assume are fixed by n in the following way:

$$l = 0, 1, 2, \ldots, n-1$$

As we saw in Chapter 10, the values m can assume are fixed by l as

$$m = -l, -l+1, \ldots, l-1, l$$

The radial function and therefore the energy do not depend on m in any way. As l ranges from $0 \rightarrow n-1$, there are

$$g_n = 2\sum_{l=0}^{n-1} 2l+1 = 2n^2$$

different states that have the same energy (therefore this is the degeneracy). In SI units, the potential V is

$$V = -\frac{1}{4\pi\varepsilon_o}\frac{q^2}{r}$$

and the ground state energy is

$$E_1 = -\frac{m^2 q^4}{32\,\pi^2\varepsilon_o^2\hbar^2}$$

The energy of the nth state is

$$E_n = -\frac{m^2 q^4}{32\,\pi^2\varepsilon_o^2\hbar^2}\frac{1}{n^2}$$

Following are a few of the first radial functions and their plots (see Figs. 12-5 through 12-9).

$$R_{10}(r) = 2a_o^{-3/2} e^{-r/a_o}$$

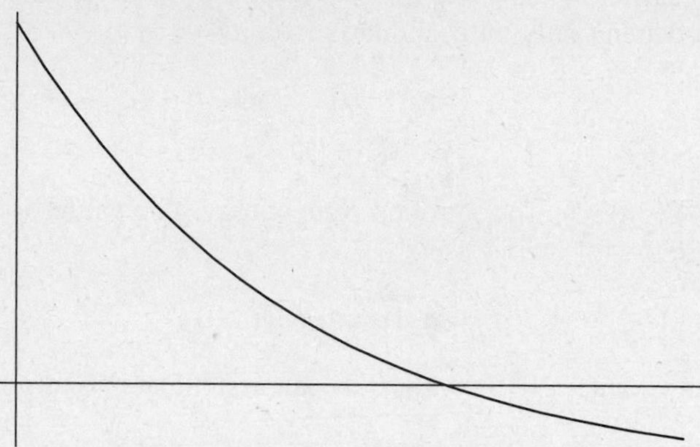

Fig. 12-5

$$R_{20}(r) = 2(2a_o)^{-3/2}\left(1 - \frac{r}{2a_o}\right) e^{-r/2a_o}$$

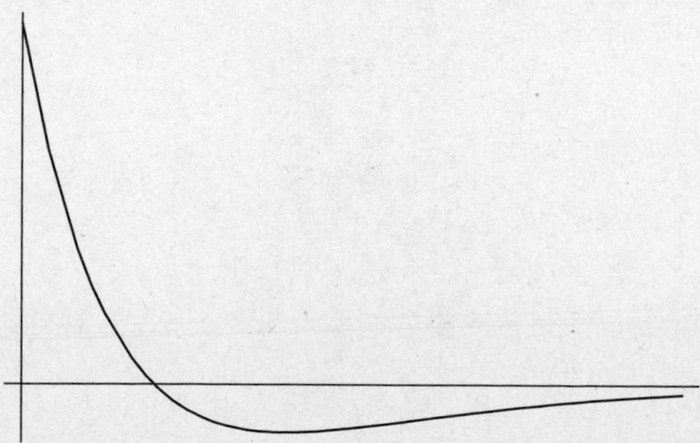

Fig. 12-6

$$R_{21}(r) = (2a_o)^{-3/2}\frac{1}{\sqrt{3}}\frac{r}{a_o}e^{-r/a_o}$$

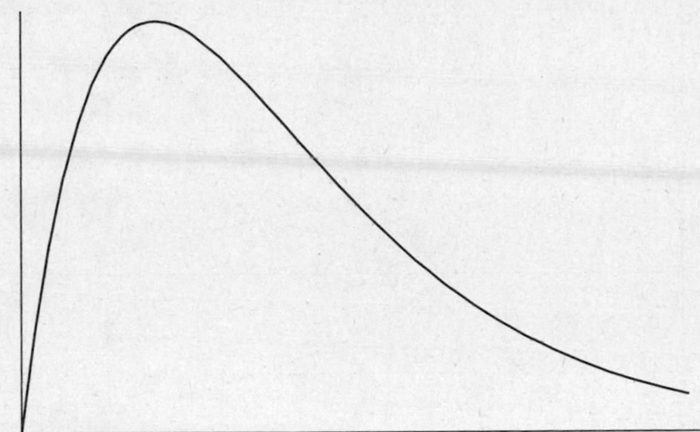

Fig. 12-7

$$R_{32}(r) = \frac{4}{81\sqrt{30}}(a_o)^{-3/2}\frac{1}{\sqrt{3}}\left(\frac{r}{a_o}\right)^2 e^{-r/3a_o}$$

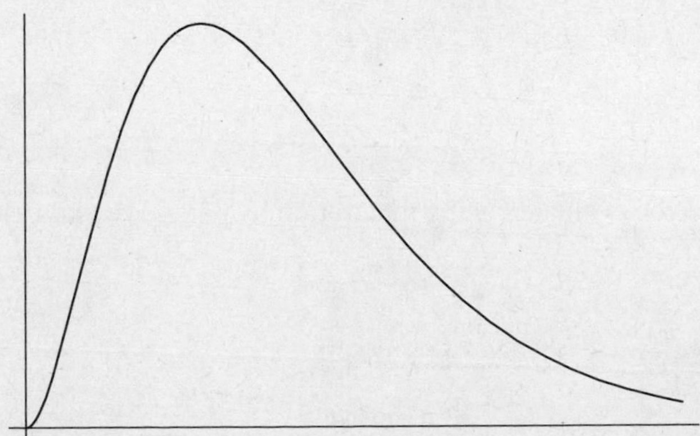

Fig. 12-8

$$R_{41}(r) = \frac{\sqrt{5}}{16\sqrt{3}}(a_o)^{-3/2}\left(1 - \frac{1}{4}\frac{r}{a_o} + \frac{1}{80}\left(\frac{r}{a_o}\right)^2\right)\left(\frac{r}{a_o}\right)e^{-r/4a_o}$$

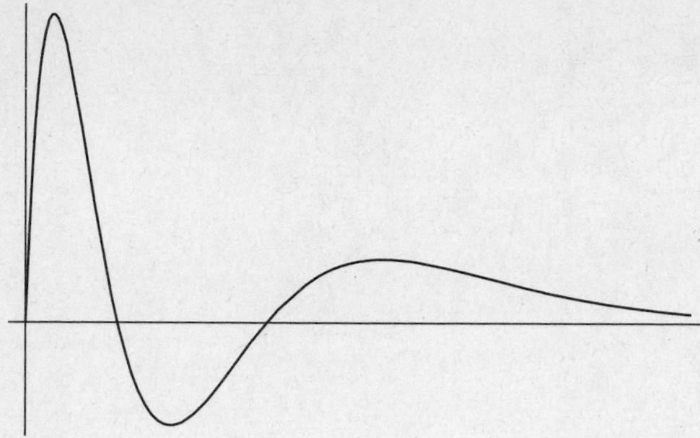

Fig. 12-9

EXAMPLE 12.2

In the ground state of hydrogen, what is the probability that the electron is found inside the Bohr radius?

SOLUTION

The ground state wavefunction is

$$R_{10}(r) = 2a_o^{-3/2}e^{-r/a_o}$$

The probability that the particle is found in the range $0 < r < a_o$ is

$$\int_0^{a_o} r^2 |R_{10}(r)|^2 \, dr = \int_0^{a_o} r^2 \left(2a_o^{-3/2}e^{-r/a_o}\right)^2 dr$$

$$= 4a_o^{-3}\int_0^{a_o} r^2 e^{-2r/a_o} \, dr$$

This integral can be evaluated using integration by parts. We make the substitutions

$$u = r^2, \Rightarrow du = 2r\,dr \qquad dV = e^{-2r/a_o}, \Rightarrow V = -\frac{a_o}{2}e^{-2r/a_o}$$

Using $\int u\,dv = uv - \int v\,du$, we obtain

$$prob = 4a_o^{-3}\left[-\frac{a_o}{2}r^2 e^{-2r/a_o} + a_o\int_0^{a_o} re^{-2r/a_o} dr\right]$$

On the second integral, we again use integration by parts. This time we have

$$u = r, \Rightarrow du = dr$$

$$dV = e^{-2r/a_o}, \quad \Rightarrow V = -\frac{a_o}{2}e^{-2r/a_o}$$

This leads us to

$$prob = 4a_o^{-3}\left[-\frac{a_o}{2}r^2 e^{-2r/a_o} + a_o\left(-\frac{a_o}{2}re^{-2r/a_o} + \frac{a_o}{2}\int_0^{a_o} e^{-2r/a_o}dr\right)\right]$$

The last integral can be done immediately

$$\int_0^{a_o} e^{-2r/a_o}dr = -\frac{a_o}{2}e^{-2r/a_o}$$

Inserting this term into our expression for the probability, we now evaluate at the limits.

$$prob = 4a_o^{-3}\left[-\frac{a_o}{2}r^2 e^{-2r/a_o} + a_o\left(-\frac{a_o}{2}re^{-2r/a_o} - \frac{a_o^2}{4}e^{-2r/a_o}\right)\right]\Bigg|_0^{a_o}$$

$$= 4a_o^{-3}\left[-\frac{a_o^3}{2}e^{-2} - \frac{a_o^3}{2}e^{-2} - \frac{a_o^3}{4}e^{-2} + \frac{a_o^3}{4}\right] = -5e^{-2} + 1 \approx 0.323$$

EXAMPLE 12.3 e.g.

Suppose that $n = 2$ and $l = 0$. Find the average radius of the electron's orbit, which is given by $\langle r \rangle$.

SOLUTION ✔

The radial wavefunction for the state is

$$R_{20}(r) = 2(2a_o)^{-3/2}\left(1 - \frac{r}{2a_o}\right)e^{-r/2a_o}$$

and so we have

$$\langle r \rangle = \int_0^\infty r^{2+1}|R_{20}(r)|^2\,dr = \int_0^\infty r^3\left[2(2a_o)^{-3/2}\left(1 - \frac{r}{2a_o}\right)e^{-r/2a_o}\right]^2 dr$$

$$= 4(2a_o)^{-3}\int_0^\infty r^3\left(1 - \frac{r}{a_o} + \frac{r^2}{4a_o^2}\right)e^{-r/a_o}\,dr$$

$$= 4(2a_o)^{-3}\int_0^\infty \left(r^3 - \frac{r^4}{a_o} + \frac{r^5}{4a_o^2}\right)e^{-r/a_o}\,dr$$

Each term can be evaluated by repeated integration by parts, using $dV = e^{-r/a_o}$. The three integrals evaluate as (check)

$$\int_0^\infty r^3 e^{-r/a_o} dr = 6\,a_o^4$$

$$\int_0^\infty \frac{r^4}{a_o} e^{-r/a_o} dr = 24\,a_o^4$$

$$\int_0^\infty \frac{r^5}{4a_o^2} e^{-r/a_o} dr = 30\,a_o^4$$

Putting these results together we obtain

$$\begin{aligned}\langle r\rangle &= 4\,(2\,a_o)^{-3}\left[6\,a_o^4 - 24\,a_o^4 + 30\,a_o^4\right]\\ &= 4\,(2\,a_o)^{-3}\left(12\,a_o^4\right) = 6a_o\end{aligned}$$

Here are some useful formulas that work when computing expectation values for the hydrogen atom:

$$\langle r\rangle = \frac{a_o}{2}\left(3n^2 - l(l+1)\right)$$

$$\left\langle r^2\right\rangle = \frac{a_o^2 n^2}{2}\left(5n^2 + 1 - 3l(l+1)\right)$$

$$\left\langle \frac{1}{r}\right\rangle = \frac{1}{a_o n^2}$$

The angular probability distributions are given by the spherical harmonics. These give the probability of finding the electron at angle (θ, ϕ). Fig. 12-10 shows the distribution for $l = 0, m = 0$.

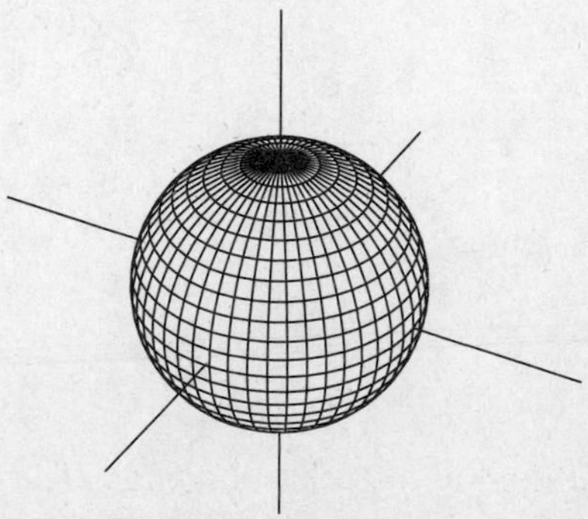

Fig. 12-10

Fig. 12-11 shows the angular distribution for $l = 1, m = 0$.

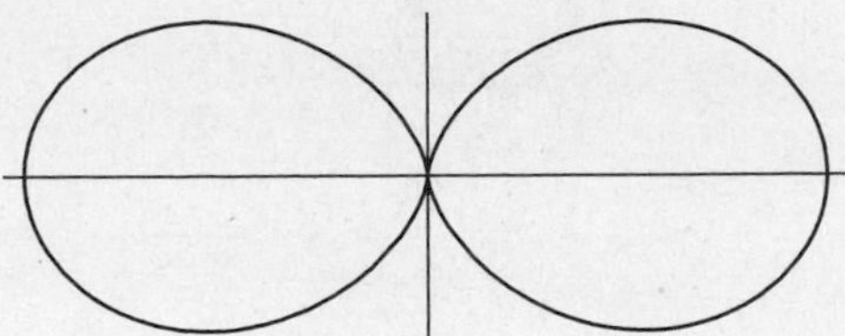

Fig. 12-11

Fig. 12-12 shows $l = 2, m = 0$.

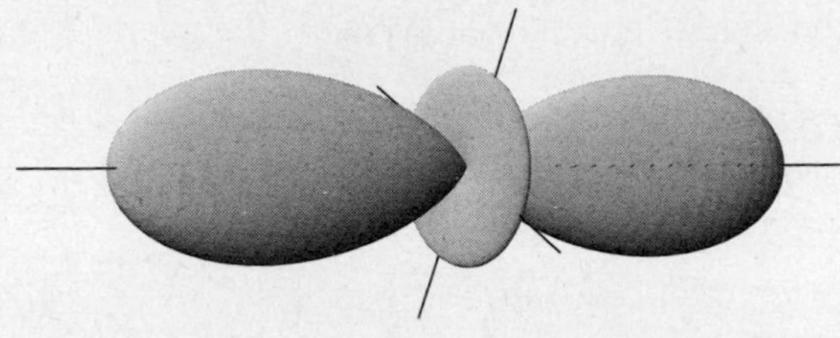

Fig. 12-12

Spectroscopic notation is sometimes used to identify the value of l. This is done with a letter label to specify the angular momentum. The first few states are shown in Table 12-1.

Table 12-1

l	Designation
$l = 0$	s
$l = 1$	p
$l = 2$	d
$l = 3$	f

Table 12-2 Hydrogen Wave Functions

States can then be identified in the following way. The $(n, l) = (2, 0)$ state can be written as Ψ_{2s}, while the $(3, 2)$ state can be written as Ψ_{3d}. Note that these states are orthonormal.

Just like any quantum state, a valid state of the hydrogen atoms can be formed by superposition states, so for example we might have

$$\phi = \frac{1}{\sqrt{2}}\psi_{1s} + \frac{1}{2}\psi_{2p} + \frac{1}{2}\psi_{3d}$$

EXAMPLE 12.4

The state of a hydrogen atoms is

$$\phi = \frac{1}{\sqrt{2}}\psi_{1s} + A\psi_{2p} + \frac{1}{\sqrt{8}}\psi_{3s}$$

Find A so that the state is normalized. What is the average energy of the state?

SOLUTION

We compute the norm of the state

$$\begin{aligned}\langle\phi\,|\phi\rangle &= \frac{1}{2}\langle\psi_{1s}\,|\psi_{1s}\rangle + |A|^2\left\langle\psi_{2p}\,|\psi_{2p}\right\rangle + \frac{1}{8}\langle\psi_{3s}\,|\psi_{3s}\rangle \\ &= \frac{1}{2} + |A|^2 + \frac{1}{8} = \frac{5}{8} + |A|^2\end{aligned}$$

For this state to be normalized, the inner product must be unity. Therefore

$$|A|^2 = 1 - \frac{5}{8} = \frac{3}{8}, \Rightarrow A = \sqrt{\frac{3}{8}}$$

The average energy of the state, which depends on n only, is given by $\langle\phi|\,H\,|\phi\rangle$. Recalling the energy dependence on n

$$E_n = -\frac{E_1}{n^2}$$

we find

$$H\,|\psi_{1s}\rangle = -E_1\,|\psi_{1s}\rangle \quad \left(\textit{probability}\ \frac{1}{2}\right)$$

$$H\,\left|\psi_{2p}\right\rangle = -\frac{E_1}{4}\,|\psi_{2p}\rangle \quad \left(\textit{probability}\ \frac{3}{8}\right)$$

$$H\,|\psi_{3s}\rangle = -\frac{E_1}{9}\,|\psi_{3s}\rangle \quad \left(\textit{probability}\ \frac{1}{8}\right)$$

The mean energy is found by summing over the possible energy values, each multiplied by the respective probability

$$\langle H \rangle = \sum p_i E_i = -E_1\left(\frac{1}{2}\right) - \frac{E_1}{4}\left(\frac{3}{8}\right) - \frac{E_1}{9}\left(\frac{1}{8}\right) = -E_1\frac{31}{288}$$

EXAMPLE 12.5

e.g.

An electron in a hydrogen atom is in the state

$$\psi_{nlm} = R_{32}\left(\sqrt{\frac{1}{6}}Y_2^1 + \sqrt{\frac{1}{2}}Y_2^0 + \sqrt{\frac{1}{3}}Y_2^{-1}\right)$$

(a) What is the energy of the electron?

(b) L^2 is measured. What values can be found?

(c) If L_z is measured, what values can be found and with what probabilities? What is the expectation value of L_z?

SOLUTION

(a) The energy level is determined from n. The radial function is, so $n = 3$. The energy for the $n = 3$ state is

$$E_3 = -\frac{E_1}{3^2} = -\frac{13.6}{9}eV = -1.51\ eV$$

(b) A measurement of L^2 gives $\hbar^2 l(l+1)$. In this case all of the spherical harmonics have $l = 2$, and so the only possible measurement result is $6\hbar^2$.

(c) To consider measurement results for L_z, we only need to worry about the angular part of the wavefunction. This is

$$\sqrt{\frac{1}{6}}Y_2^1 + \sqrt{\frac{1}{2}}Y_2^0 + \sqrt{\frac{1}{3}}Y_2^{-1}$$

Recalling that

$$L_z Y_l^m = m\hbar Y_l^m$$

we see by inspection that in this state the possible results are (see Table 12-3):

Table 12-3

Measurement Result	Probability
$\hbar$	$\left(\sqrt{\frac{1}{6}}\right)^2 = \frac{1}{6}$
0	$\left(\sqrt{\frac{1}{2}}\right)^2 = \frac{1}{2}$
$-\hbar$	$\left(\sqrt{\frac{1}{3}}\right)^2 = \frac{1}{3}$

We find the expectation value of to be

$$\langle L_z \rangle = (\hbar)\left(\frac{1}{6}\right) + (0)\left(\frac{1}{2}\right) + (-\hbar)\left(\frac{1}{3}\right)$$
$$= -\frac{\hbar}{6}$$

We can form the complete probability distribution that depends on n, l, and m for the location of the electron for a given state of hydrogen. The plots show the z-axis going from left to right along the page. The first example (see Fig. 12-13) shows $n = 2, l = 0, m = 0$.

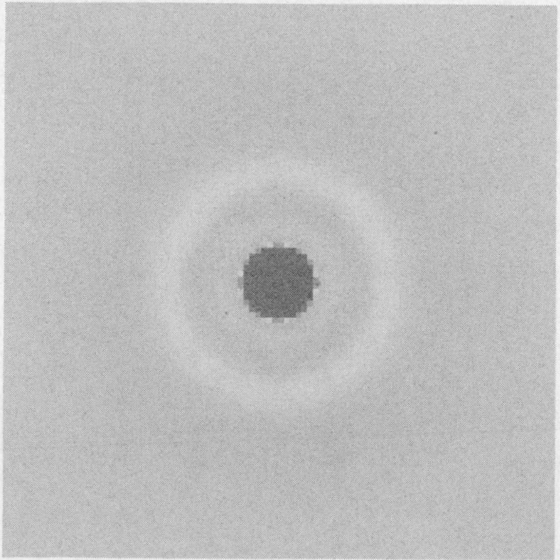

Fig. 12-13

The next example (see Fig. 12-14) shows $n = 2$, $l = 1$, and $m = -1$.

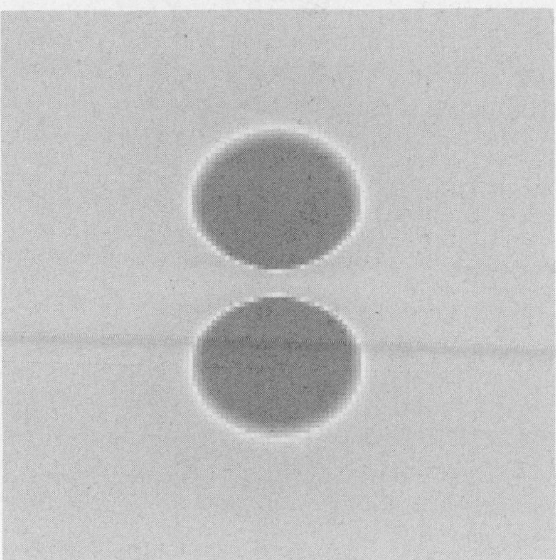

Fig. 12-14

Fig. 12-15 shows $n = 4$, $l = 2$, $m = 1$.

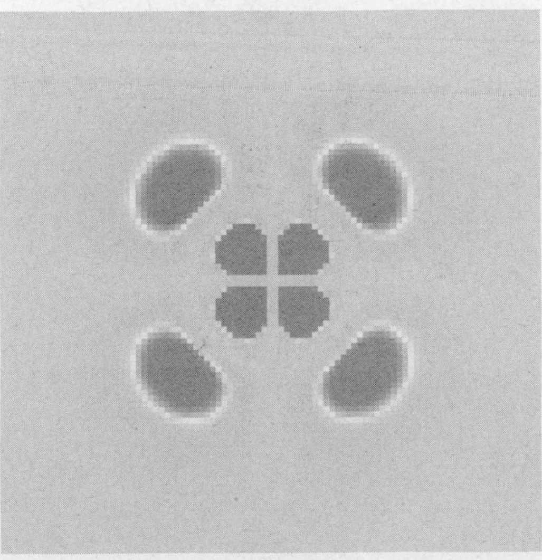

Fig. 12-15

Quiz

1. The momentum operator in multiple dimensions can be written as
   (a) $\vec{p} = -\hbar^2\nabla^2$
   (b) $\vec{p} = -\hbar\nabla \cdot \vec{v}$ where $\vec{v}$ is the particle's velocity
   (c) $\vec{p} = -i\hbar\nabla$

2. The angular momentum operator acts on a state $\psi(r, \theta, \phi)$ as
   (a) $L^2\psi = \hbar^2 l(l+1)\psi$
   (b) $L\psi = \hbar l(l+1)\psi$
   (c) $L\psi = \hbar^2 l(l+1)\psi$

3. For the hydrogen atom, the radial part of the wavefunction $R(r)$ is normalized according to which formula
   (a) $\int_0^\infty |R(r)|^2 \, dr = 1$
   (b) $\int_{a_o}^\infty r^2 |R(r)|^2 \, dr = 1$
   (c) $\int_0^\infty r^2 |R(r)|^2 \, dr = 1$

4. The angular part of the wavefunction for the hydrogen atom is written using
   (a) spherical harmonics.
   (b) Bessel functions.
   (c) Hankel functions.
   (d) Bessel functions of the second kind.

5. The average radius of an electron's orbit in the hydrogen atom can be calculated using
   (a) $\langle r \rangle = \int_0^\infty r \, |R_{nl}(r)|^2 \, dr$
   (b) $\langle r \rangle = \int_0^\infty r^3 \, |R_{nl}(r)|^2 \, dr$
   (c) $\langle r \rangle = \int_0^{a_o} r^3 \, |R_{nl}(r)|^2 \, dr$

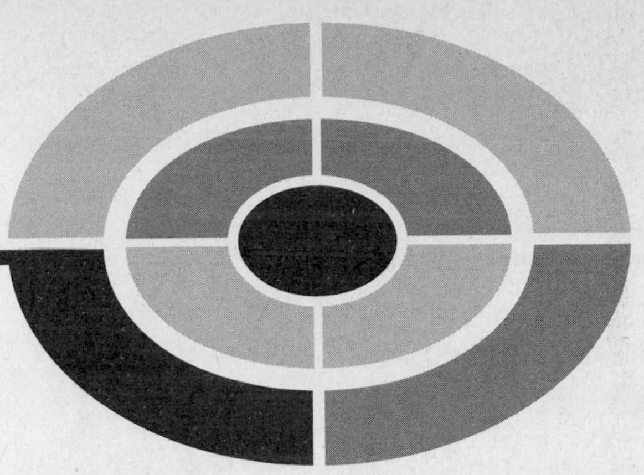

Final Exam

1. A wavefunction is defined by

$$\psi(x) = \begin{cases} A & \text{for } -a \le x \le a \\ 0 & \text{otherwise} \end{cases}$$

The best description of the Fourier transform of this function is that it includes

(a) a Hankel function.

(b) a sum of cos functions.

(c) the Dirac Delta function.

2. The time-energy uncertainty relation can be written as

(a) $\Delta E \Delta t \ge \hbar$

(b) $\Delta E \Delta t \ge \hbar/4$

(c) $\Delta E \Delta t > \hbar$

(d) $\Delta \omega \Delta t \ge \hbar$

3. The wavefunction for a particle trapped in an infinite square well defined for $0 \leq x \leq a$ can be written as
   (a) $\phi_n(x) = \sin\left(\frac{n\pi x}{a}\right)$
   (b) $\phi_n(x) = \sqrt{\frac{2}{a}}\sin\left(\frac{nx}{a}\right)$
   (c) $\phi_n(x) = \sqrt{\frac{2}{a}}\cos\left(\frac{n\pi x}{a}\right)$
   (d) $\phi_n(x) = \sqrt{\frac{2}{a}}\sin\left(\frac{n\pi x}{a}\right)$

4. In one-dimension, the full Schrödinger equation is
   (a) $-\hbar\omega\frac{\partial\psi}{\partial t} = -\frac{\hbar^2}{2m}\frac{\partial^2\psi}{\partial x^2} + V(x)\psi$
   (b) $-\hbar\frac{\partial\psi}{\partial t} = -\frac{\hbar^2}{2m}\frac{\partial^2\psi}{\partial x^2} + V(x)\psi$
   (c) $i\hbar\frac{\partial\psi}{\partial t} = -\frac{\hbar^2}{2m}\frac{\partial^2\psi}{\partial x^2} + V(x)\psi$
   (d) $-i\hbar\frac{\partial\psi}{\partial t} = -\frac{\hbar}{2m}\frac{\partial^2\psi}{\partial x^2} + V(x)\psi$

5. The *expectation value* of an operator A can be written as
   (a) $\langle A\rangle = \int_{-\infty}^{\infty}\psi^*(x)A\psi(x)\,dx$
   (b) $\langle A\rangle = \langle\psi|\,A\,|\psi\rangle$
   (c) $\langle A\rangle = \int\psi^*(x)A^2\psi(x)\,dx$
   (d) Both a & b are correct

6. A wavefunction is expanded in a set of basis states as $\psi = \sum_n c_n\phi_n$. The coefficients of the expansion must satisfy
   (a) $\sum_n c_n = 1$
   (b) $\sum_n |c_n|^2 = 1$
   (c) The coefficients c_n must be real numbers.
   (d) There are no restrictions.

7. A wavefunction is written in terms of an orthonormal basis as

$$|\psi\rangle = \frac{1}{\sqrt{3}}|u_1\rangle + \frac{1}{\sqrt{6}}|u_2\rangle + \sqrt{\frac{5}{6}}|u_3\rangle$$

   The probability that the system is found in the state $|u_2\rangle$ is
   (a) 1/3
   (b) 0.17
   (c) 0.13
   (d) 5/6

8. A wavefunction is expanded in terms of an orthonormal basis in the following way

$$|\psi\rangle = \frac{1}{\sqrt{5}}|u_1\rangle + \sqrt{\frac{3}{5}}|u_2\rangle + A|u_3\rangle$$

A must be
(a) $1/\sqrt{5}$
(b) $1/5$
(c) $2/5$
(d) $1/2$
(e) There is not enough information given.

9. The coefficients of expansion for some basis can be written as
(a) $c_n = \left(\sum_n |\phi_n\rangle\right)|\psi\rangle$
(b) $c_n = \int dx\, \phi_n^*(x)\psi^*(x)$
(c) $c_n = \int dx\, \phi_n^*(x)x\psi(x)$
(d) $c_n = \langle\phi_n \mid \psi\rangle = \int dx\, \phi_n^*(x)\psi(x)$

10. The base kets of position space satisfy
(a) $\sum_n |x_n\rangle\langle x_n| = 1$
(b) $\int dx\, |x\rangle\langle x| = 1$
(c) $\int dx\, |x\rangle\langle x \mid p\rangle = 1$
(d) $\int dx\, |p\rangle\langle x \mid p\rangle = 1$

11. The *Hermitian conjugate* of an operator A is
(a) $\langle\phi \mid A^\dagger \mid \psi\rangle = \langle\psi \mid A \mid \phi\rangle^*$
(b) $\langle\phi \mid A^\dagger \mid \psi\rangle = \langle\phi \mid A \mid \psi\rangle^*$
(c) $\langle\phi \mid A \mid \psi\rangle = \langle\psi \mid A \mid \phi\rangle^*$
(d) $\langle\phi \mid A^\dagger \mid \psi\rangle = -\langle\psi \mid A \mid \phi\rangle^*$

12. If we denote the position space operator by $\hat{x}$, which of the following best describes its action on the base kets
(a) $\langle x \mid \hat{x} \mid x'\rangle = -x'\delta(x - x')$
(b) $\langle x \mid \hat{x} \mid x'\rangle = x'$
(c) $\langle x \mid \hat{x} \mid x'\rangle = x^*\delta(x - x')$
(d) $\langle x \mid \hat{x} \mid x'\rangle = x'\delta(x - x')$

13. A unitary operator satisfies
    (a) $UU^\dagger = -U^\dagger U = I$
    (b) $UU^\dagger = 0$
    (c) $UU^\dagger = U^\dagger U = I$
    (d) $UU^\dagger = U^\dagger U = -I$

14. The energy levels of the one-dimensional harmonic oscillator are
    (a) equally spaced,
    (b) 2-fold degenerate,
    (c) 3-fold degenerate,
    (d) not degenerate,
    (e) a & d

15. In the coordinate representation, the eigenstates of the one-dimensional harmonic oscillator can be written in terms of
    (a) Hermite polynomials.
    (b) Hankel functions.
    (c) Legendre polynomials.
    (d) Bessel functions of the second kind.

16. The energy eigenvalues of the one-dimensional harmonic oscillator are
    (a) $E_n = \left(n + \frac{1}{2}\right)\hbar\omega, \quad n = 1, 2, 3, \ldots$
    (b) $E_n = \left(n + \frac{1}{2}\right)\hbar\omega, \quad n = 0, 1, 2, \ldots$
    (c) $E_n = \left(n - \frac{1}{2}\right)\hbar\omega, \quad n = 0, 1, 2, \ldots$
    (d) $E_n = \left(n - \frac{1}{2}\right)\hbar\omega, \quad n = 1, 2, 3, \ldots$

17. Consider the one-dimensional harmonic oscillator. The commutator $[H, a^\dagger]$ is equal to
    (a) $a^\dagger$
    (b) $\hbar\omega a$
    (c) $-\hbar\omega a^\dagger$
    (d) $\hbar\omega a^\dagger$

18. A particle is in an angular momentum eigenstate $|lm\rangle$. The expectation value $\langle L_x \rangle$ is equal to
    (a) $\hbar l(l+1)$
    (b) $\sqrt{\hbar^2 l^2 (l+1) - m(m+1)}$
    (c) 0
    (d) $\hbar^2 l^2 (l+1)$

19. A spin-1 particle is in the state

$$|\psi\rangle = \frac{1}{\sqrt{7}} \begin{pmatrix} 2 \\ 1 \\ \sqrt{2} \end{pmatrix}$$

If spin is measured in the z-direction, the probability that $m = +1$ is found is
    (a) 3/7
    (b) 1/2
    (c) 4/7
    (d) 1/7

20. Consider the hydrogen atom. The angular probability distribution is given by
    (a) $Y_l^m(\theta, \phi)$
    (b) $\left|\theta Y_l^m(\theta, \phi)\right|^2 \, d\Omega$
    (c) $\left|Y_l^m(\theta, \phi)\right|^2 \, d\Omega$
    (d) $\left|\cos\theta Y_l^m(\theta, \phi)\right|^2 \, d\Omega$

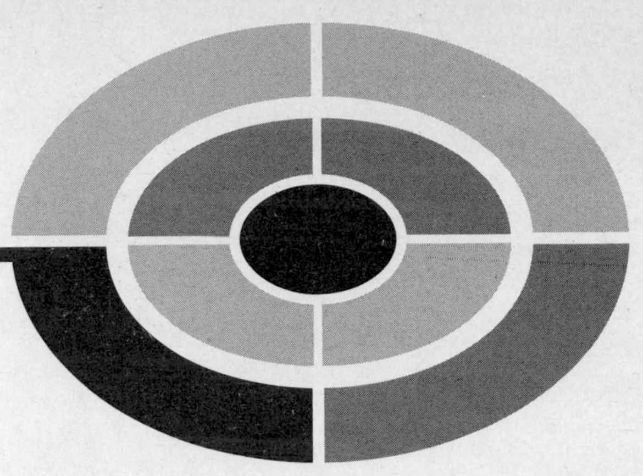

Answers to Quiz and Exam Questions

Chapter 1

1. Given that

$$f(\varepsilon) = \sum_{n=0}^{\infty} e^{-n\varepsilon/kT}$$

We see that

$$\frac{df}{d\varepsilon} = \sum_{n=1}^{\infty} \left(\frac{n}{kT}\right) e^{-n\varepsilon/kT}$$

Some manipulation of the series given for g can put it in this form. We have

$$g(\varepsilon) = \sum_{n=1}^{\infty} n\varepsilon e^{-n\varepsilon/kT} = \varepsilon \sum_{n=1}^{\infty} n e^{-n\varepsilon/kT} = -kT\,\varepsilon \sum_{n=1}^{\infty} \left(\frac{-n}{kT}\right) e^{-n\varepsilon/kT} = -kT\,\varepsilon \frac{df}{d\varepsilon}$$

Using the geometric series result, we can write

$$f(\varepsilon) = \sum_{n=0}^{\infty} e^{-n\varepsilon/kT} = \frac{1}{1 - e^{-\varepsilon/kT}}$$

Computing the derivative of f using this representation we find

$$\frac{df}{d\varepsilon} = \frac{d}{d\varepsilon}\left(\frac{1}{1 - e^{-\varepsilon/kT}}\right) = -\frac{1}{kT}\frac{e^{-\varepsilon/kT}}{\left(1 - e^{-\varepsilon/kT}\right)^2}$$

So we can write g in the following way

$$g(\varepsilon) = -kT\,\varepsilon\frac{df}{d\varepsilon} = -kT\,\varepsilon\left[\frac{e^{-\varepsilon/kT}}{\left(1 - e^{-\varepsilon/kT}\right)^2}\right] = \varepsilon\frac{e^{-\varepsilon/kT}}{\left(1 - e^{-\varepsilon/kT}\right)^2}$$

2. The lowest energy of the hydrogen atom is

$$E_1 = -\frac{2\pi^2 m\left(\frac{e^2}{4\pi\varepsilon_o}\right)^2}{2\hbar^2} = -\frac{me^4}{2\hbar^2}$$

Now explicitly including the permittivity of free space, we have

$$E_1 = -\frac{mq^4}{2\hbar^2}\left(\frac{1}{4\pi\varepsilon_o}\right)^2$$

Let's take a look at the units, to see what value we should use for Planck's constant. Since we want the final answer in electron volts, for the electron mass we use

$$m = 0.511\ \mathrm{MeV}/c^2 = \left(0.511 \times 10^6\right)\ \mathrm{eV}\left(\frac{1}{2.99 \times 10^8}\frac{\mathrm{s}^2}{\mathrm{m}^2}\right)$$

where we included the speed of light in m/s. The units of ε_o are $\mathrm{C}^2/\mathrm{N\ m}^2$, and so using the fact that a Joule is a N-m we get the right units if we use the value of Planck's constant written in terms of J-s. Writing charge in Coulombs (C) we would have

$$[E_1] = \left[\frac{mq^4}{2\hbar^2}\left(\frac{1}{4\pi\varepsilon_o}\right)^2\right] = \frac{[m]\left[q^4\right]}{\left[\hbar^2\right]}\frac{1}{\left[\varepsilon_o^2\right]} = \frac{\left(\mathrm{eV} * \frac{\mathrm{s}^2}{\mathrm{m}^2}\right)\left(\mathrm{C}^4\right)}{\left(\mathrm{N}^2\mathrm{m}^2\mathrm{s}^2\right)}\left(\frac{\mathrm{N}^2\mathrm{m}^4}{\mathrm{C}^4}\right) = \mathrm{eV}$$

So, in addition to the mass of the electron defined above we use the following values

$$\varepsilon_o = 8.85 \times 10^{-12}\text{C}^2\text{Nm}^2, \Rightarrow \frac{1}{4\pi\varepsilon_o} = 8.9918 \times 10^9 \frac{\text{Nm}^2}{\text{C}^2}$$

$$\hbar^2 = \left(1.055 \times 10^{-34}\text{J-s}\right)^2$$

And we obtain

$$E_1 = -\frac{mq^4}{2\hbar^2}\left(\frac{1}{4\pi\varepsilon_o}\right)^2$$

$$= -\frac{\left(0.511 \times 10^6\right)}{2\left(1.055 \times 10^{-34}\right)^2}\left(8.9918 \times 10^9\right)^2\left(1.602 \times 10^{-19}\right)^4 \frac{1}{\left(2.99 \times 10^8\right)}\text{eV}$$

$$= -13.6 \text{ eV}$$

3. Use $a_o = \dfrac{\hbar^2}{me^2}$

Chapter 2

1. (a) Setting $\psi(x,t) = \phi(x) f(t)$ and using the Schrödinger equation, we have

$$-i\hbar\frac{\partial\psi}{\partial t} = -i\hbar\phi(x)\frac{df}{dt}$$

For the right hand side we obtain

$$-\frac{\hbar^2}{2m}\frac{\partial^2\psi}{\partial x^2} + V\psi = -\frac{\hbar^2}{2m}f(t)\frac{d^2\phi}{dx^2} + Vf(t)\phi(x)$$

Setting this equal to the time derivative and dividing through by $\psi(x,t) = \phi(x) f(t)$ gives

$$-i\hbar\frac{1}{f}\frac{df}{dt} = -\frac{\hbar^2}{2m}\frac{1}{\phi}\frac{d^2\phi}{dx^2} + V = -E$$

We have set these terms equal to a constant because on one side we have a function of t only while on the other side we have a function of x only, and the only possible way they can be equal is if they are each constant. We call the constant E in anticipation that this is the energy. Looking at the equation involving time we have

$$i\hbar \frac{1}{f}\frac{df}{dt} = E$$

This integrates immediately to give

$$f(t) = e^{-iEt/\hbar}$$

(b) For the given wavefunction we have

$$\frac{d\psi}{dx} = kA\cos kx - kB\sin kx$$

$$\frac{d^2\psi}{dx^2} = -k^2 A\sin kx - k^2 B\cos kx = -k^2\psi$$

The result follows.

2. The complex conjugate of the wavefunction is (a)

$$\psi(x) = C\frac{1+ix}{1+ix^2}, \Rightarrow \psi^*(x) = C^*\frac{1-ix}{1-ix^2}$$

And so we find that

$$|\psi(x)|^2 = \psi^*(x)\,\psi(x) = |C|^2\,\frac{1+x^2}{1+x^4}$$

(b) To find the normalization constant we need to solve

$$\int_{-\infty}^{\infty} |\psi(x)|^2 dx = 1$$

It turns out that

$$\int_{-\infty}^{\infty} \frac{1+x^2}{1+x^4}dx = \sqrt{2}\pi$$

And so $C = \frac{1}{\sqrt{\sqrt{2}\pi}}$

(c) Using the following integral

$$\int_0^1 \frac{1+x^2}{1+x^4}dx = \frac{1}{2}\left[\sqrt{2}\arctan\left(\frac{2-\sqrt{2}}{\sqrt{2}}\right) + \sqrt{2}\arctan\left(\frac{2+\sqrt{2}}{\sqrt{2}}\right)\right]$$

we find that the probability is ≈ 0.52.

3. (a) The complex conjugate is

$$\psi^*(x) = C^* \frac{1}{x} e^{-i\omega t}$$

Therefore the normalization condition is

$$1 = |C|^2 \int_1^2 \frac{1}{x^2} dx = \frac{|C|^2}{2}, \Rightarrow C = \sqrt{2}$$

(b) The probability is given by

$$\int_{3/2}^{2} |\psi(x)|^2 dx = 2 \int_{3/2}^{2} \frac{1}{x^2} dx = \frac{1}{3}$$

4. The expectation value is

$$\langle x \rangle = \int_{-\infty}^{\infty} \psi^*(x) x \psi(x)\, dx = \frac{54}{\pi} \int_{-\infty}^{\infty} \frac{x}{(x^2+9)} dx$$

This is an odd function, as we can see by looking at the plot

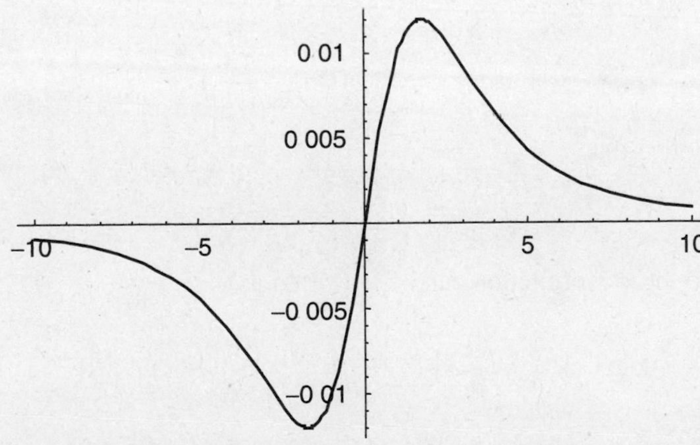

Fig. -1

Therefore the integral vanishes. Now for $\langle x^2 \rangle$, we encounter the integral

$$\int \frac{x^2}{(x^2+9)} dx = -\frac{x}{2(x^2+9)} + \frac{1}{6} \arctan\left(\frac{x}{3}\right)$$

Using L'Hopital's rule, the first term goes to zero at the limits of integration, for example

$$\lim_{x\to\infty} \frac{x}{2\left(x^2+9\right)} = \lim_{x\to\infty} \frac{1}{4x} = 0$$

For the second term, note that

$$\lim_{x\to\infty} \arctan\left(\frac{x}{3}\right) = \frac{\pi}{2}, \quad \lim_{x\to-\infty} \arctan\left(\frac{x}{3}\right) = -\frac{\pi}{2}$$

So you should find that the expectation value is

$$\langle x^2 \rangle = \frac{54}{\pi} \int_{-\infty}^{\infty} \frac{x^2}{\left(x^2+9\right)} dx = \frac{54}{\pi} \frac{\pi}{6} = 9$$

5. The function given is not square integrable, so it cannot be a valid wavefunction. So, the question makes no sense.
6. X is Hermitian but iX is not.
7. You should find that $j(x,t) = 0 = \frac{\partial \rho}{\partial t}$
8. Recall that for the infinite square well

(a)

$$\phi_n(x) = \sqrt{\frac{2}{a}} \sin\left(\frac{n\pi x}{a}\right)$$

Notice that

$$\sqrt{\frac{1}{2a}} = \sqrt{\frac{2}{2}}\sqrt{\frac{1}{2a}} = \frac{1}{2}\sqrt{\frac{2}{a}}$$

So the wavefunction can be rewritten as

$$\psi(x,t) = i\frac{\sqrt{3}}{2}\phi_1(x)\, e^{-iE_1/\hbar} + \frac{1}{2}\phi_3(x)\, e^{-iE_3/\hbar}$$

Normalization requires that $\sum_n |c_n|^2 = 1$ where the c_n are the coefficients of the expansion. In this case we have

$$\sum_n |c_n|^2 = \left| i\frac{\sqrt{3}}{2} \right|^2 + \left| \frac{1}{2} \right|^2 = \frac{3}{4} + \frac{1}{4} = 1$$

Hence the wavefunction is normalized.

(b) The possible values of the energy are

$$E_1 = \frac{\hbar^2\pi^2}{2ma^2} \text{ with probability } \left| i\frac{\sqrt{3}}{2} \right|^2 = \frac{3}{4}$$

$$E_3 = \frac{9\hbar^2\pi^2}{2ma^2} \text{ with probability } \left| \frac{1}{2} \right|^2 = \frac{1}{4}$$

(c) $\langle x \rangle = \langle p \rangle = 0$. To see this consider that integrals such as

$$\int_0^a \sin\left(\frac{\pi x}{a}\right) \sin\left(\frac{3\pi x}{a}\right) dx = 0$$

Chapter 3

1. b **2.** a **3.** b
4. c **5.** a

Chapter 4

1. The inner products are

$$(A, B) = (2)(1) + (-4i)(0) + (0)(1) + (1)(9i) + (7i)(2) = 2 + 23i$$

$$(B, A) = (1)(2) + (0)(4i) + (1)(0) + (-9i)(1) + (2)(-7i) = 2 - 23i$$

2. To see if the functions belong to L_2 we compute $\int_{-\infty}^{\infty} |f(x)|^2 dx$ and see if it is finite.

(a) In the first case we obtain

$$\int_{-\infty}^{\infty} (\operatorname{sech}(x))^2 \, dx = 2 < \infty$$

Since it is finite, the function belongs to L_2.

(b) This integral diverges, so it does not belong to L_2.

(c) In the final case we find

$$\int_0^{\infty} \left(e^{-x} \sin \pi x\right)^2 dx = \frac{\pi^2}{4 + 4\pi^2} < \infty$$

So the function belongs to L_2.

3. By the sampling property of the delta function

$$f\,(x)\,\delta\,(x-a) = f\,(a)\,\delta\,(x-a)$$

Therefore if we write $\delta\,(x)$ then we set a to zero and

$$x\delta\,(x) = (0)\delta\,(x) = 0$$

(a) Do a u substitution in the integral

$$\int f(x)\delta(ax)dx = \int f(u/a)\delta(u)du/\,|a| = \frac{1}{|a|}f\,(0)$$

Chapter 5

1. (a) The inner products are

$$\langle u|u\rangle = 13,\ \langle v|v\rangle = 52$$

And so the norms are $||u|| = \sqrt{13},\ ||v|| = \sqrt{52}$ The vectors are not normalized.
(b) $(4-2i)\,|u\rangle = \begin{pmatrix} 6+12i \\ 8-4i \end{pmatrix}$
(c) $\langle u|v\rangle = \langle v|u\rangle = 26$

2. The set of 2×2 matrices does constitute a vector space. For a basis, try the Pauli matrices and the identity matrix.

3. They are linearly independent.

4. No they are not, since we can write the third as a linear combination of the first two

$$i\,(1,1,0) + \frac{i}{2}\,(0,0,2) = (i,i,i)$$

5. We square the term on the left to obtain

$$\begin{aligned}
|z_1+z_2|^2 &= (z_1+z_2)\left(z_1^*+z_2^*\right) \\
&= z_1z_1^* + z_2z_2^* + z_1z_2^* + z_2z_1^* \\
&= |z_1|^2 + |z_2|^2 + 2Re^{(z_1z_2^*)} \\
&\le |z_1|^2 + |z_2|^2 + 2\,|z_1z_2| = (|z_1|+|z_2|)^2
\end{aligned}$$

Take the square root of both sides and the result follows.

6. The basis is

$$\{(1, 2, 1), (-3/2, 1, -1/2), (8/21 + 8/21i, 4/21 + 4/21i, -16/21 - 16/21i)\}$$

7. (a) The inner products are

$$\langle A|A\rangle = 29, \langle B|B\rangle = 113$$

And so the normalized vectors are

$$|A'\rangle = \frac{1}{\sqrt{29}}|A\rangle, |B'\rangle = \frac{1}{\sqrt{113}}|B\rangle$$

(b) The sum of the vectors is

$$|A + B\rangle = (9 + 3i)|u_1\rangle + (4 - 8i)|u_2\rangle$$

(c)

$$\langle A|A\rangle = 29, \quad \langle B|B\rangle = 113, \quad \langle A + B|A + B\rangle = 170$$

$$\Rightarrow \sqrt{\langle A|A\rangle} + \sqrt{\langle B|B\rangle} \approx 16$$

$$\sqrt{\langle A + B|A + B\rangle} \approx 13$$

$$\text{and so } \sqrt{\langle A + B|A + B\rangle} < \sqrt{\langle A|A\rangle} + \sqrt{\langle B|B\rangle}$$

(d)

$$\langle A|A\rangle\langle B|B\rangle = (29)(113) = 3277$$

$$\langle A|B\rangle = 14 - 53i, \quad \Rightarrow |\langle A|B\rangle|^2 = (14 + 53i)(14 - 53i) = 3005$$

$$\text{and so } |\langle A|B\rangle|^2 < \langle A|A\rangle\langle B|B\rangle$$

8. The dual vector is

(a) $\langle A| = (9 + 2i)\langle u_1| - 4i\langle u_2| - \langle u_3| - i\langle u_4|$

(b) The column vector representation is

$$|A\rangle = \begin{pmatrix} 9 - 2i \\ 4i \\ -1 \\ i \end{pmatrix}$$

(c) The row vector representation is

$$\langle A| = \begin{pmatrix} 9 + 2i & -4i & -1 & -i \end{pmatrix}$$

(d) $\langle A|A\rangle = 103$, to normalize divide by the square root of this quantity.

Chapter 6

1. The eigenvalues of all three matrices are ± 1. The eigenvectors are

$$\sigma_x : \left\{ \begin{pmatrix} \frac{1}{\sqrt{2}} \\ \frac{1}{\sqrt{2}} \end{pmatrix}, \begin{pmatrix} \frac{1}{\sqrt{2}} \\ -\frac{1}{\sqrt{2}} \end{pmatrix} \right\}, \quad \sigma_y : \left\{ \begin{pmatrix} \frac{i}{\sqrt{2}} \\ \frac{1}{\sqrt{2}} \end{pmatrix}, \begin{pmatrix} \frac{-i}{\sqrt{2}} \\ \frac{1}{\sqrt{2}} \end{pmatrix} \right\}, \quad \sigma_z : \left\{ \begin{pmatrix} 1 \\ 0 \end{pmatrix}, \begin{pmatrix} 0 \\ 1 \end{pmatrix} \right\}$$

$$\left[\sigma_x, \sigma_y\right] = 2i\sigma_z, \quad \left\{\sigma_x, \sigma_y\right\} = 0$$

The matrices are Hermitian and unitary.

2. To do the proof, recall that $A^\dagger = -A$ for an anti-Hermitian operator. Then we have

$$\langle a| (A|a\rangle) = \langle a| (a|a\rangle) = a\langle a|a\rangle$$

$$\langle a| (A|a\rangle) = \left(\langle a|A^\dagger\right)|a\rangle = \left(\langle a| - A^\dagger\right)|a\rangle = (\langle a| - a)\,|a\rangle = -a^*\langle aa\rangle$$

Comparison of the two equations shows that $a = -a^*$, so a must be pure imaginary.

3. Assume that we have two eigenvectors of an operator A such that $A|a\rangle = a|a\rangle$, $A|a'\rangle = a'|a'\rangle$ and $a \neq a'$. Recall that a Hermitian operator has real eigenvalues. Then

$$\langle a| \left(A\left|a'\right\rangle\right) = \langle a|\, a' \left|a'\right\rangle = a' \left\langle a|a'\right\rangle$$

$$\langle a| \left(A\left|a'\right\rangle\right) = \left(\langle a|\, A^\dagger\right)\left|a'\right\rangle = \left(\langle a|\, A^\dagger\right)\left|a'\right\rangle = (\langle a|\, a)\left|a'\right\rangle = a\left\langle a|a'\right\rangle$$

The term on the first line is equal to the term on the second line. Subtracting, we get

$$\left(a - a'\right)\left\langle a|a'\right\rangle = 0.$$

Since $a \neq a'$, then $\left\langle a|a'\right\rangle = 0$.

4. (a) $BA = (1/2)(\{A, B\} - [A, B])$.
(b)

$$\begin{aligned}[A + B, C] &= (A + B)\,C - C\,(A + B) \\ &= AC + BC - CA - CB \\ &= AC - CA + BC - CB \\ &= [A, C] + [B, C]\end{aligned}$$

5. The eigenvectors are

$$|a\rangle = \begin{pmatrix} 0 \\ 0 \\ 1 \end{pmatrix}, \quad |b\rangle = \frac{1}{\sqrt{2}} \begin{pmatrix} 1 \\ 1 \\ 0 \end{pmatrix}, \quad |c\rangle = \begin{pmatrix} 0 \\ 0 \\ 0 \end{pmatrix}$$

6. Since U is unitary we know that $U^{\dagger}U = UU^{\dagger} = I$. Also, we have $\langle \phi U| = \langle \phi |U^{\dagger}$, and so

$$\langle U\phi|U\psi\rangle = \langle \phi|U^{\dagger}U|\psi\rangle = \langle \phi|I|\psi\rangle = \langle \phi|\psi\rangle$$

7. $[A, B]^{\dagger} = (AB - BA)^{\dagger} = B^{\dagger}A^{\dagger} - A^{\dagger}B^{\dagger} = -\left(A^{\dagger}B^{\dagger} - B^{\dagger}A^{\dagger}\right) = -\left[A^{\dagger}, B^{\dagger}\right]$

8. (a) The matrix is Hermitian.
(b) A is not unitary.
(c) $Tr\,(A) = 7$.
(d) The eigenvalues are

$$\left\{4, \frac{3+\sqrt{5}}{2}, \frac{3-\sqrt{5}}{2}\right\}$$

(e) The eigenvectors are

$$|a\rangle = \begin{pmatrix} 0 \\ 0 \\ 1 \end{pmatrix}, \quad |b\rangle = \begin{pmatrix} \frac{i}{2}\left(1-\sqrt{5}\right) \\ 1 \\ 0 \end{pmatrix}, \quad |c\rangle = \begin{pmatrix} \frac{i}{2}\left(1+\sqrt{5}\right) \\ 1 \\ 0 \end{pmatrix}$$

Chapter 7

1. The eigenvalues of σ_x are $\{1, -1\}$ and the normalized eigenvectors are

$$|+_x\rangle = \frac{1}{\sqrt{2}} \begin{pmatrix} 1 \\ 1 \end{pmatrix}, |-_x\rangle = \frac{1}{\sqrt{2}} \begin{pmatrix} 1 \\ -1 \end{pmatrix}$$

To construct a unitary matrix to diagonalize σ_x, we use the eigenvectors for each column of the matrix, i.e.

$$U = \frac{1}{\sqrt{2}} \begin{pmatrix} 1 & 1 \\ 1 & -1 \end{pmatrix}$$

2. The eigenvalues are $(2, i, -i)$ and the (unnormalized) eigenvectors are

$$|X_1\rangle = \begin{pmatrix} -i\left(-2+\sqrt{5}\right) \\ 1 \\ 0 \end{pmatrix}, |X_2\rangle = \begin{pmatrix} i\left(2+\sqrt{5}\right) \\ 1 \\ 0 \end{pmatrix}, |X_3\rangle = \begin{pmatrix} 0 \\ 0 \\ 1 \end{pmatrix}$$

The matrix which diagonalizes X is constructed from these eigenvectors (but normalize them)

$$U = \left(\left|\tilde{X}_1\right\rangle \ \left|\tilde{X}_2\right\rangle \ \left|\tilde{X}_3\right\rangle \right)$$

3. (a) Noting that partial derivatives commute, we have

$$[A, B] f = (AB - BA) f = \frac{\partial}{\partial x}\left(-i\frac{\partial}{\partial x} + \frac{\partial}{\partial y}\right) f - \left(-i\frac{\partial}{\partial x} + \frac{\partial}{\partial y}\right)\frac{\partial}{\partial x} f$$

$$= -i\frac{\partial^2 f}{\partial x^2} + \frac{\partial^2 f}{\partial x \partial y} + i\frac{\partial^2 f}{\partial x^2} - \frac{\partial^2 f}{\partial y \partial x} = 0$$

(b) To find the eigenvalues, apply each operator to the given function. We see that the eigenvalue of B is $(1+i)$

$$Af = \frac{\partial}{\partial x}\left(e^{-x+y}\right) = -e^{-x+y} = -f$$

$$Bf = -i\frac{\partial}{\partial x}\left(e^{-x+y}\right) + \frac{\partial}{\partial y}\left(e^{-x+y}\right) = (1+i)e^{-x+y} = (1+i) f$$

4. Write down the Taylor series expansions of the exponentials.

5. Will work if F is a real function.

6. (a) Use $p = -i\hbar\frac{d}{dx}$ and $[x, p] = xp - px = i\hbar$ to find

$$\begin{aligned}
\left[p, x^n\right] &= px^n - x^n p = -i\hbar\frac{dx^n}{dx} - x^n p \\
&= -i\hbar n x^{n-1} - x^{n-1}(xp) \\
&= -i\hbar n x^{n-1} - x^{n-1}(px + i\hbar) \\
&= -i\hbar n x^{n-1} - x^{n-1}(-i\hbar + i\hbar) \\
&= -i\hbar n x^{n-1}
\end{aligned}$$

(b) Use $[p, f(x)] = -i\hbar\frac{df}{dx} - f(x)p$ then expand $f(x)$ in a Taylor series, using $[x, p] = xp - px = i\hbar$ to show that the second term is zero.

7. (a) Noting that partial derivatives commute, we have

$$[A, B] f = (AB - BA) f = \frac{\partial}{\partial x}\left(-i\frac{\partial}{\partial x} + \frac{\partial}{\partial y}\right) f - \left(-i\frac{\partial}{\partial x} + \frac{\partial}{\partial y}\right)\frac{\partial}{\partial x} f$$

$$= -i\frac{\partial^2 f}{\partial x^2} + \frac{\partial^2 f}{\partial x \partial y} + i\frac{\partial^2 f}{\partial x^2} - \frac{\partial^2 f}{\partial y \partial x} = 0$$

(b) To find the eigenvalues, apply each operator to the given function. We see that the eigenvalue of B is $(1 + i)$

$$Af = \frac{\partial}{\partial x}\left(e^{-x+y}\right) = -e^{-x+y} = -f$$

$$Bf = -i\frac{\partial}{\partial x}\left(e^{-x+y}\right) + \frac{\partial}{\partial y}\left(e^{-x+y}\right) = (1 + i)e^{-x+y} = (1 + i) f$$

8. (a) The matrix representations of each operator are

$$|+\rangle\langle+| = \begin{pmatrix} 1 & 0 \\ 0 & 0 \end{pmatrix}, \quad |-\rangle\langle-| = \begin{pmatrix} 0 & 0 \\ 0 & 1 \end{pmatrix}$$

(b) These are projection operators, as can be seen by checking the requirements projection operators must meet. For example, consider the square of each operator. Noting that the set is orthonormal, we have

$$(|+\rangle\langle+|)^2 = |+\rangle\langle+|+\rangle\langle+| = |+\rangle\langle+|$$

$$(|-\rangle\langle-|)^2 = |-\rangle\langle-|-\rangle\langle-| = |-\rangle\langle-|$$

Since $P^2 = P$ in both cases, they are projection operators.
(c) First we check the normalization

$$\langle\psi|\psi\rangle = (-2i)(2i) + (-4)(-4) = 4 + 16 = 20$$

Since the inner product does not evaluate to 1, the state is not normalized. We divide by the square root of this quantity to get a normalized state, which we denote with a tilde. The matrix representation of the state is

$$\left|\tilde{\psi}\right\rangle = \frac{1}{\sqrt{20}}\begin{pmatrix} 2i \\ -4 \end{pmatrix}$$

(d) Using the outer product notation, we have

$$(|+\rangle\langle+|)|\tilde{\psi}\rangle = \frac{1}{\sqrt{20}}\left[(2i)(|+\rangle\langle+|)|+\rangle + (-4)(|+\rangle\langle+|)|-\rangle\right] = \frac{2i}{\sqrt{20}}|+\rangle = \frac{i}{\sqrt{5}}|+\rangle$$

$$(|-\rangle\langle-|)|\tilde{\psi}\rangle = \frac{1}{\sqrt{20}}\left[(2i)(|-\rangle\langle-|)|+\rangle + (-4)(|-\rangle\langle-|)|-\rangle\right] = \frac{-4}{\sqrt{20}}|-\rangle = \frac{-2}{\sqrt{5}}|-\rangle$$

Using matrices

$$\begin{pmatrix} 1 & 0 \\ 0 & 0 \end{pmatrix} \begin{pmatrix} \frac{2i}{\sqrt{20}} \\ \frac{-4}{\sqrt{20}} \end{pmatrix} = \begin{pmatrix} \frac{2i}{\sqrt{20}} \\ 0 \end{pmatrix} = \frac{i}{\sqrt{5}} \begin{pmatrix} 1 \\ 0 \end{pmatrix}$$

$$\begin{pmatrix} 0 & 0 \\ 0 & 1 \end{pmatrix} \begin{pmatrix} \frac{2i}{\sqrt{20}} \\ \frac{-4}{\sqrt{20}} \end{pmatrix} = \begin{pmatrix} 0 \\ \frac{-4}{\sqrt{20}} \end{pmatrix} = \frac{-2}{\sqrt{5}} \begin{pmatrix} 0 \\ 1 \end{pmatrix}$$

Chapter 8

1. Applying the matrix to each basis vector we obtain

$$H|0\rangle = \frac{1}{\sqrt{2}} \begin{pmatrix} 1 & 1 \\ 1 & -1 \end{pmatrix} \begin{pmatrix} 1 \\ 0 \end{pmatrix} = \frac{1}{\sqrt{2}} \begin{pmatrix} 1 \\ 1 \end{pmatrix} = \frac{|0\rangle + |1\rangle}{\sqrt{2}}$$

$$H|1\rangle = \frac{1}{\sqrt{2}} \begin{pmatrix} 1 & 1 \\ 1 & -1 \end{pmatrix} \begin{pmatrix} 0 \\ 1 \end{pmatrix} = \frac{1}{\sqrt{2}} \begin{pmatrix} 1 \\ -1 \end{pmatrix} = \frac{|0\rangle - |1\rangle}{\sqrt{2}}$$

2. The expectation value is found by calculating $\langle Z \rangle = Tr(\rho Z)$

$$\rho Z = \begin{pmatrix} 63/80 & \frac{24+5\sqrt{3}}{80} \\ \frac{24+5\sqrt{3}}{80} & 17/80 \end{pmatrix} \begin{pmatrix} 1 & 0 \\ 0 & -1 \end{pmatrix} = \begin{pmatrix} 63/80 & -\frac{24+5\sqrt{3}}{80} \\ \frac{24+5\sqrt{3}}{80} & -17/80 \end{pmatrix}$$

$$\Rightarrow \langle Z \rangle = Tr(\rho Z) = 63/80 - 17/80 = 23/40$$

3. The density matrices are formed from

   (a) $|+\rangle \langle +|$ and $|-\rangle \langle -|$

$$\rho_+ = |+\rangle \langle +| = \left(\frac{|0\rangle + |1\rangle}{\sqrt{2}} \right) \left(\frac{\langle 0| + \langle 1|}{\sqrt{2}} \right) = \frac{1}{2} (|0\rangle \langle 0| + |1\rangle \langle 0| + |0\rangle \langle 1| + |1\rangle \langle 1|)$$

$$= \frac{1}{2} \begin{pmatrix} 1 & 1 \\ 1 & 1 \end{pmatrix}$$

$$\rho_- = |-\rangle \langle -| = \left(\frac{|0\rangle - |1\rangle}{\sqrt{2}} \right) \left(\frac{\langle 0| - \langle 1|}{\sqrt{2}} \right) = \frac{1}{2} (|0\rangle \langle 0| - |1\rangle \langle 0| - |0\rangle \langle 1| + |1\rangle \langle 1|)$$

$$= \frac{1}{2} \begin{pmatrix} 1 & -1 \\ -1 & 1 \end{pmatrix}$$

(b) In this case the density operator is

$$\rho = \frac{1}{2}\left|+\right\rangle\left\langle+\right| + \frac{1}{2}\left|-\right\rangle\left\langle-\right| = \frac{1}{2}\begin{pmatrix}1 & 0\\ 0 & 1\end{pmatrix}$$

It is a good idea to verify that the trace of each matrix representing these density operators is 1, as it should be.

(c) To see if the state is pure or mixed we square each matrix and compute the trace

$$\rho_+^2 = \frac{1}{4}\begin{pmatrix}1 & 1\\ 1 & 1\end{pmatrix}^2 = \frac{1}{2}\begin{pmatrix}1 & 1\\ 1 & 1\end{pmatrix}$$

$Tr(\rho_+^2) = 1/2 + 1/2 = 1 \Rightarrow \rho_+$ represents a pure state.

$$\rho_-^2 = \frac{1}{4}\begin{pmatrix}1 & -1\\ -1 & 1\end{pmatrix}^2 = \frac{1}{2}\begin{pmatrix}1 & -1\\ -1 & 1\end{pmatrix}$$

$Tr(\rho_-^2) = 1/2 + 1/2 = 1 \Rightarrow \rho_-$ represents a pure state.

$$\rho_-^2 = \frac{1}{4}\begin{pmatrix}1 & 0\\ 0 & 1\end{pmatrix}^2 = \frac{1}{4}\begin{pmatrix}1 & 0\\ 0 & 1\end{pmatrix}$$

$Tr(\rho_-^2) = 1/4 + 1/4 = 1/2 < 1 \Rightarrow \rho_-$ represents a mixed state.

The probabilities are found by calculating $\left\langle+\right|\rho_+\left|+\right\rangle$, $\left\langle+\right|\rho_-\left|+\right\rangle$, and $\left\langle+\right|\rho_-\left|+\right\rangle$. For example

$$\left\langle+\right|\rho_-\left|+\right\rangle = \begin{pmatrix}1/\sqrt{2} & 1/\sqrt{2}\end{pmatrix}\begin{pmatrix}1/2 & 0\\ 0 & 1/2\end{pmatrix}\begin{pmatrix}1/\sqrt{2}\\ 1/\sqrt{2}\end{pmatrix}$$
$$= \begin{pmatrix}1/2 & 1/2\end{pmatrix}\begin{pmatrix}1/2\\ 1/2\end{pmatrix} = 1/4 + 1/4 = 1/2$$

4. (a) To determine if this is a pure state, we square the density operator and compute the trace

$$\rho^2 = \begin{pmatrix}3/4 & -i/4\\ i/4 & 1/4\end{pmatrix}\begin{pmatrix}3/4 & -i/4\\ i/4 & 1/4\end{pmatrix} = \begin{pmatrix}5/8 & -i/4\\ i/4 & 1/8\end{pmatrix}$$

Therefore we find that

$$Tr\left(\rho^2\right) = 5/8 + 1/8 = 3/4 < 1$$

This means that this density operator represents a mixed state.

(b) First we calculate

$$\rho X = \begin{pmatrix} 3/4 & -i/4 \\ i/4 & 1/4 \end{pmatrix}\begin{pmatrix} 0 & 1 \\ 1 & 0 \end{pmatrix} = \begin{pmatrix} -i/4 & 3/4 \\ 1/4 & i/4 \end{pmatrix}$$

$$\rho Y = \begin{pmatrix} 3/4 & -i/4 \\ i/4 & 1/4 \end{pmatrix}\begin{pmatrix} 0 & -i \\ i & 0 \end{pmatrix} = \begin{pmatrix} 1/4 & -i3/4 \\ i/4 & 1/4 \end{pmatrix}$$

$$\rho Z = \begin{pmatrix} 3/4 & -i/4 \\ i/4 & 1/4 \end{pmatrix}\begin{pmatrix} 1 & 0 \\ 0 & -1 \end{pmatrix} = \begin{pmatrix} 3/4 & i/4 \\ i/4 & -1/4 \end{pmatrix}$$

We obtain the components of the Bloch vector by taking the trace of each of these matrices

$$r_x = Tr(\rho X) = -i/4 + i/4 = 0$$

$$r_y = Tr(\rho Y) = 1/4 + 1/4 = 1/2$$

$$r_z = Tr(\rho Z) = 3/4 - 1/4 = 1/2$$

The magnitude of the Bloch vector is

$$\|\vec{r}\| = \sqrt{r_x^2 + r_y^2 + r_z^2} = \sqrt{0 + 1/4 + 1/4} = \sqrt{1/2} < 1$$

Since $\|\vec{r}\| < 1$ this is a mixed state.

(c) The probability is found by calculating $Tr\,(p_1 \rho)$

$$p_1 \rho = \begin{pmatrix} 0 & 0 \\ 0 & 1 \end{pmatrix}\begin{pmatrix} 3/4 & -i/4 \\ i/4 & 1/4 \end{pmatrix} = \begin{pmatrix} 0 & 0 \\ i/4 & 1/4 \end{pmatrix}$$

$$\Rightarrow Tr\,(p_1 \rho) = 0 + 1/4 = 1/4$$

5. (a) $\langle \psi | = A^* \langle u_1 | + 1/2 \langle u_2 | + i/\sqrt{3} \langle u_3 |$

(b) We compute the inner product and set it equal to 1, and solve for A

$$1 = \langle \psi | \psi \rangle = |A|^2 + 3/12 + 4/12 = |A|^2 + 7/12$$

$$\Rightarrow A = \sqrt{5/12}$$

(c) The matrix representation is given by

$$\begin{pmatrix} \hbar\omega & 0 & 0 \\ 0 & 3\hbar\omega & 0 \\ 0 & 0 & 5\hbar\omega \end{pmatrix}$$

(d) The column vector is

$$|\psi\rangle = \begin{pmatrix} \sqrt{5/12} \\ 1/2 \\ -i/\sqrt{3} \end{pmatrix}$$

(e) Using the Born rule the probabilities are

$$prob(\hbar\omega) = |\langle u_1|\psi\rangle|^2 = 5/12$$

$$prob(3\hbar\omega) = |\langle u_2|\psi\rangle|^2 = 1/4$$

$$prob(5\hbar\omega) = |\langle u_3|\psi\rangle|^2 = 1/3$$

Notice that these probabilities sum to one

$$5/12 + 1/4 + 1/3 = 1$$

(f) We can find the average value of the energy we can use $\langle H\rangle = \sum_n E_n p_n$ where p_n is the probability of finding energy E_n. This gives

$$\langle H\rangle = (\hbar\omega)\,(5/12) + (3\hbar\omega)\,(1/4) + (5\hbar\omega)\,(1/3) = \frac{17}{6}\hbar\omega$$

(g) The state changes with time according to $|\psi\rangle = \sum_n e^{-iE_nt/\hbar} c_n\,|u_n\rangle$ and so we have

$$|\psi\rangle = e^{-i\omega t}\sqrt{5/12}\,|u_1\rangle + \frac{e^{-i3\omega t}}{2}\,|u_2\rangle - i\frac{e^{-i5\omega t}}{\sqrt{3}}\,|u_3\rangle$$

Chapter 9

1. (a) Using the fact that the basis states are orthonormal, we calculate $\langle\psi|\psi\rangle$ giving

$$1 = \langle\psi|\psi\rangle = \frac{1}{8} + \frac{1}{2} + A^2$$

Solving we find that $A = \sqrt{\frac{3}{8}}$

(b) The energy for the state $|\phi_n\rangle$ of the harmonic oscillator is $E_n = \left(n + \frac{1}{2}\right)\hbar\omega$. The possible energies that can be found in the given state and their respective probabilities are

$$E_0 = \frac{\hbar\omega}{2} \text{ with probability } 1/8$$

$$E_1 = \frac{3\hbar\omega}{2} \text{ with probability } 1/2$$

$$E_2 = \frac{5\hbar\omega}{2} \text{ with probability } 3/8$$

(c) The state at a later time t can be written as

$$\psi(x,t) = \frac{1}{\sqrt{8}}e^{-i\omega t/2}\phi_0(x) + \frac{1}{\sqrt{2}}e^{-i3\omega t/2}\phi_1(x) + \frac{1}{\sqrt{8}}e^{-i5\omega t/2}\phi_2(x)$$

2. Follow the procedure used in example 9.6. In particular, use $\left[a, a^\dagger\right] = aa^\dagger - a^\dagger a = 1$ to write

$$H = \frac{\hbar\omega}{2}\left(aa^\dagger + a^\dagger a\right) = \frac{\hbar\omega}{2}\left(aa^\dagger + aa^\dagger - 1\right) = \hbar\omega\left(aa^\dagger - \frac{1}{2}\right)$$

3. Starting by writing the Hamiltonian operator in terms of the ladder operators,

$$H|n\rangle = \hbar\omega\left(aa^\dagger - \frac{1}{2}\right)|n\rangle = \hbar\omega\left(aa^\dagger|n\rangle\right) - \frac{\hbar\omega}{2}|n\rangle$$

However, we know that

$$H|n\rangle = \hbar\omega\left(n + \frac{1}{2}\right)|n\rangle$$

We can equate this to $\hbar\omega\left(aa^\dagger|n\rangle\right) - \frac{\hbar\omega}{2}|n\rangle$ and divide through by $\hbar\omega$. This allows us to write

$$aa^\dagger|n\rangle = \left(n + \frac{1}{2} + \frac{1}{2}\right)|n\rangle = (n+1)|n\rangle$$

4. Write the Hamiltonian in terms of the number operator and consider the fact that $\|a|n\rangle\|^2 = \langle n| a^\dagger a |n\rangle \geq 0$. Then consider the ground state.
5. We write the number operator explicitly and then use $[A, B] = -[B, A]$ and $[A, BC] = [A, B]C + B[A, C]$. In the first case this gives

$$[N, a] = \left[a^\dagger a, a\right] = -\left[a, a^\dagger a\right] = -\left[a, a^\dagger\right]a - a^\dagger[a, a] = -\left[a, a^\dagger\right]a = -a$$

For the other commutator we obtain

$$\left[N, a^\dagger\right] = \left[a^\dagger a, a^\dagger\right] = -\left[a^\dagger, a^\dagger a\right] = -\left[a^\dagger, a^\dagger\right]a - a^\dagger\left[a^\dagger, a\right] = a^\dagger\left[a, a^\dagger\right] = a^\dagger$$

6. In the coordinate representation, we have

$$\psi_n(x) = \frac{1}{\sqrt{2^n n!}}\left(\frac{m\omega}{\pi\hbar}\right)^{1/4} H_n\left(\sqrt{\frac{m\omega}{\hbar}}x\right) e^{-m\omega x^2/2\hbar}$$

The expectation value of x in any state is going to be

$$\langle x\rangle = \int_{-\infty}^{\infty} \psi_n^*(x) x \psi_n(x)\, dx$$

$$= \frac{1}{2^n n!}\left(\frac{m\omega}{\pi\hbar}\right)^{1/2} \int_{-\infty}^{\infty} H_n\left(\sqrt{\frac{m\omega}{\hbar}}x\right) x H_n\left(\sqrt{\frac{m\omega}{\hbar}}x\right) e^{-m\omega x^2/\hbar} dx$$

This integral can be rewritten using

$$uH_n(u) = \frac{1}{2}H_{n+1}(u) + nH_{n-1}(u)$$

Then we recall that the Hermite polynomials satisfy

$$\int_{-\infty}^{\infty} H_m(u)H_n(u)e^{-u^2} du = \sqrt{\pi}2^n n!\delta_{mn}$$

Note the presence of the Kronecker delta δ_{mn} in this formula. This means that if $m \neq n$ the integral vanishes. In particular, you will find that $\langle x\rangle = 0$ in the ground state (or consider that the integral you will obtain is of an odd function over a symmetric interval—so it must vanish).

For momentum, write the momentum operator in coordinate representation (i.e. as a derivative) and consider the relation

$$\frac{dH_n}{du} = 2nH_{n-1}(u)$$

together with the integral formula for the Hermite polynomials.

Chapter 10

1. First write down the basis states

$$|1_y\rangle = \frac{1}{2}\begin{pmatrix} 1 \\ i\sqrt{2} \\ -1 \end{pmatrix}, |0_y\rangle = \frac{1}{\sqrt{2}}\begin{pmatrix} 1 \\ 0 \\ 1 \end{pmatrix}, |-1_y\rangle = \frac{1}{2}\begin{pmatrix} 1 \\ -i\sqrt{2} \\ -1 \end{pmatrix}$$

The possible results of measurement are $+\hbar$, 0, and $-\hbar$. The probabilities of obtaining each measurement result are obtained by applying the Born rule, and are given in turn as $|\langle 1_y|\psi\rangle|^2$, $|\langle 0_y|\psi\rangle|^2$, and $|\langle -1_y|\psi\rangle|^2$. Proceeding, using the particle state given in the problem the first inner product is

$$\langle 1_y|\psi\rangle = \frac{1}{2\sqrt{14}}\begin{pmatrix} 1 & -i\sqrt{2} & -1 \end{pmatrix}\begin{pmatrix} 3 \\ 1 \\ 2 \end{pmatrix} = \frac{1-i\sqrt{2}}{2\sqrt{14}}$$

Therefore the probability of obtaining measurement result $+\hbar$ is

$$|\langle 1_y|\psi\rangle|^2 = \left|\frac{1-i\sqrt{2}}{2\sqrt{14}}\right|^2 = \left(\frac{1-i\sqrt{2}}{2\sqrt{14}}\right)\left(\frac{1+i\sqrt{2}}{2\sqrt{14}}\right) = \frac{3}{4(14)} = \frac{3}{56}$$

Next, we find that

$$\langle 0_y|\psi\rangle = \frac{1}{\sqrt{2}\sqrt{14}}\begin{pmatrix} 1 & 0 & 1 \end{pmatrix}\begin{pmatrix} 3 \\ 1 \\ 2 \end{pmatrix} = \frac{5}{\sqrt{28}}$$

So the probability of obtaining measurement result 0 is

$$|\langle 0_y|\psi\rangle|^2 = \left|\frac{5}{\sqrt{28}}\right|^2 = \frac{25}{28} = \frac{50}{56}$$

Finally, we have

$$\langle -1_y|\psi\rangle = \frac{1}{2\sqrt{14}}\begin{pmatrix} 1 & i\sqrt{2} & -1 \end{pmatrix}\begin{pmatrix} 3 \\ 1 \\ 2 \end{pmatrix} = \frac{1+i\sqrt{2}}{2\sqrt{14}}$$

So the probability of finding measurement result $-\hbar$ is

$$|\langle -1_y|\psi\rangle|^2 = \left|\frac{1+i\sqrt{2}}{2\sqrt{14}}\right|^2 = \left(\frac{1-i\sqrt{2}}{2\sqrt{14}}\right)\left(\frac{1+i\sqrt{2}}{2\sqrt{14}}\right) = \frac{3}{4(14)} = \frac{3}{56}$$

The reader should verify that these probabilities sum to one.

2. (a) It is easier to find N if we write the state in Dirac notation

$$|\psi\rangle = N\frac{2}{\sqrt{5}}|0\rangle + \frac{N}{2}|-1\rangle - \frac{N}{\sqrt{5}}|1\rangle$$

Using $\langle\psi|\psi\rangle = 1$ we find that $N = \frac{2}{\sqrt{5}}$.

(b) Incorporating the normalization constant into the state we have

$$|\psi\rangle = \frac{4}{5}|0\rangle + \frac{1}{\sqrt{5}}|-1\rangle - \frac{2}{5}|1\rangle$$

Then using the Born rule the probabilities are

$$prob\,(+\hbar) = |\langle 1|\psi\rangle|^2 = \frac{4}{25}$$

$$prob\,(0) = |\langle 0|\psi\rangle|^2 = \frac{16}{25}$$

$$prob\,(-\hbar) = |\langle -1|\psi\rangle|^2 = \frac{5}{25}$$

(c) The ladder operators act on the basis states according to

$$L_+|+1\rangle = 0,\ L_+|0\rangle = \sqrt{2}\hbar\,|+1\rangle\,,\ L_+|-1\rangle = \sqrt{2}\hbar\,|0\rangle$$

$$L_-|+1\rangle = \sqrt{2}\hbar\,|0\rangle\,,\ L_-|0\rangle = \sqrt{2}\hbar\,|-1\rangle\,,\ L_-|-1\rangle = 0$$

So we find that

$$L_+|\psi\rangle = \frac{4\sqrt{2}}{5}\hbar\,|+1\rangle + \sqrt{\frac{2}{5}}\hbar\,|0\rangle$$

$$L_-|\psi\rangle = \frac{4\sqrt{2}}{5}\hbar\,|-1\rangle - \frac{2\sqrt{2}}{5}\hbar\,|0\rangle$$

3. With $J = 2$ we can have $m = -2, -1, 0, 1, 2$.

4. We begin by applying a famous trig identity

$$\sin 2\theta \cos\phi = 2\sin\theta\cos\theta\cos\phi$$

Now use Euler's formula

$$2\sin\theta\cos\theta\cos\phi = \sin\theta\cos\theta e^{i\phi} + \sin\theta\cos\theta e^{-i\phi}$$

Using $Y_2^{\pm 1} = \mp\sqrt{\frac{15}{8\pi}}e^{\pm i\phi}\sin\theta\cos\theta$ we arrive at the final result

$$\sin 2\theta\cos\phi = -\sqrt{\frac{8\pi}{15}}Y_2^1 + \sqrt{\frac{8\pi}{15}}Y_2^{-1}$$

5. c

Chapter 11

1. c	**2.** d	**3.** c
4. b	**5.** c	**6.** a

Chapter 12

1. c	**2.** a	**3.** c
4. a	**5.** b	

Final Exam Solutions

1. c	**2.** a	**3.** d	**4.** c	**5.** d
6. b	**7.** b	**8.** a	**9.** d	**10.** b
11. a	**12.** d	**13.** c	**14.** e	**15.** a
16. b	**17.** d	**18.** c	**19.** c	**20.** b

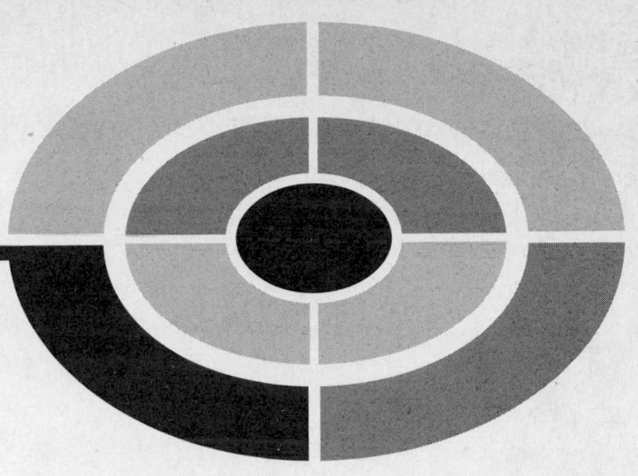

References

Cohen-Tannoudj, C., B. Dui, and F. Lahöe. *Quantum Mechanics,* Volume I. New York, John Wiley, 1977.

Deutsch, Ivan. Online course notes for Physics 521 at the University of New Mexico, *Pure vs. mixed states—density operators,* available at http://info.phys.unm.edu/%7Edeutschgroup/Classes/Phys521F02/Phys521.html

Griffiths, D. *Introduction to Quantum Mechanics.* Englewood Cliffs, N.J., Prentice-Hall, 1995.

Goswami, A. *Quantum Mechanics,* 2nd ed. New York, McGraw-Hill, 1996.

Liboff, R. L. *Introductory Quantum Mechanics,* 3rd ed. Reading, Mass., Addison-Wesley, 1997.

Peleg, Y., R. Pnini, and E. Zaarur. *Schaum's Outlines: Quantum Mechanics.* New York, McGraw-Hill, 1998.

Wikipedia online Encyclopedia, Density Matrix entry, found at http://en.wikipedia.org/wiki/Density_operator

Zettilli, N. *Quantum Mechanics, Concepts and Applications.* New York, John Wiley, 2001.

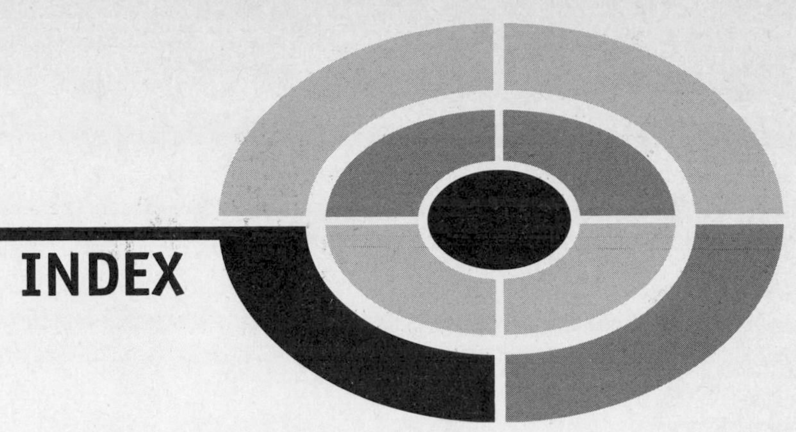

INDEX

A

B

C

D

E

F

G

H

N

O

P

Q

R

S

ABOUT THE AUTHOR

David McMahon works as a researcher in the national labs on nuclear energy. He has advanced degrees in physics and applied mathematics, and has written several titles for McGraw-Hill.